Cellular Agriculture for Revolutionized Food Production

Ahmed M. Hamad
Benha University, Egypt

Tanima Bhattacharya
Lincoln University College, Malaysia

A volume in the Advances in Environmental Engineering and Green Technologies (AEEGT) Book Series

Published in the United States of America by
IGI Global
Engineering Science Reference (an imprint of IGI Global)
701 E. Chocolate Avenue
Hershey PA, USA 17033
Tel: 717-533-8845
Fax: 717-533-8661
E-mail: cust@igi-global.com
Web site: http://www.igi-global.com

Copyright © 2024 by IGI Global. All rights reserved. No part of this publication may be reproduced, stored or distributed in any form or by any means, electronic or mechanical, including photocopying, without written permission from the publisher.
Product or company names used in this set are for identification purposes only. Inclusion of the names of the products or companies does not indicate a claim of ownership by IGI Global of the trademark or registered trademark.

Library of Congress Cataloging-in-Publication Data

CIP DATA PENDING

ISBN: 9798369341155
EISBN: 9798369341162

British Cataloguing in Publication Data
A Cataloguing in Publication record for this book is available from the British Library.

All work contributed to this book is new, previously-unpublished material.
The views expressed in this book are those of the authors, but not necessarily of the publisher.

For electronic access to this publication, please contact: eresources@igi-global.com.

Table of Contents

Preface xii

Chapter 1
Introduction to Cellular Agriculture 1
    Wasiya Farzana, Karunya Institute of Technology and Sciences, India
    Gintu Sara George, Karunya Institute of Technology and Sciences, India

Chapter 2
Unveiling the Science From Cells to Cultivated Food 24
    Jayasree S. Kanathasan, Lincoln University College, Malaysia
    Devi Nallappan, Lincoln University College, Malaysia
    Gunavathy Selvarajh, Lincoln University College, Malaysia
    Zaliha Binti Harun, Lincoln University College, Malaysia

Chapter 3
The Environmental Impact of Cellular Agriculture 50
    Gunavathy Selvarajh, Lincoln University College, Malaysia
    Farzana Yasmin, Lincoln University College, Malaysia
    Udugalage Isuru Harsha Kumara, Lincoln University College, Malaysia
    Jayasree Kanathasan, Lincoln University College, Malaysia
    Devi Nallapan, Lincoln University College, Malaysia

Chapter 4
Navigating Ethics and Animal Welfare 67
    Devi Nallappan, Lincoln University College, Malaysia
    Jayasree S. Kanathasan, Lincoln University College, Malaysia

Chapter 5
The Lab Grown Plate: A Deep Dive Into Cellular Agriculture 92
    Rathimalar Ayakannu, Lincoln University College, Malaysia
    Hemapriyaa Vijayan, Lincoln University College, Malaysia
    Asita Elengoe, Lincoln University College, Malaysia

Chapter 6
In Vitro Harvest 109
    Zaliha Harun, Lincoln University College, Malaysia
    Gunavathy Selvarajh, Lincoln University College, Malaysia
    Norhashima Abd Rashid, Lincoln University College, Malaysia
    Nik Nurul Najihah Nik Mat Daud, Lincoln University College, Malaysia

Chapter 7
Cellular Milk and Dairy Products 128
    Dipali Saxena, Shri Vaishnav Vidyapeeth Vishwadyalaya, India
    Rolly Mehrotra, Gorakhpur University, India
    Manisha Trivedi, Shri Vaishnav Vidyapeeth Vishwavidyalaya, India

Chapter 8
In Vitro Cultured Meat 149
    Shimaa N. Edris, Benha University, Egypt
    Aya Tayel, Benha University, Egypt
    Ahmed M. Alhussaini Hamad, Benha University, Egypt
    Islam I. Sabeq, Benha University, Egypt

Chapter 9
Future of Cellular Agriculture 208
    Idris Adewale Ahmed, Lincoln University College, Malaysia
    Ibrahim Bello, North Dakota State University, USA
    Abdullateef Akintunde Raji, Federal University of Kashere, Nigeria
    Maryam Abimbola Mikail, Mimia Sdn Bhd, Malaysia

Chapter 10
Addressing Global Food Security Through Cellular Agriculture 231
    Ahmed M. Alhussaini Hamad, Benha University, Egypt

Chapter 11
The Role of Cellular Agriculture in Mitigating Climate Change 255
    Ahmed Hamad, Benha University, Egypt
    Dina A. B. Awad, Benha University, Egypt

Chapter 12
Regulatory Landscape and Public Perception 288
　Peter Yu, APAC Society for Cellular Agriculture, Singapore
　Calisa Lim, APAC Society for Cellular Agriculture, Singapore

Compilation of References 323

About the Contributors 413

Index 419

Detailed Table of Contents

Preface xii

Chapter 1
Introduction to Cellular Agriculture 1
    Wasiya Farzana, Karunya Institute of Technology and Sciences, India
    Gintu Sara George, Karunya Institute of Technology and Sciences, India

Today, microbes grown in bioreactors provide leather and fibers for clothing, bags, and shoes, as well as flavors, sweeteners, and proteins from eggs and milk for human use. Additionally, components from plant cell and tissue cultures boost immunity and enhance skin texture; in vitro meat, a different precommercial cellular agriculture product, is currently the subject of a lot of interest. All of these strategies could help traditional agriculture continue to meet the nutritional needs of an expanding global population while releasing valuable resources like arable land. It will be extremely difficult to feed a growing population, and there is demand to lessen the damaging effects of traditional agriculture on the environment. The prospect of consuming plant cells presents a very alluring substitute for obtaining wholesome, high-protein, and nutritionally balanced food raw materials. Moreover, a variety of biosynthetic compounds can be produced from cultured bacteria.

Chapter 2
Unveiling the Science From Cells to Cultivated Food 24
    Jayasree S. Kanathasan, Lincoln University College, Malaysia
    Devi Nallappan, Lincoln University College, Malaysia
    Gunavathy Selvarajh, Lincoln University College, Malaysia
    Zaliha Binti Harun, Lincoln University College, Malaysia

The food crisis has been a crucial concern to meet the rising food demand worldwide. Cellular agriculture could be a prominent solution for sustainable food production. One of the main aims for cellular agriculture is to increase the capacity of the food to feed the increasing world population, whilst maintaining the quality of the environment and food. Therefore, this chapter introduces cellular agriculture as one of the latest food bioengineering technologies. Then, the principles and factors of cultivating cultured meat and plant tissues are discussed. Furthermore, applications of the cultivated food are explored. Following this, the ethical aspects and the challenges that will direct the future perspectives were discussed. Being at an early stage of venture in the food industry, the benefits and significant improvements of cellular agriculture will make it the most effective alternative in the food industry.

Chapter 3
The Environmental Impact of Cellular Agriculture 50
    Gunavathy Selvarajh, Lincoln University College, Malaysia
    Farzana Yasmin, Lincoln University College, Malaysia
    Udugalage Isuru Harsha Kumara, Lincoln University College, Malaysia
    Jayasree Kanathasan, Lincoln University College, Malaysia
    Devi Nallapan, Lincoln University College, Malaysia

Cellular agriculture, known as cultured or lab-grown meat production, involves the cultivation of animal cells in vitro to generate meat products without the need for traditional animal rearing and slaughtering. There are numerous environmental issues stemming from traditional animal agriculture. These include land degradation, water scarcity, greenhouse gas emissions, and biodiversity loss, which collectively contribute to environmental degradation and climate change. Against this backdrop, cellular agriculture offers a promising solution by leveraging biotechnological advancements to produce animal-derived products without the need for intensive livestock farming. This chapter evaluates the environmental impact of cellular agriculture across multiple dimensions. Further, this chapter also evaluates the water conservation, followed by biodiversity conservation. The final topic is on the reduced energy consumption due to the implementation of cellular agriculture.

Chapter 4
Navigating Ethics and Animal Welfare 67
    Devi Nallappan, Lincoln University College, Malaysia
    Jayasree S. Kanathasan, Lincoln University College, Malaysia

Animal ethics encompasses moral, legal, and ethical standards governing human interactions with animals, including pre-clinical research and industrial settings. Animal welfare assesses animals' well-being and adaptability. Scientific evaluation of welfare has been significant, with scientists and philosophers working together to define the appropriate relationship between humans and animals. This chapter focuses on navigation ethics and animal welfare. It begins with an introduction to animal research ethics. Then, it delves into the general application, issues and development of animal ethics. The subsequent part highlights animal ethics in cellular agriculture and food production. It then discusses the connection between animal ethics and animal welfare. The chapter concludes with an overview of the challenges, limitations and prospects of integrating animal ethics and animal welfare. This review aims to provide new perspectives on ethics and animal welfare for researchers, policymakers, regulators, and others involved in animal research and welfare fields.

Chapter 5
The Lab Grown Plate: A Deep Dive Into Cellular Agriculture 92
    Rathimalar Ayakannu, Lincoln University College, Malaysia
    Hemapriyaa Vijayan, Lincoln University College, Malaysia
    Asita Elengoe, Lincoln University College, Malaysia

This chapter highlights the cutting-edge field of cellular agriculture, emphasizing the creation, principles, and applications of lab-grown food. The lab-grown plate refers to a concept representing food that is produced through cellular agriculture techniques in a laboratory setting. It covers the idea of creating food products, such as meat, dairy, and other animal products, through cell culture techniques, without the need for conventional animal husbandry. Instead, cells are cultured and grown in controlled settings to produce edible tissue, which is subsequently processed into a variety of food products. However, there are still challenges to overcome in scaling up cellular agriculture to commercial levels, including technological hurdles, regulatory considerations, and consumer acceptance. Nevertheless, continued research and investment in this field hold promise for transforming the way animal products are produced and consumed in the future.

Chapter 6
In Vitro Harvest 109
    Zaliha Harun, Lincoln University College, Malaysia
    Gunavathy Selvarajh, Lincoln University College, Malaysia
    Norhashima Abd Rashid, Lincoln University College, Malaysia
    Nik Nurul Najihah Nik Mat Daud, Lincoln University College, Malaysia

In vitro harvesting is a technique involving growing animal cells in controlled laboratories. This technology provides sustainable, ethical, and scalable options to meet the world's protein needs. This chapter addresses the environmental advantages and diminished impact of traditional farming methods. This chapter explains the cell culture techniques, different cell sources, nutritional media formulations, and bioreactor technologies in meat production. The ethical implications of the topic are discussed, with a specific focus on the well-being of animals and concerns related to the slaughter process. Successfully integrating in vitro harvests into mainstream markets depends on overcoming regulatory barriers and gaining customer acceptability. The chapter finishes by providing a concise overview of the significant impact that cellular agriculture can have on redefining sustainability, ethics, and nutritional diversity in the global food system. This transformation is made possible through the collaboration of many fields of study and the application of advanced technology.

Chapter 7
Cellular Milk and Dairy Products 128
    Dipali Saxena, Shri Vaishnav Vidyapeeth Vishwadyalaya, India
    Rolly Mehrotra, Gorakhpur University, India
    Manisha Trivedi, Shri Vaishnav Vidyapeeth Vishwavidyalaya, India

Cellular agriculture could develop foods with changed macro- and micronutrient content to promote optimal health or flavor, potentially producing a new class of superfoods. This multi-disciplinary area focuses on resulting in existing agricultural products, particularly animal-based products, using cell culture rather than living organisms. Conventional dairy farming has a major environmental impact, including greenhouse gas emissions, water use, land use, and animal welfare challenges. Cell-based dairy, also known as lab-grown or cultured dairy, is a promising alternative that produces milk directly from cell cultures, eliminating the need for cows. This technique can significantly minimize the environmental footprint. This chapter focuses on cellular agriculture, which has the potential to reduce the number of animals required for food production and thereby minimize the environmental implications of extensive animal husbandry.

Chapter 8
In Vitro Cultured Meat 149
    Shimaa N. Edris, Benha University, Egypt
    Aya Tayel, Benha University, Egypt
    Ahmed M. Alhussaini Hamad, Benha University, Egypt
    Islam I. Sabeq, Benha University, Egypt

The advent of in vitro cultured meat represents a groundbreaking advancement in food technology and sustainable agriculture. This chapter delves into the intricacies of lab-grown meat, exploring its potential to revolutionize the meat industry by offering a viable alternative to traditional livestock farming. In vitro cultured meat is produced by culturing animal cells in a controlled environment, allowing for the creation of muscle tissue that mirrors conventional meat without the need for animal slaughter. This method addresses a myriad of concerns related to environmental sustainability, animal welfare, and food security. In conclusion, in vitro cultured meat has the potential to transform the meat industry by offering a sustainable, ethical, and safe alternative to traditional meat. As research and technology continue to advance, cultured meat could play a pivotal role in addressing some of the most pressing issues facing global food systems today.

Chapter 9
Future of Cellular Agriculture 208
    Idris Adewale Ahmed, Lincoln University College, Malaysia
    Ibrahim Bello, North Dakota State University, USA
    Abdullateef Akintunde Raji, Federal University of Kashere, Nigeria
    Maryam Abimbola Mikail, Mimia Sdn Bhd, Malaysia

Cellular agriculture, a transformative field at the intersection of biotechnology and food production, is poised to revolutionize the global food system. This review delves into the multifaceted landscape of cellular agriculture, examining key aspects such as technological advancements driving innovation, emerging market trends and lucrative opportunities, regulatory frameworks shaping industry development, and the environmental impact of transitioning to cellular agriculture. Moreover, it explores consumer acceptance and perception, crucial for mainstream adoption, alongside economic viability considerations and the evolving supply chain and infrastructure requirements. Ethical considerations, particularly concerning animal welfare, are scrutinized, highlighting the importance of addressing these concerns for industry sustainability. Furthermore, the review also evaluates the roles of international collaboration and partnerships in fostering growth and overcoming challenges.

Chapter 10
Addressing Global Food Security Through Cellular Agriculture 231
    Ahmed M. Alhussaini Hamad, Benha University, Egypt

In the face of an ever-growing global population and the escalating impacts of climate change, ensuring food security has become a critical challenge of our time. This chapter explores the transformative potential of cellular agriculture as a viable solution to address global food security. Cellular agriculture, which involves producing animal products from cell cultures rather than traditional livestock farming, offers a sustainable and efficient alternative to conventional agriculture. The chapter also examines the socio-economic implications of cellular agriculture, such as its potential to create new job opportunities, support rural economies, and contribute to food equity by making nutritious food accessible to diverse populations. Through case studies and expert insights, this chapter provides a comprehensive overview of how cellular agriculture can play a pivotal role in building a resilient, sustainable, and secure global food system for the future.

Chapter 11
The Role of Cellular Agriculture in Mitigating Climate Change 255
   Ahmed Hamad, Benha University, Egypt
   Dina A. B. Awad, Benha University, Egypt

Cellular agriculture, a revolutionary approach to food production, holds significant potential for mitigating climate change. Cellular agriculture addresses the environmental burdens associated with conventional animal agriculture by cultivating animal products such as meat, dairy, and eggs from cell cultures rather than traditional livestock farming. The chapter explores cellular agriculture's technological advancements, economic implications, and regulatory challenges, emphasizing its role in decreasing deforestation, methane emissions, and water usage. Furthermore, it highlights the potential for cellular agriculture to reduce the reliance on antibiotics, thereby contributing to public health. Through a comprehensive analysis of current research, case studies, and future projections, this chapter underscores the critical importance of cellular agriculture in the global effort to combat climate change and foster a sustainable food system.

Chapter 12
Regulatory Landscape and Public Perception 288
   Peter Yu, APAC Society for Cellular Agriculture, Singapore
   Calisa Lim, APAC Society for Cellular Agriculture, Singapore

The current food system faces global disruptions from climate change and population growth, destabilising the state of food security in many countries. In this new global context, cultivated meat (CM) presents an innovative solution to provide a sustainable source of protein. There are two essential aspects of the CM industry: development of clear and transparent regulatory frameworks and positive consumer perception. Regulatory frameworks allow companies to demonstrate food safety, a pre-market requirement that most countries mandate before commercialisation. Post-commercialisation success will be measured by how well the CM products are received by the public. This chapter explores both topics by providing a comprehensive overview of the current state of each area, with further insights on key drivers that facilitate progressive development, such as international coordination and harmonised labelling guidelines. The chapter concludes with a combined outlook of the CM industry going forward.

Compilation of References 323

About the Contributors 413

Index 419

Preface

Traditional food production methods are increasingly proving unsustainable in a world confronted with the dual crises of climate change and food insecurity. Industrial agriculture, which has dominated the global food landscape for decades, is not only resource-intensive but also a significant contributor to environmental degradation. The urgent need for a paradigm shift in how we produce food has never been more apparent. This book, *Cellular Agriculture for Revolutionized Food Production*, explores a revolutionary approach that promises to address these pressing challenges: cellular agriculture.

Cellular agriculture, the practice of producing agricultural products from cell cultures rather than whole plants or animals, represents a radical departure from conventional farming methods. It holds the potential to produce meat, dairy, and other essential food products in a way that is more efficient, sustainable, and humane. By cultivating cells in controlled environments, we can dramatically reduce the environmental footprint of food production, eliminate the need for vast tracts of farmland, and mitigate the ethical concerns associated with livestock farming.

This book delves into the science and technology underpinning cellular agriculture, presenting a comprehensive overview of its current and prospects. It highlights the innovative research and pioneering companies driving this field forward, transforming what once seemed like science fiction into tangible reality. From lab-grown meat to precision-fermented dairy, we explore the diverse applications of cellular agriculture and their implications for the global food system.

However, the transition to cellular agriculture is not without its challenges. This book also addresses the economic, regulatory, and societal hurdles that must be overcome to achieve widespread adoption. It discusses the need for supportive policies, public acceptance, and the development of robust supply chains. Furthermore, it considers the ethical questions surrounding this new technology and its potential impact on rural economies and traditional farming communities.

Cellular Agriculture for Revolutionized Food Production is not just a technical manual or an academic treatise but a call to action. It aims to inspire researchers, policymakers, entrepreneurs, and consumers to embrace this transformative approach to food production. By thoroughly understanding cellular agriculture's potential and the steps needed to realize it, this book seeks to catalyze the change required to ensure a sustainable and secure food future.

As you embark on this journey through the pages ahead, I hope you will be as captivated by the promise of cellular agriculture as I am. Together, we can envision and build a future where food production is both innovative and sustainable, meeting the needs of a growing population without compromising the health of our planet.

Ahmed M. Hamad
Benha University, Egypt

Tanima Bhattacharya
Lincoln University College, Malaysia

Chapter 1
Introduction to Cellular Agriculture

Wasiya Farzana
 https://orcid.org/0000-0003-2974-9771
Karunya Institute of Technology and Sciences, India

Gintu Sara George
 https://orcid.org/0009-0002-1082-011X
Karunya Institute of Technology and Sciences, India

ABSTRACT

Today, microbes grown in bioreactors provide leather and fibers for clothing, bags, and shoes, as well as flavors, sweeteners, and proteins from eggs and milk for human use. Additionally, components from plant cell and tissue cultures boost immunity and enhance skin texture; in vitro meat, a different precommercial cellular agriculture product, is currently the subject of a lot of interest. All of these strategies could help traditional agriculture continue to meet the nutritional needs of an expanding global population while releasing valuable resources like arable land. It will be extremely difficult to feed a growing population, and there is demand to lessen the damaging effects of traditional agriculture on the environment. The prospect of consuming plant cells presents a very alluring substitute for obtaining wholesome, high-protein, and nutritionally balanced food raw materials. Moreover, a variety of biosynthetic compounds can be produced from cultured bacteria.

DOI: 10.4018/979-8-3693-4115-5.ch001

INTRODUCTION

One of the major difficulties of the modern era is to simultaneously transform the food system to provide nutritious foods and environmental sustainability. Reaching these goals is essential to achieving the UN Sustainable Development Goals (SDGs). Even though, up until now, the world's food production has kept up with population growth, there remains a significant disparity in the amount of food produced. Additionally, there are problems with inadequate diets that result in micronutrient deficiencies as well as diet-related obesity (Willett et al., 2019). On its alone, conventional agriculture might not be able to handle the enormous challenges that lie ahead. By 2050, according to current predictions, the world's food needs will increase by 60%, but there will only be 2% additional arable agricultural land, or 40% of the overall land area, to meet this demand. Contrary to popular assumption, human activity has actually lowered the overall biomass on World approximately two instead of increased it. This includes domesticating cattle, adapting agriculture, and generating an enormous rise in the human population through the industrial revolution (Bar-On et al., 2018). Grazing and forest management are primarily to blame for this (Erb et al., 2017). Only about 2% of total plant biomass worldwide is made up of crops (Erb et al., 2017). "Cellular agriculture" refers to production of the agricultural commodities using cell cultures of all host species rather than raising livestock or crops (Mattick, 2018).

The challenges facing traditional agriculture are enormous right now. The demand for food along with the other agricultural goods are rising ultimately resulting the world's population growth, expected to increase to 9–11 billion people by 2050 (Röös et al., 2017). Additionally, forms a threat from climate change and a shortage of arable land. The Traditional agricultural productivity must increase dramatically in the near future in order to meet this goal. Cellular agriculture is a more environmentally friendly and sustainable method of producing agricultural products (Rischer et al., 2020). Compared to conventional farming methods, it uses less water and land and emits less greenhouse gases, making it a possible solution to this issue.

Producing agricultural goods that are molecularly comparable to those produced using conventional agricultural techniques is the main goal of cellular agriculture. In order to achieve this, cells from plants and animals and tissue cultures as well as microorganism cultures (such those of bacteria, yeasts, fungi, and algae). The end products can be either cellular or acellular; the latter is composed of cells from plants or animals, either living or dead, and is often not genetically altered (Rischer et al., 2020), the former is made up of organic molecules like protein from the proteins found in milk, silk, and eggs, as well as lipids and is usually generated using genetically modified microorganisms.

Regardless of the aforementioned factors, the initial stage of production is always conducted in a bioreactor, a closed, temperature-controlled glass, steel, as well as the plastic container that is agitated and aerated while cells and nutrients are combined. This in vitro production technique allows the use of productivity-maximizing parameters. However, since some production use organisms that grow relatively slowly and may be grown out of by contaminating microbes, it is imperative to make sure the entire process is carried out aseptically, especially when performing the transfer of the culture medium and the properly chosen producing strain and cell line to the bioreactor. Once a desirable level of biomass of cell and material titer is reached, the bioreactor's contents are removed, and the intended product which may include cells, tissue, protein, and secondary metabolites is then separated, purified, as well as if required, created. Complete monitoring of the production process is ensured by this closed production approach, leading to consistent and repeatable product quality. Furthermore, custom goods can be created by modifying the producing organism's metabolism.

EVOLUTION OF DEVELOPMENT OF CELLULAR AGRICULTURE

The method of creating agricultural goods is called cellular agriculture through the use of cell and tissue cultures, has seen significant advancements, some of which are summarized in Figure 1. The field of cellular agriculture has roots dating back to the early twentieth century, despite Isha Datar coining the phrase in 2015, (Stephens & Ellis, 2020). The discoveries of plant-based totipotency (v. Guttenberg, 1943), animal tissue and cell cultivation in a laboratory (Carrel, 1912), sterile fermentation methods, and recombinant bacterial DNA production (Cohen et al., 1973) established the scientific as well as technological basis of cellular agriculture.

Research on instead of using plants to manufacture secondary metabolites, use plant tissue as well as cell cultures had already begun to appear in the 1970s, and the first experiments in the lab using permanent lines of embryonic stem cells had started in 1981 (Evans & Kaufman, 1981). Quorn is from Marlow Foods, the initial single-cell protein, has been approved for consumption by humans along with released in the UK in 1985. Another significant outcome of research conducted in the 1980s was the acceptance of ginsenosides, the first secondary metabolites, and the red pigment shikonin (Malik et al., 2014), which are generated by plant cell and tissues cultures. Global tissue engineering efforts began around the same time to produce organ and cell substitutes for regenerative medicine. The early 2000s saw the initiation of tissue cultivation projects for food production in the US (Benjaminson et al., 2002) as well as Netherlands (Wilschut et al., 2008). The introduction of methods for tissue engineering was a key prerequisite for these projects.

Figure 1. The history of cellular agriculture (Eibl et al. 2021)

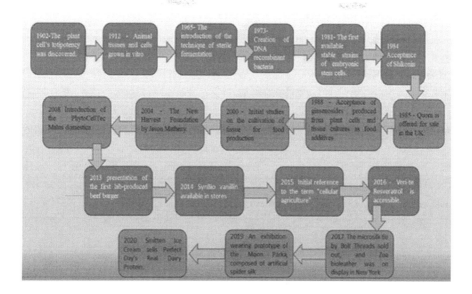

Another significant event occurred in 2008 when the Swiss company Mibelle biochemistry introduced PhytoCellTec Malus domestica, a product whose anti-aging efficacy was demonstrated supposedly inspired by plant stem cells and created utilizing human skin cells. Originating from plant-based cell cultures, it was the initial commercial extract and it completely changed the cosmetics industry. Later on, there was a fresh wave of uses for plant cells and tissue culture that produced a plethora of novel cosmetic goods (Eibl et al., 2018).

Furthermore, the market was introduced to yeast-based tastes (SynBio Vanillin) and sweeteners (Ever-Sweet), as well as milk proteins like those made by Perfect Day Foods, Bolt Thread along with the Spiber's spider-silk proteins, as well as the Modern Meadow's organic leather (Zoa) (García & Prieto, 2018). Furthermore, there has been and continues to be a great deal of research done on new strategies for bacterial, fungal, and algal single-cell protein production (Ritala et al., 2017).

Cell as well as tissue cultures from plants and animals are used as producing organisms by a large number of enterprises now creating products for cellular agriculture. Our evaluation emphasises plant tissue as well as cell extracts along with in vitro meat, as well as the related production procedures, for the food and cosmetics industries.

CELLULAR AGRICULTURE BASED ON PLANT CELL AND TISSUE CULTURE

Plant Tissue Culture and Cell Extractions on a Commercial Level

The goods, which are produced utilizing plant tissue and cell cultures rather than whole plants, comprise both acellular as well as cellular products. All product examples, with the exception of CBN Biotech's extract made from ginseng root (Murthy et al., 2014), have all been created to a level of market maturity in the last twelve years. The specific types of plants that are difficult to cultivate in the wild, endangered, or protected are among those used as source material; as a result, the direct advantage of cellular agriculture is that it shields these plant species from future exploitation.

The list includes extracts made in Europe that are utilised by the cosmetic industry to make medium- and high-priced personal care products, in along with extracts like ACETOS 10P, TEUPOL 10P, 50P, ECHINAN 4P, ECHIGENA PluS, as well as the TEOSIDE 10, utilized as ingredients in supplements for food and the supplements contain phenylpropanoids, demonstrated to have an immune system-stimulating impact and have received European approval as new foods (Dal Toso & Melandri, 2011). Following the Korean and US Food as well as the authorization of CBN Biotech's ginseng-derived extracts by the Drug Administration (Murthy et al., 2015), ginseng powder has been used in a range of goods.

Suspension cultures employing cells that have dedifferentiated (DDCs) along with preferred over root cultures for the production of plant cell as well as the tissue extracts is cambial meristematic cells (CMCs) because it is simpler to grow and handle. Additionally, there is a wider range of suitable bioreactors currently accessible for suspension cultures, and the procedure management is simpler to accomplish.

DEVELOPMENT OF PLANT CELL AS WELL AS THE EXTRACTS OF THE TISSUE CULTURE: DEVELOPING THE MANUFACTURING METHOD

The developmental steps for plant cell as well as the tissue culture extract products derived from nongenetically modified cells are depicted in Figure 2. The strong emphasis on nongenetically altered plant tissue and cells cultured commodities utilized by the food as well as cosmetics business is due to consumers' rejection of genetically engineered products, particularly in European countries. When creating procedures for plant cells as well as the tissue culture, the following assumptions must be followed:

- The plant variety has already been chosen;
- The necessary material from the plant (parent material containing a substantial amount of the intended result) is easily accessible;
- The tissue of explant/cellular origins (e.g., leaf, roots, etc.) has been selected;
- The part of the plant where the desired item collects is known;
- It has been determined what kind of plant-cultured tissues and cells will be created.

Figure 2. Significant advancements in the methods for producing extract of plant cell and tissue culture (Eibl et al. 2021)

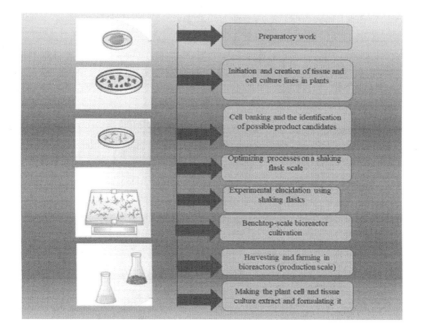

Establishing Cultures and Cell Banking

Utilizing sterile tweezers, scalpels, as well as a three-stage surface sterilization procedure that includes four washing stages along with chemical sterilization with a solution of hypochlorite, the isolation and fragmentation of explants are done under aseptic circumstances. Before sterilization, a brief ultrasonic therapy may be administered to increase the effectiveness of the sterilizing process. The sterilized explants are then used to induce a number of plant cultured cell as well as the tissue lines.

Plant tissue and cell lines are mass-propagated at temperatures ranging from 23°C to 27°C in either a liquid environment (shaking flasks) or in a solid environment (Petri dishes). Plant tissue and cell culture candidates can be identified for further optimization operations in shaking flasks as well as bioreactors by tracking their growth and product creation behavior. The likelihood of somaclonal changes in DDC-based suspension cultures is decreased and laboratory cell maintenance is made simpler by creating cell banks from plant tissue and cell culture candidates. Additionally, cell banking generally streamlines product approval and procedure patenting.

The preferred technique for further cryopreservation is a gradual, controlled-rate freezing procedure, which is also applied to animal cells. The procedure, that depends on the cryoprotectant chemicals to permeate cells, works best for CMC-based suspension cultures because they grow as single cells and have fewer vacuoles than DDC- suspension cultures (Lee et al., 2010).

Optimizing Processes, Such as Elicitation

Shaking flasks are typically used for the process optimization for plant tissue and cell lines, which aims to boost cell biomass as well as product concentrations. It comprises studies on the effects light (Yue et al., 2014) and/or elicitors (Halder et al., 2019) on cell growth and product formation (Tassoni et al., 2012), as well as media composition (phytohormones, phosphate, nitrogen, and sucrose) and optimization of parameters (temperature, pH value, mixing, aeration). Elicitation is thought to be the best technique for boosting the accumulation of secondary metabolite products, as it can even cause product secretion in root and suspension cultures. The activation of gene expression, which is commonly induced by the addition of foreign molecules or elicitors, aids in stimulating the creation of secondary metabolites. Multiparametric outcome analyses and Design of Experiment (DoE) techniques help the design as well as the evaluation of trials for the optimization of process (Rasche et al., 2016). The production of biomass of a pear suspension of cells culture was increased by more than twofold in just five weeks (Rasche et al., 2016) by utilizing solely a DoE technique for each of the four cultivating factors (light, temperatures, incubating time, and inoculum density). Furthermore, they achieved a more than 50% reduction in the expenses of items sold per kilogram of biomass.

Regarding product safety, it is crucial to make sure that no elicitors or hazardous media components are utilized. Since their usage is prohibited in the food as well as cosmetic industries, they are used in the synthesis of plant tissue as well as cell culture extracts. Similarly, natural phytohormones like zeatin should be added into the growth medium in place of synthetic phytohormones. When it comes to elicitors, it is advised to use salicylic acid or methyl jasmonate, or to expose the cells to light sources like UV light (Murthy et al., 2015).

Cultivation of Bioreactors

Following the development and description of the procedure in shaken flasks, it might be ramped up to production size after being characterized in benchtop-scale bioreactors, like those with a 1–10 L operating volume. The optimal kind of bioreactor is selected based on the type of plant tissue as well as cell culture, the amount of biomass and target product to be generated at production scale, along with additional considerations. The production scale used in food as well as cosmetic applications is the two-digit liter range (such as., product on the gramme level) as opposed to a single-digit cubic meter level (such as., products on a kilogram scale). Stirred bioreactors are the preferable method for suspension cultures when manufacturing at the cubic meter level is necessary. The height to diameter ratio of these stirred bioreactors ought to be 2:1 or 3:1. Usually, they have distributed power input impellers, axial impellers, or a combination of an axial impeller as well as a Rushton turbine. Baffles and a ring sparger are also added to the bioreactor tank.

Due to their higher height: diameter ratios such as 6:1 or 14:1pneumatically driven bubble-column bioreactors tend to produce substantially stronger wall-growth phenomena during plant cell cultivation than stirred bioreactors. By altering the shape of a bubble column bioreactor such that it looks like a pear-shaped container, to reduce these unwanted effects; this type of bioreactor is known as a balloon-type bubble bioreactor (Paek et al., 2005) For the purpose of producing ginsenoside for commercial use using adventitious root cultures, a balloon-type bubbles bioreactor with a capacity of 10 m^3 is now in operation.

When working quantities up to 100 L are sufficient, the suspension cultures can be grown in wave-mixed bioreactors which are stirred along a single axis. The bioreactor is favored by producers of plant cultured cells and tissues extract, like Mibelle Biochemistry, because of how simple it is to use. CerCell, GE Healthcare (now Cytiva), Sartorius, as well as ThermoFisher Scientific all have versions of this sort of bioreactor available. They consist of a sterile, usable polyethylene bag, a control device, and a rocker. The pouch is decontaminated as well as thrown away after one usage.

The single-use bag is fastened to a rocker, agitates it along one axis that is, moves it in both directions after being filled with culture media and injected with cells. This creates a wave that moves through the bag back and forth, allowing for surface aeration without bubbles and mixing. The filling volume, rocking angle, and rocking rate can all be used to modify the power input. The medium surface is constantly renewed by this rocking motion, which also continually dissolves any foam generated during cultivation. Thus, the use of an antifoam agent for foam control is usually not required, unlike cultivations in stirred and also bubble-column bioreactors, where the antifoam agent inhibits further downstream processing.

A suspension of cultures can be developed in wave-mixed bioreactors that are swirled along a single axis when working capacities up to 100 L are adequate. Because of its ease of use, this type of bioreactor is preferred by companies that make extract from plant cultivated cells and tissues, such as Mibelle Biochemistry. This type of bioreactor is offered from CerCell, GE Healthcare (formerly Cytiva), Sartorius, and ThermoFisher Scientific. They are made up of a rocker, a control device, as well as a sterile, reusable polyethylene bag. After one use, the pouch is disposed of and decontaminated.

Harvesting, Formulating, and Downstreaming-Processing

Harvesting the content is the last step in the mass growth of plants tissue and cell culture in industrial bioreactors. The creation of a liquid or powder extract comes next. Basic process engineering procedures can be used, based on the type of plant tissue as well as cell culture, along with the product's planned usage involving plant cells and tissue culture. These include, for example, extraction processes (such as the soxhlet extraction, hydro distillation, the maceration process pressurized fluid extraction, ultrasonic-assisted extraction, filtration and separation, and the drying processes (such as microwave-assisted drying, vacuum drying, forced air drying, and freeze drying) (Fierascu et al., 2020).

Furthermore, it might be required to perform chromatographic purification procedures that, in case of intracellular products, disrupt cells (Barbulova et al., 2014). Manufacturer-specific protocols and patents, certain extracts from plant cells and tissue cultures are utilized as components in cosmetic products. For example, the PhytoCellTec product line's liquid extract is made by homogenizing cells under high pressure and then incorporating liposomes, phenoxyethanol, as well as antioxidants into the broth of the bioreactor that contains the cells.

CHALLENGES AND NOVEL APPROACHES

The list of brands that sell plant tissue and cell culture extracts aside of those used for a cosmetic products is still rather small, despite the fact that the functioning techniques for creating and managing plant cell as well as tissue cultures, widely known and commercially accessible, as are the required instruments and chemically defined culture medium. This is probably because the circumstances in the food and cosmetics industries are not exactly the same. Operational effectiveness is the most crucial factor for product commercialization in the food manufacturing industry. Cell growth as well as product titer have a significant impact on process productivity.

It's also important to note that research on DDC-based suspension cultures revealed that the product titer decreased as the culture grew older (Qu et al., 2005), plant cultured cells and tissues herbal extracts can work synergistically, and levels utilized in final products for cosmetics are relatively low, making a low product titer for these extracts seem less problematic (Barbulova et al., 2014). The more stringent product approval procedures in the food industry undoubtedly contribute to the greater percentage of plant tissue and cell extracts utilized in cosmetic products. While product safety in the manufacture of cosmetic ingredients is solely the responsibility of the manufacturing company (Zappelli et al., 2016), approval from a national authority with the necessary expertise (such as the U.S. FDA) is necessary for manufacturing food or food additives. The planned use (e.g., food, dietary supplements, taste component) must be approved beforehand is taken into consideration during the biosafety and toxicological evaluations, which additionally encompass pharmacokinetic as well as clinical studies.

If plant tissue and cell culture-based food ingredients are utilized to create luxury goods or high-value superfoods that consumers are prepared to pay more for, then the greater cost of manufacture and/or approval of these ingredients may not be as troublesome. For instance, luxury cuisines have a unique flavor or texture. Theobroma cacao suspensions grown within a wave-mixed bioreactor using DDC technology were utilized to create cocoa powder and, in the end, a prototype chocolate with an original flavor (Eibl et al., 2018).

In the future, a customer's home kitchen can create 500 g of fresh weight of cell biomass in less than a week using a specifically constructed bioreactor called VTT's Home Bioreactor, which functions similarly to a Nespresso machines. Research on the nutritional makeup of berry biomass generated in vitro revealed that the biomass had a fresh flavor and aroma, and that its composition was ideal for human nutrition, with 21–37% fiber in the diet, an appropriate lipid quality, an equilibrium amino acid profile, 18–33% sugars as those 14–19% protein molecules, and 0.3–1% starch (Nordlund et al., 2018).

CELLULAR AGRICULTURE BASED ON ANIMAL CELL AS WELL AS TISSUE CULTURE

Products in Progress

The surrogate meat, including fish meat, is currently the main output of cellular agriculture using animal tissue as well as cell culture (Rubio et al., 2019). Skeletal muscle makes up the majority of edible meat. Other cell types that enhance the flavour, texture, and overall sensory experience of food of edible meat include fibroblasts, blood vessels, RBC, adipocytes, endothelial cells, leukocytes, and connective tissue. Creating a three-dimensional structure is the aim of animal cell as well as the tissue cultured meat, known to be artificial meat, unadulterated meat, as well as in vitro meat. that has a sensory and nutritional profile that is identical to the original.

MesTech, an Israeli business, is now pursuing a 3D or 4D bioprinting method that might produce highly organised meat by using a hydrogel that contains cells and a bioprinter. While the field of regenerative medicine has reported the first results with 3D-printed tissues/organs (Rana Khalid et al., 2019), it seems as though the industry of 3D printing food will be the pioneer. There might be a more genuine flavor experience and fewer obstacles to overcome.

Similar to organ printing, extremely three-dimensionally organized meat, such steaks, can be produced via tissue culture or self-organizing techniques enlarge the muscles that were taken out of the goldfish skeleton explants by combining it with different compositions of culture medium (Benjaminson et al., 2002). The authors were able to increase the surface area by 13% to 79% in just one week. The cultured tissue looked like a fillet of fish, and it received high ratings from a sensory panel. Meat made using the self-organizing process does, however, lack characteristics like fat marbling and vascularization. The same holds true for meat derived from scaffolding or cell culture, despite the fact that scaling up the production procedure is simpler.

Meat Derived From Animal Cell Culture (Scaffolding Technique)

Although the scaffolding technique approach is more suited for processed meat-based goods like sausages, hot dogs, burgers, and meatballs, numerous developers of meat from culture prefer it. Because the final culture cells are barely 0.5 mm thick, meat grown in this manner does not have a hard feel. Figure 3 demonstrates a standard process (with six phases) for creating in vitro meat utilising the scaffolding approach. The following processes are involved: (a) biopsy; (b) isolation and banking of cells; (c) cell proliferation; (d) cell differentiation; (e) harvesting myofibers; as well as (f) processing of meat.

Figure 3. Diagram illustrating the scaffolding technique used to produce myofibers from satellite cells (Eibl et al., 2021)

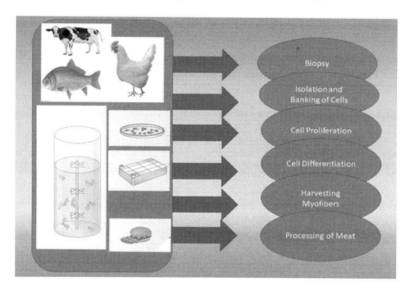

The Production Procedure

Muscle stem cells that are commonly used are called mononuclear satellite cells, or myosatellite cells (Bhat et al., 2015). Satellite cells can be successfully separated using techniques that are accessible, and they were taken out of numerous animal species' muscle tissue, including fish, poultry, pigs, lambs, and turkeys. Extracellular cells are created while retaining their stemness after cell isolation and banking, and they differentiate from myotubes into myofibers. Static culturing methods or dynamic systems for cultivation that promote satellite cell adhesion can be used for cell isolation, banking, and expansion (Bodiou et al., 2020).

It is important to remember that Petri dishes as well as multitray methods like CellFactories, which are static culture systems, have the significant drawback of having less surface area for differentiation and growth of cells, which might lead to oxygen limits, and they also lack instrumentation. Stated differently, myofiber output cannot be efficiently produced at a commercial scale using static culture technologies. For this use, mechanically as well as hydraulically powered bioreactors having integrated scaffolds that have controllable temperature, pH levels, and dissolved oxygen content are preferable (also refer to the section on dynamic bioreactor types as well as scaffolds). The best scaffolds for satellite cell growth or differentiation should mimic in vivo settings.

Large surface areas are important, but so are flexible, non-animal materials that permit contractions and maximize medium diffusion, additionally make it possible for the myofiber product to be easily separated during harvest (Engler et al., 2004). The presence of edible as well as dissolvable scaffolds in the final myofiber product reduces the requirement for the myofibers to be mechanically or enzymatically separated in order to remove the scaffold. Furthermore, edible scaffolds could be incorporated into the product to offer extra texture. The processing of meat is the final stage of the meat-producing process, which yields the finished product.

Different Kinds of Stem Cells

To create cultured flesh, besides the varieties of satellite cells already described. Additionally, embryonic stem cells, pluripotent stem cells that have been induced, as well as mesenchymal stem cells may be used. Unlike cell satellites, which have a finite potential, embryonic stem cells have an endless ability for multiplication (Bach et al., 2003). But when cell generations increase, mutations may happen, and particular stimulation is required for the required differentiation into muscle cells (Amit et al., 2000). Additionally, it is well acknowledged that creating embryonic stem lines is challenging, particularly for ungulates (Keefer et al., 2007), as well as is doubt about whether or not consumers will accept the use of these cells to produce food.

Utilizing pluripotent stem cells that have been induced, which are created through transforming somatic cells of adults in a lab, is a promising and more modern method (Stanton et al., 2018). . Less research has been done on the scaling-up of generated pluripotent stem cell procedures than on mesenchymal stem cells produced from adipose tissue, which are also thought to be viable options for producing meat in vitro. According to Nogueira et al. (2019), the upper limit of the operating volume for cultures that contain generated pluripotent stem cells is currently limited to three-digit millilitres. Conversely, mesenchymal stem cells generated from adipose tissue have been grown to a performing volume of 35 L with success (Schirmaier et al., 2014).

It is simple to separate multipotent mesenchymal stem cells generated from subcutaneous fat in adipose tissue and induce them to develop into adipogenic, myogenic, osteogenic, as well as chondrogenic cells (Kim et al., 2006). As opposed to how muscle tissue is separated from satellite cells, theirs is thought to be a less intrusive process. Mesenchymal lines of stem cells produced from adipose tissue are thought to be harmless as long as they do not develop for lengthy periods of time (4-5 months), as doing so may result in undesirable alterations (Lazennec & Jorgensen, 2008).

Media of Culture

A proper medium for culture free of antibiotics as well as components of animal is necessary cell multiplication as well as differentiation, regardless of the kind of stem cell utilized. Since sterile workbenches, sterilization procedures, and since the last 25 years, filtering techniques have greatly evolved, including the use of streptomycin and penicillin. to avoid bacterial contamination is no longer necessary. The usage of media compositions lacking animal ingredients, like serum, is more challenging. It has been demonstrated that the application of horse serum or foetal calf serum is beneficial for both phases of the development of skeletal muscle (Stephens et al., 2018). Additionally, these serum includes components that shield the cells from extreme stress and promote cell development and differentiation. Despite this, there are still safety and ethical questions about the usage, consistency, and manufacture of serum (Brunner, 2010).

Expertise has been gathered in the serum-free production as well as the chemically specified media formulation for the cell therapies and biopharmaceutical manufacturing processes (Sagaradze et al., 2019). The commercial medium and additives used in these formulations, however, are not impacted by the same budgetary constraints as food applications based on animal tissues and cells in culture. Biotherapeutics comprise human mesenchymal stem cell-based cell therapies or monoclonal antibodies made from Chinese hamster ovary cells, are valuable products that occasionally become blockbuster medications. In this case, higher cultural media costs are probably going to be approved, particularly given the higher standards for the product and related processes (pharma grade versus food grade). The Culture medium is thought to be the primary cost factor in commercial vitro meat manufacturing.

Types and Forms of Dynamic Bioreactors

In in vitro meat production efficiency is also greatly enhanced by the expansion as well as the differentiation of cells using bioreactors. The growth, differentiation, and morphology of the type of stem cell being used should be taken into consideration while choosing the right kinds of bioreactors. Because not all bioreactor types are available without size restrictions, working volume as well as scaffold type should additionally be considered. Moreover, not all bioreactor types can be used with all types of scaffolds without creating new difficulties (Cesarz & Tamama, 2016)

The stirred reactors (Rafiq et al., 2018), rotating wall as well as the bed bioreactors (Varley et al., 2017), wave-mixed bioreactors (Shekaran et al., 2016), hollow fiber bioreactors (Lechanteur, 2014), and fixed bed bioreactors are some examples of bioreactor types. With the exception of stirred reactors, when it comes to executing manufacturing procedures on a cubic metre scale, all one of these bioreactor models have scaled-up constraints. For the competitive production of in vitro meat, the stirred bioreactor serves as the preferred system. Microcarriers, which are spherical substances that resemble beads and provide cell growth surfaces, are useful scaffolds for stirred reactors. The Bovine satellite cells successfully multiplied on Synthemax, CellBind, and Cytodex 1 microcarriers in spinner flasks (Verbruggen et al., 2017). Furthermore, scientists were able to expand the cells' growth surface by only introducing fresh microcarriers as well as Utilising transmission between beads. The mesenchymal stem cells and satellite cells exhibit comparable behaviors (Verbruggen et al., 2017). Specifically, mesenchymal stem cells are employed to generate cell therapies in suspension bioreactors, can be cultured in a variety of microcarriers (Jossen et al., 2018).

The surface charge, diameter, density, and core material of these microcarriers vary. The microcarrier that is most appropriate for the culture task depends, in part, on the media composition, based on our experience with microcarrier-based procedures for cell therapies based on mesenchymal stem cells. This means that in order to create novel culture medium compositions for in vitro meat, it is also necessary to incorporate new scientifically specified microcarriers that comply with food legislation. In order to save capital expenditure, performing differentiation and proliferation of cells in the exact same bioreactor with the same microcarrier (Bodiou et al., 2020). This is a problem because the two process phases have different needs for the medium and physical environment, necessitating a Alteration of the medium (between differentiation and growth media). Additionally, shear stress must be used to alter the culturing settings in order to promote cell differentiation. Using the microcarrier that's edible, that is a substitute option for the two-phase bioreactor process and has no detrimental sensory effects on the finished product. Therefore, the texture and flavor might also be enhanced by using the microcarrier design. Alternatively, For a single-phase procedure with two bioreactors, the proliferation step requires only one separable and biodegradable microcarrier (Bodiou et al., 2020).

CONCLUSION

Numerous prospects exist for the environmentally friendly manufacture of agricultural goods through cellular agriculture. Western markets are the target of the initial items as well as continuous product developments. As producing organisms, tissue and cell cultures of plants and animals are crucial. The requisite technical equipment and expertise in process optimization, product approval, and culture establishment are provided. It is likely that newly developed products continue in having an influence on the cosmetics industry specifically, given the frequently low levels of secondary metabolites in this context. If novel food components derived from plants cell as well as the tissue culture are to compete with those conventionally made from or from plants, they should prioritize improving product titer. As an alternative, high-value goods with advantageous features catered to the user might be pursued, such as upscale cuisine, customized cuisine, or entirely novel techniques that enable food preparation using plant tissue and cell cultures at home. On the other hand, animal tissue and cell culture-based cellular agriculture is still at its infancy.

It is evident that the production of meat including fish is the current priority. This is to be expected, since by 2050, it is anticipated that meat consumption would quadruple globally (Gaydhane et al., 2018). Nowadays, the scaffolding approach and satellite cells are frequently used in the vitro meat production process. The first in vitro meat product prototypes have been shown thus far, and the first commercial items are anticipated to be released within the next two years. Many basic questions remain to be addressed if animal tissue and cell culture-based agriculture using cells will be competing with conventional production of meat within the years to come. Chemically specified culture media, scaffolds and bioreactors, various process modes and controls (feed batches or continuous, expansion as well as differentiation, regeneration of medium elements and microcarriers, within the same bioreactor or separately, and regulatory concerns are the primary topics of concern. There could be valuable insights for the food industry from the biopharmaceutical sector. Even though the food business operates on far bigger scales and with fewer rigorous regulations than the pharmaceutical industry, It will be essential for its activities to start making more financial sense. New educational initiatives and training programmes should be developed, experts in cell culture technology for agricultural products should be produced, and facility design should incorporate the fundamental knowledge of biochemical engineering, food processing, tissue as well as cell culture technology, quality control, and legislation. Beyond all technical constraints, consumer acceptance which may ultimately come down to taste is what determines the viability for in vitro meat along with related goods.

REFERENCES

Amit, M., Carpenter, M. K., Inokuma, M. S., Chiu, C. P., Harris, C. P., Waknitz, M. A., Itskovitz-Eldor, J., & Thomson, J. A. (2000, November). Clonally Derived Human Embryonic Stem Cell Lines Maintain Pluripotency and Proliferative Potential for Prolonged Periods of Culture. *Developmental Biology*, 227(2), 271–278. 10.1006/dbio.2000.991211071754

Bach, A., Stern-Straeter, J., Beier, J., Bannasch, H., & Stark, G. (2003, October). Engineering of muscle tissue. *Clinics in Plastic Surgery*, 30(4), 589–599. 10.1016/S0094-1298(03)00077-414621307

Bar-On, Y. M., Phillips, R., & Milo, R. (2018, May 21). The biomass distribution on Earth. *Proceedings of the National Academy of Sciences of the United States of America*, 115(25), 6506–6511. 10.1073/pnas.171184211529784790

Barbulova, A., Apone, F., & Colucci, G. (2014, April 22). Plant Cell Cultures as Source of Cosmetic Active Ingredients. *Cosmetics*, 1(2), 94–104. 10.3390/cosmetics1020094

Barbulova, A., Apone, F., & Colucci, G. (2014, April 22). Plant Cell Cultures as Source of Cosmetic Active Ingredients. *Cosmetics*, 1(2), 94–104. 10.3390/cosmetics1020094

Benjaminson, M., Gilchriest, J., & Lorenz, M. (2002, December). In vitro edible muscle protein production system (mpps): Stage 1, fish. *Acta Astronautica*, 51(12), 879–889. 10.1016/S0094-5765(02)00033-412416526

Benjaminson, M., Gilchriest, J., & Lorenz, M. (2002, December). In vitro edible muscle protein production system (mpps): Stage 1, fish. *Acta Astronautica*, 51(12), 879–889. 10.1016/S0094-5765(02)00033-412416526

Bhat, Z. F., Kumar, S., & Bhat, H. F. (2015, May 5). In vitro meat: A future animal-free harvest. *Critical Reviews in Food Science and Nutrition*, 57(4), 782–789. 10.1080/10408398.2014.92489925942290

Bodiou, V., Moutsatsou, P., & Post, M. J. (2020, February 20). Microcarriers for Upscaling Cultured Meat Production. *Frontiers in Nutrition*, 7.32154261

Brunner, D. (2010). Serum-free cell culture: The serum-free media interactive online database. *ALTEX*, 53–62. 10.14573/altex.2010.1.5320390239

Campuzano, S., & Pelling, A. E. (2019, May 17). Scaffolds for 3D Cell Culture and Cellular Agriculture Applications Derived From Non-animal Sources. *Frontiers in Sustainable Food Systems*, 3, 3. 10.3389/fsufs.2019.00038

Carrel, A. (1912, May 1). On the permanent life of tissues outside of the organism. *The Journal of Experimental Medicine*, 15(5), 516–528. 10.1084/jem.15.5.51619867545

Cesarz, Z., & Tamama, K. (2016). Spheroid Culture of Mesenchymal Stem Cells. *Stem Cells International*, 2016(1), 1–11. 10.1155/2016/917635726649054

Cohen, S. N., Chang, A. C. Y., Boyer, H. W., & Helling, R. B. (1973, November). Construction of Biologically Functional Bacterial Plasmids In Vitro. *Proceedings of the National Academy of Sciences of the United States of America*, 70(11), 3240–3244. 10.1073/pnas.70.11.32404594039

Dal Toso, R., & Melandri, F. (2011, January). Echinacea angustifolia cell culture extract. *NUTRAfoods : International Journal of Science and Marketing for Nutraceutical Actives, Raw Materials, Finish Products*, 10(1), 19–24. 10.1007/BF03223351

Eibl, R., Meier, P., Stutz, I., Schildberger, D., Hühn, T., & Eibl, D. (2018, August 11). Plant cell culture technology in the cosmetics and food industries: Current state and future trends. *Applied Microbiology and Biotechnology*, 102(20), 8661–8675. 10.1007/s00253-018-9279-830099571

Engler, A. J., Griffin, M. A., Sen, S., Bönnemann, C. G., Sweeney, H. L., & Discher, D. E. (2004, September 13). Myotubes differentiate optimally on substrates with tissue-like stiffness. *The Journal of Cell Biology*, 166(6), 877–887. 10.1083/jcb.20040500415364962

Erb, K. H., Kastner, T., Plutzar, C., Bais, A. L. S., Carvalhais, N., Fetzel, T., Gingrich, S., Haberl, H., Lauk, C., Niedertscheider, M., Pongratz, J., Thurner, M., & Luyssaert, S. (2017, December 20). Unexpectedly large impact of forest management and grazing on global vegetation biomass. *Nature*, 553(7686), 73–76. 10.1038/nature2513829258288

Evans, M. J., & Kaufman, M. H. (1981, July). Establishment in culture of pluripotential cells from mouse embryos. *Nature*, 292(5819), 154–156. 10.1038/292154a07242681

Fierascu, R. C., Fierascu, I., Ortan, A., Georgiev, M. I., & Sieniawska, E. (2020, January 12). Innovative Approaches for Recovery of Phytoconstituents from Medicinal/Aromatic Plants and Biotechnological Production. *Molecules (Basel, Switzerland)*, 25(2), 309. 10.3390/molecules2502030931940923

García, C., & Prieto, M. A. (2018, August 22). Bacterial cellulose as a potential bioleather substitute for the footwear industry. *Microbial Biotechnology*, 12(4), 582–585. 10.1111/1751-7915.1330630136366

vGuttenberg, H. (1943, July). Kulturversuche mit isolierten Pflanzenzellen. *Planta*, 33(4), 576–588. 10.1007/BF01916543

Halder, M., Sarkar, S., & Jha, S. (2019, July 25). Elicitation: A biotechnological tool for enhanced production of secondary metabolites in hairy root cultures. *Engineering in Life Sciences*, 19(12), 880–895. 10.1002/elsc.20190005832624980

Jossen, V., van den Bos, C., Eibl, R., & Eibl, D. (2018, March 22). Manufacturing human mesenchymal stem cells at clinical scale: Process and regulatory challenges. *Applied Microbiology and Biotechnology*, 102(9), 3981–3994. 10.1007/s00253-018-8912-x29564526

Keefer, C., Pant, D., Blomberg, L., & Talbot, N. (2007, March). Challenges and prospects for the establishment of embryonic stem cell lines of domesticated ungulates. *Animal Reproduction Science*, 98(1–2), 147–168. 10.1016/j.anireprosci.2006.10.00917097839

Kim, M., Choi, Y. S., Yang, S. H., Hong, H. N., Cho, S. W., Cha, S. M., Pak, J. H., Kim, C. W., Kwon, S. W., & Park, C. J. (2006, November). Erratum to "Muscle regeneration by adipose tissue-derived adult stem cells attached to injectable PLGA spheres" [Biochem. Biophys. Res. Commun. 348 (2006) 386–392]. *Biochemical and Biophysical Research Communications*, 350(2), 499. 10.1016/j.bbrc.2006.09.05115369779

Lazennec, G., & Jorgensen, C. (2008, April 3). Concise Review: Adult Multipotent Stromal Cells and Cancer: Risk or Benefit? *Stem Cells (Dayton, Ohio)*, 26(6), 1387–1394. 10.1634/stemcells.2007-100618388305

Lechanteur, C. (2014). Large-Scale Clinical Expansion of Mesenchymal Stem Cells in the GMP-Compliant, Closed Automated Quantum® Cell Expansion System: Comparison with Expansion in Traditional T-Flasks. *Journal of Stem Cell Research & Therapy*, 04(08). Advance online publication. 10.4172/2157-7633.1000222

Lee, E. K., Jin, Y. W., Park, J. H., Yoo, Y. M., Hong, S. M., Amir, R., Yan, Z., Kwon, E., Elfick, A., Tomlinson, S., Halbritter, F., Waibel, T., Yun, B. W., & Loake, G. J. (2010, October 24). Cultured cambial meristematic cells as a source of plant natural products. *Nature Biotechnology*, 28(11), 1213–1217. 10.1038/nbt.169320972422

Malik, S., Bhushan, S., Sharma, M., & Ahuja, P. S. (2014, October 16). Biotechnological approaches to the production of shikonins: A critical review with recent updates. *Critical Reviews in Biotechnology*, 36(2), 327–340. 10.3109/07388551.2014.96100325319455

Mattick, C. S. (2018, January 2). Cellular agriculture: The coming revolution in food production. *Bulletin of the Atomic Scientists*, 74(1), 32–35. 10.1080/00963402.2017.1413059

Murthy, H. N., Georgiev, M. I., Kim, Y. S., Jeong, C. S., Kim, S. J., Park, S. Y., & Paek, K. Y. (2014, May 25). Ginsenosides: Prospective for sustainable biotechnological production. *Applied Microbiology and Biotechnology*, 98(14), 6243–6254. 10.1007/s00253-014-5801-924859520

Murthy, H. N., Georgiev, M. I., Park, S. Y., Dandin, V. S., & Paek, K. Y. (2015, June). The safety assessment of food ingredients derived from plant cell, tissue and organ cultures: A review. *Food Chemistry*, 176, 426–432. 10.1016/j.foodchem.2014.12.07525624252

Nogueira, D. E. S., Rodrigues, C. A. V., Carvalho, M. S., Miranda, C. C., Hashimura, Y., Jung, S., Lee, B., & Cabral, J. M. S. (2019, September 14). Strategies for the expansion of human induced pluripotent stem cells as aggregates in single-use Vertical-Wheel™ bioreactors. *Journal of Biological Engineering*, 13(1), 74. 10.1186/s13036-019-0204-131534477

Nordlund, E., Lille, M., Silventoinen, P., Nygren, H., Seppänen-Laakso, T., Mikkelson, A., Aura, A. M., Heiniö, R. L., Nohynek, L., Puupponen-Pimiä, R., & Rischer, H. (2018, May). Plant cells as food – A concept taking shape. *Food Research International*, 107, 297–305. 10.1016/j.foodres.2018.02.04529580489

Paek, K. Y., Chakrabarty, D., & Hahn, E. J. (2005, June). Application of bioreactor systems for large scale production of horticultural and medicinal plants. *Plant Cell, Tissue and Organ Culture*, 81(3), 287–300. 10.1007/s11240-004-6648-z

Qu, J., Zhang, W., Yu, X., & Jin, M. (2005, April). Instability of anthocyanin accumulation in Vitis vinifera L. var. Gamay Fréaux suspension cultures. *Biotechnology and Bioprocess Engineering; BBE*, 10(2), 155–161. 10.1007/BF02932586

Rafiq, Q. A., Ruck, S., Hanga, M. P., Heathman, T. R., Coopman, K., Nienow, A. W., Williams, D. J., & Hewitt, C. J. (2018, July). Qualitative and quantitative demonstration of bead-to-bead transfer with bone marrow-derived human mesenchymal stem cells on microcarriers: Utilising the phenomenon to improve culture performance. *Biochemical Engineering Journal*, 135, 11–21. 10.1016/j.bej.2017.11.005

Rana Khalid, I., Darakhshanda, I., & Rafi, R. (2019, July 5). 3D Bioprinting: An attractive alternative to traditional organ transplantation. *Archive of Biomedical Science and Engineering, 5*(1), 7–18.

Rasche, S., Herwartz, D., Schuster, F., Jablonka, N., Weber, A., Fischer, R., & Schillberg, S. (2016, March 18). More for less: Improving the biomass yield of a pear cell suspension culture by design of experiments. *Scientific Reports, 6*(1), 23371. 10.1038/srep2337126988402

Rischer, H., Szilvay, G. R., & Oksman-Caldentey, K. M. (2020, February). Cellular agriculture — Industrial biotechnology for food and materials. *Current Opinion in Biotechnology, 61*, 128–134. 10.1016/j.copbio.2019.12.00331926477

Ritala, A., Häkkinen, S. T., Toivari, M., & Wiebe, M. G. (2017, October 13). Single Cell Protein—State-of-the-Art, Industrial Landscape and Patents 2001–2016. *Frontiers in Microbiology, 8*.

Röös, E., Bajželj, B., Smith, P., Patel, M., Little, D., & Garnett, T. (2017, November). Greedy or needy? Land use and climate impacts of food in 2050 under different livestock futures. *Global Environmental Change, 47*, 1–12. 10.1016/j.gloenvcha.2017.09.001

Rubio, N., Datar, I., Stachura, D., Kaplan, D., & Krueger, K. (2019, June 11). Cell-Based Fish: A Novel Approach to Seafood Production and an Opportunity for Cellular Agriculture. *Frontiers in Sustainable Food Systems, 3*, 3. 10.3389/fsufs.2019.00043

Sagaradze, G., Grigorieva, O., Nimiritsky, P., Basalova, N., Kalinina, N., Akopyan, Z., & Efimenko, A. (2019, April 3). Conditioned Medium from Human Mesenchymal Stromal Cells: Towards the Clinical Translation. *International Journal of Molecular Sciences, 20*(7), 1656. 10.3390/ijms2007165630987106

Schirmaier, C., Jossen, V., Kaiser, S. C., Jüngerkes, F., Brill, S., Safavi-Nab, A., Siehoff, A., van den Bos, C., Eibl, D., & Eibl, R. (2014, March 20). Scale-up of adipose tissue-derived mesenchymal stem cell production in stirred single-use bioreactors under low-serum conditions. *Engineering in Life Sciences, 14*(3), 292–303. 10.1002/elsc.201300134

Shekaran, A., Lam, A., Sim, E., Jialing, L., Jian, L., Wen, J. T. P., Chan, J. K. Y., Choolani, M., Reuveny, S., Birch, W., & Oh, S. (2016, October). Biodegradable ECM-coated PCL microcarriers support scalable human early MSC expansion and in vivo bone formation. *Cytotherapy, 18*(10), 1332–1344. 10.1016/j.jcyt.2016.06.01627503763

Stanton, M. M., Tzatzalos, E., Donne, M., Kolundzic, N., Helgason, I., & Ilic, D. (2018, September 24). Prospects for the Use of Induced Pluripotent Stem Cells in Animal Conservation and Environmental Protection. *Stem Cells Translational Medicine*, 8(1), 7–13. 10.1002/sctm.18-004730251393

Stephens, N., Di Silvio, L., Dunsford, I., Ellis, M., Glencross, A., & Sexton, A. (2018, August). Bringing cultured meat to market: Technical, socio-political, and regulatory challenges in cellular agriculture. *Trends in Food Science & Technology*, 78, 155–166. 10.1016/j.tifs.2018.04.01030100674

Stephens, N., & Ellis, M. (2020, October 12). Cellular agriculture in the UK: A review. *Wellcome Open Research*, 5, 12. 10.12688/wellcomeopenres.15685.132090174

Tassoni, A., Durante, L., & Ferri, M. (2012, May). Combined elicitation of methyl-jasmonate and red light on stilbene and anthocyanin biosynthesis. *Journal of Plant Physiology*, 169(8), 775–781. 10.1016/j.jplph.2012.01.01722424571

Varley, M. C., Markaki, A. E., & Brooks, R. A. (2017, June). Effect of Rotation on Scaffold Motion and Cell Growth in Rotating Bioreactors. *Tissue Engineering. Part A*, 23(11–12), 522–534. 10.1089/ten.tea.2016.035728125920

Verbruggen, S., Luining, D., van Essen, A., & Post, M. J. (2017, May 3). Bovine myoblast cell production in a microcarriers-based system. *Cytotechnology*, 70(2), 503–512. 10.1007/s10616-017-0101-828470539

Willett, W., Rockström, J., Loken, B., Springmann, M., Lang, T., Vermeulen, S., Garnett, T., Tilman, D., DeClerck, F., Wood, A., Jonell, M., Clark, M., Gordon, L. J., Fanzo, J., Hawkes, C., Zurayk, R., Rivera, J. A., De Vries, W., Majele Sibanda, L., & Murray, C. J. L. (2019, February). Food in the Anthropocene: The EAT–Lancet Commission on healthy diets from sustainable food systems. *Lancet*, 393(10170), 447–492. 10.1016/S0140-6736(18)31788-430660336

Wilschut, K. J., Jaksani, S., Van Den Dolder, J., Haagsman, H. P., & Roelen, B. A. (2008, September 26). Isolation and characterization of porcine adult muscle-derived progenitor cells. *Journal of Cellular Biochemistry*, 105(5), 1228–1239. 10.1002/jcb.2192118821573

Yue, W., Ming, Q. L., Lin, B., Rahman, K., Zheng, C. J., Han, T., & Qin, L. P. (2014, June 25). Medicinal plant cell suspension cultures: Pharmaceutical applications and high-yielding strategies for the desired secondary metabolites. *Critical Reviews in Biotechnology*, 36(2), 215–232. 10.3109/07388551.2014.92398624963701

Zappelli, C., Barbulova, A., Apone, F., & Colucci, G. (2016, November 18). Effective Active Ingredients Obtained through Biotechnology. *Cosmetics*, 3(4), 39. 10.3390/cosmetics3040039

Chapter 2
Unveiling the Science From Cells to Cultivated Food

Jayasree S. Kanathasan
https://orcid.org/0000-0001-7509-6746
Lincoln University College, Malaysia

Devi Nallappan
https://orcid.org/0000-0003-2673-5614
Lincoln University College, Malaysia

Gunavathy Selvarajh
https://orcid.org/0000-0002-6041-6660
Lincoln University College, Malaysia

Zaliha Binti Harun
Lincoln University College, Malaysia

ABSTRACT

The food crisis has been a crucial concern to meet the rising food demand worldwide. Cellular agriculture could be a prominent solution for sustainable food production. One of the main aims for cellular agriculture is to increase the capacity of the food to feed the increasing world population, whilst maintaining the quality of the environment and food. Therefore, this chapter introduces cellular agriculture as one of the latest food bioengineering technologies. Then, the principles and factors of cultivating cultured meat and plant tissues are discussed. Furthermore, applications of the cultivated food are explored. Following this, the ethical aspects and the challenges that will direct the future perspectives were discussed. Being at an early stage of venture in the food industry, the benefits and significant improvements of

DOI: 10.4018/979-8-3693-4115-5.ch002

cellular agriculture will make it the most effective alternative in the food industry.

INTRODUCTION

According to the Sustainable Development Goals (SDGs) Report in 2023, it was projected that more than 600 million people will be facing food crisis by 2030. With increasing world population, it is no surprise that this projected value will increase in the coming decades. Hence, sustainable agricultural practices were listed as one of the coordinated action plans to promote food security (Maccuzato, 2023). An important problem faced by conventional agriculture and livestock farming is limited land and water resources. Agricultural expansion accounts for 90% of global deforestation, having 38.5% being used for livestock grazing (Wali et al., 2024). In addition to that, one of the major drawbacks of traditional farming is the environmental impacts caused by the emission of greenhouse gases (GHG), especially methane gas from livestock animals which leads to global warming and eventually to the consequences arise by climate change. Despite actions taken to meet the world's food demand, the expansion of traditional farming has been a questionable step to achieve a sustainable environment (Chodkowska et al., 2022). Therefore, it is imperative to find a sustainable solution that can reduce GHG emission, land and water footprints with growing population and food demands.

Meat has been one of the commonly consumed human diet which is packed with protein, fats and micronutrients. Over the last several decades, the needs of the global meat have increased more than 200% (OECD-FAO Agricultural Outlook, 2022). Even though the COVID-19 pandemic, concerns on healthy eating and maintenance of animal welfare indicate a reduction in meat markets, plant agriculture is also greatly affected by the consequences of climate change (Chodkowska et al., 2022). Cellular agriculture is one of the recently discovered technologies that has been a central focus in the scientific world of agriculture as it produces molecularly similar agricultural products as the ones produced in traditional farms. Cellular agriculture is regarded as one of the best alternatives for traditional farming (Sexton et al., 2019). Cellular agriculture field involves manufacturing of tissue products from cell cultures of both animal and plant origins.

The scientific history of cellular agriculture was dated back in 1902 during the discovery of plant cell pluripotency (Haberlandt, 1902). The discovery of plant tissue culture as a replacement of secondary metabolite of plants has stated in 1970s which yield the production of the first embryonic plant stem cell lines in 1981 (Misawa, 1977; Evans & Kaufman, 1981). The developing tissue engineering technologies led to the plant tissue cultivation for food industry in 1991, producing the first commercial extract from plant tissue culture using ginseng product (Ushiyama, 1991).

Later, the production of plant stem cell-based products for cosmetic industry was first launched in 2008 (Imseng et al., 2014). Culture of animal cells and tissues was later discovered in laboratories (Carrel, 1912; Gey, 1958). Tissue engineering using animal cells was first started for regenerative medicine focusing on the production of tissue and organ replacement, cosmetics and then food industry (Langer, 2007). The moment which captures the transformation of cellular agriculture from laboratory to the first lab-grown beef burger has occurred in 2013 (Post, 2014).

There are several types of cellular agriculture. Fermentation-based cellular agriculture uses engineered microorganisms for production of recombinant proteins and flavoured products, while tissue-engineering-based cellular agriculture uses donor cells from animal or plant origin at the beginning to produce tissues in culture media which does not require extraction of animal cells to become the final product (Ong et al., 2020). Cultivated or lab-grown meat is identified as the most promising alternative to the real meat product, thanks to the combination of tissue engineering and cell culture of minimal extraction of original tissue. Cultured meat production is a more sustainable alternative which has reduced bacterial contamination and zoonotic diseases and it can be produced without involvement of animals' lives, enhancing animal welfare (Arshad et al., 2017). Therefore, this chapter introduces cultivated agriculture as one of the emerging bioengineering technologies in the food industry. Further aspects include principles in cultivated food process and factors involved in cultivation of cultured food are discussed in this chapter. Then, the chapter focuses on the applications of the cultivated meat and cultivated plant. Following this, the ethical considerations in the cultivation of cellular agriculture, followed by challenges faced in cellular agriculture and recommendations to enhance the cellular agriculture industry were discussed. This chapter which covers the unveiling science and basic principles and applications of cultivated foods is aimed to be a great source of information for academia, industry professionals and policymakers who are interested to get information on cellular agriculture and its prospects in future food industry.

PRINCIPLES AND MECHANISMS OF CELLULAR AGRICULTURE

(a) Cultured Meat

In vitro meat production uses advanced tissue engineering to culture stem cells from farm animals in a bioreactor to produce meat. It eliminates the need for conventional livestock farming (Bhat et al. 2015). Traditional production of meat uses a lot of resources, including land, feed and water. It also significantly contributes to deforestation, biodiversity loss and greenhouse gas emissions. Conversely, cultured meat can be produced with minimum impact to the environment. For instance, lab-grown beef has proven its potential to lower GHG emissions by up to 96% and require 45% less energy, 99% less land and 96% less water (Tuomisto et al., 2011).

The process of *in vitro* meat manufacturing often involves multiple stages. These procedures encompass cell acquisition, derivation of muscle-resident cells and cell sorting for the identification of progenitor cells. The primary cells are procured from tissue of origin from the live animal and isolated for initial cell proliferation and differentiation. Specific types of adult stem cells are optimal for the creation of meat, which comprises muscle satellite cells, fibroadipogenic progenitors (FAP) and mesenchymal stem/stromal cells (MSCs) (Guan et al., 2021). These stem cells can be isolated from various meat origins, including cattle, sheep, chickens and hogs (Lee et al., 2024). One of the most commonly used primary cells is muscle stem cells owing to its outstanding features in facilitating growth, maintenance and repair that enable it to be used in food processing (Guan et al., 2021). The anchor-dependant nature expressed by the proteins in these cells such as integrin, adhesin and cadherin enables the formation of suspension culture system in bioreactor for adherent cell growth and expansion (Bodiou et al., 2020). The second phase is all about the expansion of the cell growth in a large scale. The cells are cultured in a bioreactor to achieve proliferation. In the third phase, the muscle or fat tissue matures into 3D scaffolds. Cells within a biocompatible tissue scaffold undergo maturation, resulting in the desired shape of the meat product (Reiss et al., 2021). The final stage is the processing of the tissue into food products such as burger patties, nuggets, steaks, sausages, fillets and other meat products (Humbird et al., 2021).

Cultivated animal tissues include cultured meat, cultured seafood and cultured fat which are the alternatives to the conventionally produced animal meat and fisheries. These protein-rich products are manufactured using tissue engineering technique for food production. The growth of these tissues is done based on the differentiation of myogenic and adipogenic cells for the production of muscle and fat tissues (Kneži´c et al., 2022). Cell proliferation and differentiation depend on culture medium nutrients, growth factors and chemical specifications. Ideal meat mediums are free from xenobiotic substances and chemically specified, with considerations for cost reduction. Adult stem cells require a basal medium, L-glutamine and low fibroblast growth factor (FGF)-2 levels, while pluripotent stem cells require growth hormones (Mimura et al., 2011; Das et al., 2009).

FACTORS IN THE CULTIVATION OF CULTURED FOOD

a) Source of Cells

Stem cells or progenitor cells are used for the production of cultivated meat, which is extracted from live donor animals and the *ex-vivo* replication process of isolated cells (Kneži´c et al., 2022). Examples of cells used in the production of cultivated meat include pluripotent stem cells such as embryonic stem cells, induced pluripotent stem cells, adult stem cells such as mesenchymal stem cells, adipose tissue stem cells, fibro-adipogenic stem cells, resident muscle stem cells, muscle satellite cells and proliferating activated stem cells (Kneži´c et al., 2022). Muscle satellite cell is one of the most commonly used types of cells as it functions to develop the regeneration of the skeletal muscle tissue. These myogenic cells have been successfully isolated and differentiated into various species of domestic animals other than cattle such as sheep (Wang et al., 2021), goat (Wang et al., 2020), rabbits, chicken (Jankowski et al., 2020), duck, turkey and fisheries. Embryonic stem cells are also suitable for growth owing to their ability to multiply easily and their pluripotent nature of differentiation (Choi et al., 2020a). They can be obtained via *in vitro* fertilization (Navarro et al., 2020).

b) Scaffolding

In production of meat culture, scaffolding is necessary for tissue development and cell differentiation (O'Brien et al., 2011). Production of cultivated meat with the same nutritional content and organoleptic properties that mimic the real meat structure is imperative in animal tissue agriculture (Nayek et al., 2020). Growth of the culture on a 3D scaffolding recreates the natural microenvironment of extracellular matrix, which supports the growth of the tissue culture into a more *in vivo* like structure of the musculature fibre cells and the connective tissues (Vergeer et al., 2021). Moreover, scaffolding is important for efficient transport of oxygen, nutrients and waste products throughout the culture whilst maintaining their mechanical properties which provide sufficient surface area for vertically expansive growth (Gholobova et al., 2020). Despite the advantages exhibited by the protein-based scaffolds including extracellular proteins and self-assembling peptides, the cost effectiveness is a major drawback of using protein-based scaffolding (Orellana et al., 2020). The requirement for the scaffolding material to be biodegradable and abiding the food safety automatically eliminates the use of the synthetic polymer based scaffolding materials such as polyethylene glycol (PEG), poly-(lactic-co-glycolic acid) (PLGA) and polylactic acid (PLA) (Bartnikowski et al., 2019). On the other hand, plant and fungus-based scaffolds have been explored as a cost-effective alternative for protein-based scaffolding materials owing to their inherently vascularised structure (Rubio et al., 2019). Research is currently being conducted on hydrogels and scaffolds that are generated from plants (Ben-Arye et al., 2020). Scaffolds like gelatine fibre are opted to grow bovine aortic smooth muscle into cultivated meat using cellular agriculture (MacQueen et al., 2019). Microcarriers are also employed for a variety of purposes in scaffolding (Li et al., 2015). The utilization of 3D scaffolding in conjunction with extrusion, laser, and inkjet bioprinters appears to hold great potential (Fedorovich et al.,2011; Saunders & Derby, 2014; Dababneh & Ozbolat, 2014). Scaffolding architecture is essential for constructing a durable and supportive framework for cultured meat products, which are also safe for consumption.

c) Tools and Technologies

Usage of certain tools like spinner flasks enhances the growth of tissue culture in a 3D environment in initial stage of culture preparation for large scale production (Verbruggen et al., 2018). For lab-scale production, commercially available tools such as roller bottle, rocking bed, fluidized bed, stirred-tank and vertical bioreactors are used to enhance the growth rate of the suspension culture (Odeleye et al., 2020). Bioreactors meticulously regulate the growth of cells and the transportation of nutrients in order to cultivate meat cells. Bioreactors are found in various types including batch, fed-batch and semi-continuous operation which are available in static mode as well as stirred mode (Spier et al., 2011). Different types of bioreactors are used in different phases of growth to maintain the cellular adherence of the muscle cells and to ensure the equivalent distribution of nutrients and efficient removal of wastes and by-products (Bellani et al., 2020). One of the recent approaches in meat cultivation is bioprinting, the production of structured meat using 3D or 4D bioprinter. The first animal product made using bioprinting technology is the muscle explants of goldfish (Benjaminson et al., 2002). Bioink include amino acid based-RADA16 and methylcellulose have been used to prepare 3D-printed murine mesenchymal stem cells (Cofiño et al., 2019).

d) Growth Factor and Gene Expression

Nutrients and tissues are introduced by the use of permeability, biodegradability, and mechanical reinforcement (Ghasemi-Mobarakeh et al., 2015). Some of the growth factors used in the cultivation of cultured meat in muscle tissue culture include basic fibroblast growth factor (bFGF), interferon γ-induced protein 10 (IP-10), neural cell adhesion molecule 1 (NCAM-1) and decorin (O'Neill et al., 2022). A study by Okamoto et al. (2022) and Haraguchi et al. (2021) showed that nutrients from microalgae *Chlorella vulgaris* have improved the proliferation and differentiation of bovine fibroblasts. There are certain genes that are involved in the development of cultivated meat. Muscle satellite cells express genes including myoblast determination protein 1 (MyoD), myogenic regulatory factor (MRF) genes and Myogenin which are responsible for cellular differentiation and expansion for skeletal muscle to be formed (Cornelison et al., 2000). Some biochemical gene expression pathways also contribute to the growth and maintenance of cellular agriculture. The *in vitro* inhibition of the p38 mitogen-activated protein kinase (MAPK) cell signalling pathway helps in maintenance of the proliferative and non-differentiation for the purpose of upscaling a specific type of cells (Choi et al., 2020b). The factors involved in cultivation of cultured food is depicted in Figure 1.

Figure 1. Factors involved in cultivation of cultured food

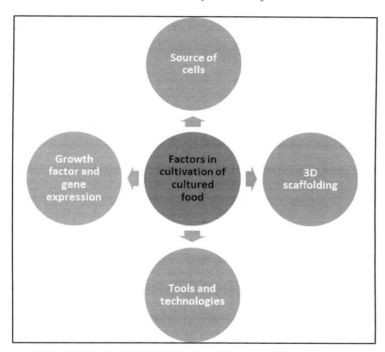

APPLICATIONS OF CULTIVATED MEAT IN CULTURED FOOD

Research on cultured meat primarily focuses on the acquisition of muscle satellite cells and the development of elements that aid in the formation of its structure. Scientists have been researching in vitro meat since the 1900s, leading to the creation of the first hamburger-like *in vitro* burger in 2013. The world's first laboratory-grown burger was discovered by a research group led by Mark Post in 2013 (Goodwin & Shoulders et al., 2013; Kulus et al., 2023). Seven years later, the growth of the cellular agriculture in food industry was proven when the regulatory approval received by company 'Eat Just', to sell the cell-cultured chicken (Helliwell & Burton, 2021). Since then, within almost half a decade, cellular agriculture has begun to transform its application in various food products include meat from livestock animals, eggs, dairy and other animal by products (Hann et al., 2023). The technical and commercial success of cellular agriculture has removed the burden caused by synthetic animal protein production which also imposes environmental and health impacts (Helliwell & Burton, 2021). However, these technologies are still in the developmental stage and are not yet suitable for mass production (Lee et al., 2023).

Advanced cell culture allows food manufacturers to use bovine, chicken, shellfish, and other animal cells. The United States, Singapore, and Israel have approved the use of lab-grown meat. Other countries are anticipated to endorse this practice once the intricate process of generating it is comprehended (Lee et al., 2024). The United States Food Drug Administration (FDA) and the United States Department of Agriculture's Food Safety and Inspection Service (USDA-FSIS) work together to ensure the safety and labelling of cultured animal cell foods. While manufacturers must promote foods that meet FDA rules, the FDA and USDA-FSIS work together to ensure compliance (U.S. Food and Drug Administration, 2023). Thus far, few companies have been working on commercializing their culture's meat. For example, UPSIDE Foods and GOOD Meat have offered their cultivated chicken products in the USA (Good Food Institute, 2023). The applications of cultivated meat that have been in practice in food industry is summarised in Table 1.

Table 1. Applications of cultivated meat as an alternative choice of meat

Cultivated food product	Founder/ Production company	References
First laboratory grown beef burger	Dr. Mark Post's research group (Mosa Meat), Netherlands	Goodwin & Shoulders et al., 2013
Cell-cultured chicken	'Eat Just' company, Singapore	Helliwell & Burton, 2021
Cultivated chicken	UPSIDE Food and GOOD Meat company, USA	Good Food Institute, 2023

PLANT TISSUE CULTURE IN CELLULAR AGRICULTURE

Plant cellular agriculture involves the production of plant-based products from parts obtained from plant cells grown into culture for cosmetic, food and medication purposes. Plant-tissue-based products are utilized to manufacture plant products without growing the whole plant, which is a concern for endangered and protected plants. Plant tissue culture (PTC) techniques involve separating, sterilizing and incubating cells, tissues, or organs from certain plants in a controlled, sterile environment to promote growth and produce many plantlets (Hasnain et al., 2022). PTC generally utilizes techniques such as organogenesis and callogenesis (Morales-Rubio et al. 2016). Organogenesis is the process by which roots or branches are formed from meristems or dedifferentiated cells, while callogenesis results in the formation of an irregular mass. The cultures promote the development of organs and the multiplication of microorganisms. Cell suspension culture involves the generation of amorphous cells by callogenesis, which can produce essential compounds or regenerate plants. In order to prepare a plant for use, it is necessary to harvest it, render it chemically sterile and then incubate it for a brief period.

Several steps are involved in all plant tissue culture techniques, which begins with selecting the appropriate plant or pre-propagation step. The growth process of plant tissue culture starts with the isolation of plant explants or fragmentation without contamination using sterile tweezers and scalpels. This procedure is conducted in a sterile condition using UV-irradiated tools and a biosafety cabinet environment (Schumacher et al., 2015). This step is followed by the initiation of the *in vitro* culture. In this step, the sterile explant or seeds are placed in appropriate culture media and incubated for a short period. The plant cells or tissues will be grown either on the petri dish onto a solid medium or shaking flask in a liquid medium at room temperature. Plant cells and tissues will be preserved in a cell bank using cryopreservation technique using liquid nitrogen for next use. The newly revived cells will take around several months to optimize its growth. Then, the next step involves propagation, in which culture will be continued with appropriate culture media for shoot, root or callus multiplication. The major objective of the second stage is to increase the cell biomass and product concentration, which is highly influenced by media composition of phytohormones, nitrogen, phosphate and sucrose. Following shoot or root multiplication, micropropagated plants undergo a hardening process in culture media to develop plants capable of photosynthesis (Morales-Rubio et al. 2016; Espinosa-Leal et al. 2018). The culture will be transferred to benchtop-scale bioreactors that can contain up to 10 litres. The growth is further developed using suspension culture in stirred bioreactors. When the growth volume reaches up to 100 litres, the suspension culture is grown in wave-mixed bioreactors along one axis, facilitating mixing and bubble free surface aeration (Eibl et al., 2021). In the final stage, the plant cell or tissue culture products in the bioreactors will be harvested. The final operation methods consist of drying, extraction and filtration steps (Fierascu et al., 2020).

Temperature, pH, mixing, aeration, light, incubation time, inoculum density and use of elicitors are some of the components which effects the growth of the plant (Halder et al., 2019). The elicitors and media components are made of non-toxic components include naturally procured indole-3-acetic acid, methyl jasmonate and salicylic acid (Murthy et al., 2015).

Traditional plant breeding has enhanced food production in developing nations; nevertheless, maintaining these gains and responding to climate change is critical (Pathira & Carimi, 2024). Micropropagation or PTC, which involves the process of growing plant cells *in vitro* might overcome this issue. PTC offers advantages over traditional field culture, including independence from variations, uniform quality and short growth cycles (Gulzar et al. 2020). PTC is essential for the bulk production of genetically identical and pest-resistant plantlets of various important plant species and has solved numerous phytosanitary concerns and made high-quality plants available to farmers worldwide (Cardoso et al., 2018). For example, plant tissue culture in propagating bananas has generated robust, pathogen-free banana plants (Justine et al., 2022). Similarly, the tissue culture approach has been used in potato farming, making it more efficient, profitable and sustainable (Hajare et al. 2021). In addition, plant tissue culture has successfully produced morphologically and genetically stable plants of various strawberry cultivars (Naing et al. 2019). Another extensive application of plant tissue culture is the production of ginseng. Culture based ginseng extract is mostly used in manufacturing many ginseng-based products (Murthy et al., 2014). In addition to that, renowned plant-based extract is used in development of anti-cancer drug Paclitaxel which is extracted from the bark of the Pacific yew tree known as *Taxus brevifolia* (Camidge, 2001).

ETHICAL CONSIDERATION IN CELLULAR AGRICULTURE

Despite the benefits of cellular agriculture, which aims to produce animal products such as meat, eggs, and milk without raising and slaughtering livestock animals, there are some key ethical considerations related to cellular agriculture that must be taken into account. One of the most pressing ethical challenges that society faces is the treatment of animals. While cellular agriculture offers a potential solution by reducing the need for traditional animal farming, it still relies on using animal cells. Despite being regarded as a more humane alternative, critics express concerns about the welfare of the cells used and the potential for animal suffering during the cell extraction process (Froggart & Wellesley, 2019). According to the Food Information to Consumers Regulation, meat is defined as the skeletal muscles of mammalian and bird species recognised as fit for human consumption with naturally included or adherent tissues. Consequently, the existing definition appears to exclude cultured meat (European Commission, 2016; Broucke et al., 2023).

Cell-based meat, also known as lab-grown meat, has the potential to address environmental concerns associated with traditional meat production. However, it's important to note that there are some complexities to consider. In certain cases, the water usage and energy consumption involved in the production of cell-based meat may be higher than those associated with traditional beef farming. Hence, the environmental benefits of meat alternatives, including plant-based substitutes and cell-based meat, may not be as significant as compared to conventional animal products such as poultry, eggs and certain types of seafood. These nuances should be openly discussed when considering the environmental impact of meat alternatives (Tuomisto, 2019; Santo et al., 2020).

Additionally, reducing animal suffering is a significant reason to endorse research and development of cultured meat. It's noteworthy that not all modern farming practices cause significant harm to animals. There has been a recent trend towards less-intensive farming, with many marketing animal products using 'free-range' and 'cage-free' labels. Supporting cultured meat is consistent with the coexistence of such humane farms, whether through direct demand for 'natural' meat or potential state subsidies if necessary (Matheny & Leahy, 2007; Kollenda et al., 2020; Schaefer & Savulescu, 2014).

Besides, the acceptance of cell-based products varies greatly across different cultures and societies. Ethical considerations should take into account the cultural norms, beliefs and preferences of the communities involved. Therefore, it is essential to engage with diverse communities and gain an understanding of their perspectives on cellular agriculture (Sapontzis, 2004). For instance, the use of animal serum in cultured meat production has implications for the *halal* status of the product while the scalability of production processes has implications for the development of the cultured meat industry (Bryant, 2020). It is reported that consuming meat emphasises the value of human relationships with the natural world. The respect for natural relationships is also highlighted in various religious traditions such as Confucianism (Fan, 2005), Hinduism (Engel & Engel, 1990) and Native American traditions (Dussias, 1998). Consuming meat can be seen as ethically acceptable when viewed as part of a broader interconnectedness with the natural world. This perspective emphasizes the importance of maintaining a delicate balance between utilizing technology and respecting the natural environment without undue intrusion.

The regulation for genetically modified food and feed, Regulation (EC) No 1829/2003 applies to cultured meat products if they contain genetically modified ingredients. Decisions made under this regulation consider risk assessments, public acceptance and economic factors. In Europe, due to strict restrictions on genetically modified foods and negative public perceptions of the technology, cultured meat with genetically modified components is unlikely to be approved (Verzijden, 2019; Bryant, 2020). Therefore, European producers must understand both European and national legislation. Nonetheless, a significant shift toward cultured meat production as well as getting away from conventional animal agriculture is likely to result in unemployment in the animal agriculture field. Ethical discussions should address the impact on farmers, labourers and rural communities (Bryant, 2020).

Some experts predict that initially, cultured meat will be offered at a high price and only available in restaurants (Purdy, 2019). During this phase, cultured meat might be considered a luxury that only affluent individuals or those with access to exclusive establishments can enjoy. As a result, consuming cultured meat at this stage could be associated with wealth and social status. However, in the long run, the production of cultured meat is expected to become more cost-effective and could even be cheaper than traditional meat if the process becomes more efficient (Fountain, 2013). Consequently, the linkage of social status to consuming cultured meat might likely to decline if it becomes more widely accessible.

Lab-grown meat presents complex moral dilemmas within our food systems, particularly concerning the loss of autonomy. The reliance on advanced technology and artificial intervention in food production may limit the natural flourishing of living organisms used as food. Moreover, there is a risk that this technology could fall under the control of multinational corporations due to patents and high technological, economic and legal barriers to entry. This could result in the creation of exclusive, high-end food products and further contribute to the proliferation of food deserts (Mahoney, 2022). It is essential to recognize the multifaceted ethical considerations at play and engage in ongoing dialogue to responsibly navigate the complexities of cellular agriculture.

CHALLENGES AND FUTURE RECOMMENDATIONS

Cultivated food from cells is the production of cultured meat and cultivated plant which can provide sufficient food supply for a growing world population. However, there are so many challenges exist in developing foods from cultivated cells. The first challenge comes in the point of scaling up cultivated meat production. The increment in global population increased the demand for meat consumption demand, while cultivated meat production is still small-scale production. The global population is expected to reach nearly more than 10 billion people by 2050 (Searchinger et al., 2018). Despite the advancements in cell culture techniques and bioprocessing, achieving large-scale production of cultivated meat remains a formidable task (Chen et al., 2022). To increase scalability, the development of cost-effective and efficient bioreactor systems, optimization of cell culture media and scaling of tissue engineering processes is needed. Bioreactor systems are used for the cultivation of animal cells at an industrial scale. Recent research has emphasized the need for innovative bioreactor designs capable of supporting large volumes of cell cultures while maintaining optimal growth conditions (Wang et al., 2005). By addressing these critical aspects of scalability and production efficiency, the cellular agriculture industry can move closer to fulfilling its potential as a sustainable cultivated meat production.

Besides, the scaling of tissue engineering processes for producing structured tissues that mimic the texture and flavour of conventional meat remains as a challenge. It had been stated that the advances in scaffold development and 3D bioprinting technology have paved the way for creating complex tissue structures with precise control over cell placement and biomaterial composition (Bomkamp et al., 2022). However, the bio-fabrication of cultured meat using 3D bioprinting techniques is not easy as being described because it takes longer time and needs a huge effort. The 3D bioprinting uses techniques include bio-fabrication of scaffolds loaded with cells are extrusion, jetting, and Vat photopolymerization. 3D bioprinting faces numerous challenges in the production of cultured meat involving cell lines, replacement of fetal bovine serum (FBS), selection of ideal biomaterials regulation and difficulty in achieving a texture and nutritional value similar to the conventional meat (Barbosa et al., 2023). The first challenge in 3D bioprinting lies in the selection of an appropriate species of animal for the cell biopsy. The selection criteria of the donor animal include sex, age, and rearing conditions (Lanzoni et al., 2022) should be done cautiously, as it can affect the growth of satellite cells in the transformation of adult skeletal muscle stem cells. This is a tremendous work that needs a longer time to obtain ideal cell source which is highly essential to promote cell proliferation and differentiation. Not only that, another challenge that arises in the 3D bioprinting is the cell viability. In one of the types of bioprinting, known as extrusion-based bioprinting process, the bioink will extrude through a needle under pressure, which can cause shear stresses between the bioink and the wall of the printing nozzle. If these stresses exceed a specific limit, they potentially can rupture cell membranes, which results in decline of cell viability (Malekpour & Chen, 2022). Hence, it can be stated that 3D printing in the bio-fabrication of meat is still in an ongoing process to meet consumer preferences.

Cultured food production considers other social issues such as consumer appeal and acceptance. Consumer acceptance of the cultivated food is important in determining the viability and success of cellular agriculture as a sustainable alternative to conventional animal farming. It can be summed up that some consumers are enthusiastic about the potential environmental and ethical benefits offered by cultivated meat, while others harbour aversion toward lab-grown foods. To overcome barriers of market adoption and building trust in cells to cultivate food products, the information on the consumer concerns regarding taste, texture, safety and naturalness must be delivered through transparent communication, education and product innovation (Frewer et al., 2011). Recent research underscores the significance of consumer perception and acceptance in shaping the future of cultivated food products. Studies have highlighted the importance of transparent communication and education initiatives in increasing consumer awareness and acceptance of cells to cultivated foods (Hubbard, 2022). Moreover, product innovation aimed at enhancing the sensory attributes of cultivated meat, such as taste and texture, is essential for winning over sceptical consumers (Bryant & Barnett, 2022). To achieve this, progress in food engineering and nutrition plays a major role in order to make a final cultivated product sensorially and nutritionally acceptable (Fish et al., 2020).

Moreover, the high production cost of cultivated food hinders the market competitiveness with conventional animal-derived foods. The existing manufacturing methods use expensive cell culture media and bioreactor equipment, contributing to elevated production costs. To reduce cost, optimization of process and minimal use of resources is needed. The increased costs of cultured meat are due to the cell culture media usage, estimated for 55% - 95% of the total product's costs (Park et al., 2013). The growth of the cells in animal tissue culture medium also requires supplementation of nutrients such as carbohydrates, proteins, lipids, sugars, amino acids, minerals, vitamins, serum, antibiotics, and hormones to induce proliferation, differentiation, and maturation. Serum is very expensive and a media without serum is highly preferable to reduce costs. Studies also have shown promising results in the development of serum-free media formulations, which eliminate the need for expensive and ethically contentious serum components (O'Neill et al., 2021). Besides, an additional cost incurs when the growth medium used for cell culture need to be arranged in a highly sanitized and sterile environment. However, despite all the sterile conditions, the in vitro cell cultures face a high risk of contamination. The contamination increases the cost of production because a new culture supplemented with the expensive nutrients need to be prepared again. Previous studies found that optimizing culture media and bioprocessing parameters such as temperature, pH, and agitation speed enhances cell growth and productivity while minimizing input costs and contamination (Gomez & Boyle, 2023). Furthermore, innovation in bioprocessing technologies will help in driving down production costs. Advanced bioreactor designs, such as perfusion bioreactors and single-use systems, offer opportunities for reducing capital investment, operational costs and contamination risks (Suntornnond et al., 2024). If cost reduction strategies are well emphasized, the cultivated food can overcome barriers by establishing itself as a competitive player in the global food market.

Despite of all the challenges in cultivated food field, the cultivated food can be established and adopted successfully if research initiatives are developed between academic, industry and government bodies. Research funding in the field of cultivated food could support and initiate different cultivated food product innovations with reduced cost. Then, the output of the research can be potentially used for technology transfer and commercialization of the cultivated food product among consumers all over the world. The commercialization process needs to be started with education and outreach programs among communities so that they are aware of the cultivated food product and accept it for consumption.

In conclusion, cellular agriculture has demonstrated the credibility and has obtained the approval of becoming a sustainable alternative to traditional farming. In this chapter, the principles and conditional requirements for production of cultivated meat and plant tissue culture have been discussed, as well as their applications in current food industry. Management of ethical implications of cellular agriculture is crucial to ensure consumer acceptance. Transformation of cellular agriculture from lab scale to large scale commercialisation holds a great future and potential benefit for the nature, environment and food supply to combat the rising food demand.

REFERENCES

Arshad, M. S., Javed, M., Sohaib, M., Saeed, F., Imran, A., & Amjad, Z. (2017). Tissue engineering approaches to develop cultured meat from cells: A mini review. *Cogent Food & Agriculture*, 3(1), 1320814. 10.1080/23311932.2017.1320814

Barbosa, W., Correia, P., Vieira, J., Leal, I., Rodrigues, L., Nery, T., Barbosa, J., & Soares, M. (2023). Trends and Technological Challenges of 3D Bioprinting in Cultured Meat: Technological Prospection. *Applied Sciences (Basel, Switzerland)*, 13(22), 12158. 10.3390/app132212158

Bartnikowski, M., Dargaville, T. R., Ivanovski, S., & Hutmacher, D. W. (2019). Degradation mechanisms of polycaprolactone in the context of chemistry, geometry and environment. *Progress in Polymer Science*, 96, 1–20. 10.1016/j.progpolymsci.2019.05.004

Bellani, C. F., Ajeian, J., Duffy, L., Miotto, M., Groenewegen, L., & Connon, C. J. (2020). Scale-Up Technologies for the Manufacture of Adherent Cells. *Frontiers in Nutrition*, 7, 575146. 10.3389/fnut.2020.57514633251241

Ben-Arye, T., Shandalov, Y., Ben-Shaul, S., Landau, S., Zagury, Y., Ianovici, I., Lavon, N., & Levenberg, S. (2020). Textured soy protein scaffolds enable the generation of three-dimensional bovine skeletal muscle tissue for cell-based meat. *Nature Food*, 1(4), 210–220. 10.1038/s43016-020-0046-5

Benjaminson, M. A., Gilchriest, J. A., & Lorenz, M. (2002). In vitro edible muscle protein production system (MPPS): Stage 1, fish. *Acta Astronautica*, 51(12), 879–889. 10.1016/S0094-5765(02)00033-412416526

Bhat, Z. F., Kumar, S., & Fayaz, H. (2015). In vitro meat production: Challenges and benefits over conventional meat production. *Journal of Integrative Agriculture*, 14(2), 241–248. 10.1016/S2095-3119(14)60887-X

Bodiou, V., Moutsatsou, P., & Post, M. J. (2020). Microcarriers for Upscaling Cultured Meat Production. *Frontiers in Nutrition*, 7, 10. 10.3389/fnut.2020.0001032154261

Bomkamp, C., Skaalure, S. C., Fernando, G. F., Ben-Arye, T., Swartz, E. W., & Specht, E. A. (2022). Scaffolding Biomaterials for 3D Cultivated Meat: Prospects and Challenges. *Advanced Science (Weinheim, Baden-Wurttemberg, Germany)*, 9(3), e2102908. 10.1002/advs.20210290834786874

Broucke, K., Van Pamel, E., Van Coillie, E., Herman, L., & Van Royen, G. (2023). Cultured meat and challenges ahead: A review on nutritional, technofunctional and sensorial properties, safety and legislation. *Meat Science*, 195, 109006. 10.1016/j.meatsci.2022.10900636274374

Bryant, C., & Barnett, J. (2020). Consumer acceptance of cultured meat: An updated review (2018–2020). *Applied Sciences (Basel, Switzerland)*, 10(15), 5201. 10.3390/app10155201

Bryant, C. J. (2020). Culture, meat, and cultured meat. *Journal of Animal Science*, 98(8), skaa172. Advance online publication. 10.1093/jas/skaa17232745186

Camidge, R. (2001). The Story of Taxol: Nature and politics in the pursuit of an anti-cancer drug. *BMJ (Clinical Research Ed.)*, 323(7304), 115. 10.1136/bmj.323.7304.115/a

Cardoso, J. C., Sheng Gerald, L. T., & Teixeira da Silva, J. A. (2018). Micropropagation in the Twenty-First Century. *Methods in Molecular Biology (Clifton, N.J.)*, 1815, 17–46. 10.1007/978-1-4939-8594-4_229981112

Carrel, A. (1912). On the permanent life of tissues outside of the organism. *The Journal of Experimental Medicine*, 15(5), 516–528. 10.1084/jem.15.5.51619867545

Chen, L., Guttieres, D., Koenigsberg, A., Barone, P. W., Sinskey, A. J., & Springs, S. L. (2022). Large-scale cultured meat production: Trends, challenges and promising biomanufacturing technologies. *Biomaterials*, 280, 121274. 10.1016/j.biomaterials.2021.12127434871881

Chodkowska, K. A., Wódz, K., & Wojciechowski, J. (2022). Sustainable Future Protein Foods: The Challenges and the Future of Cultivated Meat. *Foods*, 11(24), 4008. 10.3390/foods1124400836553750

Choi, K. H., Lee, D. K., Oh, J. N., Kim, S. H., Lee, M., Woo, S. H., Kim, D. Y., & Lee, C. K. (2020a). Pluripotent pig embryonic stem cell lines originating from in vitro-fertilized and parthenogenetic embryos. *Stem Cell Research*, 49, 102093. 10.1016/j.scr.2020.10209333232901

Choi, K. H., Yoon, J. W., Kim, M., Jeong, J., Ryu, M., Park, S., Jo, C., & Lee, C. K. (2020b). Optimization of Culture Conditions for Maintaining Pig Muscle Stem Cells In Vitro. *Food Science of Animal Resources*, 40(4), 659–667. 10.5851/kosfa.2020.e3932734272

Cofiño, C., Perez-Amodio, S., Semino, C. E., Engel, E., & Mateos-Timoneda, M. A. (2019). Development of a self-assembled peptide/methylcellulose-based bioink for 3D bioprinting. *Macromolecular Materials and Engineering*, 304(11), 1900353. 10.1002/mame.201900353

Cornelison, D. D., Olwin, B. B., Rudnicki, M. A., & Wold, B. J. (2000). MyoD(-/-) satellite cells in single-fiber culture are differentiation defective and MRF4 deficient. *Developmental Biology*, 224(2), 122–137. 10.1006/dbio.2000.968210926754

Dababneh, A. B., & Ozbolat, I. T. (2014). Bioprinting technology: A current state-of-the-art review. *Journal of Manufacturing Science and Engineering*, 136(6), 061016. 10.1115/1.4028512

Das, M., Rumsey, J. W., Bhargava, N., Gregory, C., Reidel, L., Kang, J. F., & Hickman, J. J. (2009). Developing a novel serum-free cell culture model of skeletal muscle differentiation by systematically studying the role of different growth factors in myotube formation. *In Vitro Cellular & Developmental Biology. Animal*, 45(7), 378–387. 10.1007/s11626-009-9192-719430851

Dussias, A. M. (1998). Asserting a traditional environmental ethic: Recent developments in environmental regulation involving Native American tribes. *New England Review*, 33, 653. https://heinonline.org/HOL/License

Eibl, R., Senn, Y., Gubser, G., Jossen, V., van den Bos, C., & Eibl, D. (2021). Cellular Agriculture: Opportunities and Challenges. *Annual Review of Food Science and Technology*, 12(1), 51–73. 10.1146/annurev-food-063020-12394033770467

El Wali, M., Rahimpour Golroudbary, S., Kraslawski, A., & Tuomisto, H. L. (2024). Transition to cellular agriculture reduces agriculture land use and greenhouse gas emissions but increases demand for critical materials. *Communications Earth & Environment*, 5(1), 1–17. 10.1038/s43247-024-01227-8

Engel, J. R., & Engel, J. G. (1990). Ethics of environment and development. United States. https://www.osti.gov/biblio/5476890

Espinosa-Leal, C. A., Puente-Garza, C. A., & García-Lara, S. (2018). In vitro plant tissue culture: Means for production of biological active compounds. *Planta*, 248(1), 1–18. 10.1007/s00425-018-2910-129736623

Espinosa-Leal, C. A., Puente-Garza, C. A., & García-Lara, S. (2018). In vitro plant tissue culture: Means for production of biological active compounds. *Planta*, 248(1), 1–18. 10.1007/s00425-018-2910-129736623

Evans, M. J., & Kaufman, M. H. (1981). Establishment in culture of pluripotential cells from mouse embryos. *Nature, 292*(5819), 154-156. 10.1038/292154a0

Fan, R. (2005). A reconstructionist confucian account of environmentalism: Toward a human sagely dominion over nature. *Journal of Chinese Philosophy*, 32(1), 105–122. 10.1111/j.1540-6253.2005.00178.x

Fedorovich, N. E., Alblas, J., Hennink, W. E., Oner, F. C., & Dhert, W. J. (2011). Organ printing: The future of bone regeneration? *Trends in Biotechnology*, 29(12), 601–606. 10.1016/j.tibtech.2011.07.00121831463

Fierascu, R. C., Fierascu, I., Ortan, A., Georgiev, M. I., & Sieniawska, E. (2020). Innovative Approaches for Recovery of Phytoconstituents from Medicinal/Aromatic Plants and Biotechnological Production. *Molecules (Basel, Switzerland)*, 25(2), 309. 10.3390/molecules2502030931940923

Fish, K. D., Rubio, N. R., Stout, A. J., Yuen, J. S. K., & Kaplan, D. L. (2020). Prospects and challenges for cell-cultured fat as a novel food ingredient. *Trends in Food Science & Technology*, 98, 53–67. 10.1016/j.tifs.2020.02.00532123465

Fountain, H. (2013). Engineering the $325,000 In Vitro Burger. *The New York Times, 12*.

Frewer, L. J., Bergmann, K., Brennan, M., Lion, R., Meertens, R., Rowe, G., Siegrist, M., & Vereijken, C. M. J. L. (2011). Consumer response to novel agri-food technologies: Implications for predicting consumer acceptance of emerging food technologies. *Trends in Food Science & Technology*, 22(8), 442–456. 10.1016/j.tifs.2011.05.005

Froggart, A., & Wellesley, L. (2019). Meat analogues: considerations for the EU. Chatham House. https://apo.org.au/node/222731

Gey, G. O. (1958). Normal and malignant cells in tissue culture. *Annals of the New York Academy of Sciences*, 76(3), 547–549. https://pubmed.ncbi.nlm.nih.gov/13627881/13627881

Ghasemi-Mobarakeh, L., Prabhakaran, M. P., Tian, L., Shamirzaei-Jeshvaghani, E., Dehghani, L., & Ramakrishna, S. (2015). Structural properties of scaffolds: Crucial parameters towards stem cells differentiation. *World Journal of Stem Cells*, 7(4), 728–744. 10.4252/wjsc.v7.i4.72826029344

Gholobova, D., Terrie, L., Gerard, M., Declercq, H., & Thorrez, L. (2020). Vascularization of tissue-engineered skeletal muscle constructs. *Biomaterials*, 235, 119708. 10.1016/j.biomaterials.2019.11970831999964

Gomez Romero, S., & Boyle, N. (2023). Systems biology and metabolic modeling for cultivated meat: A promising approach for cell culture media optimization and cost reduction. *Comprehensive Reviews in Food Science and Food Safety*, 22(4), 3422–3443. 10.1111/1541-4337.1319337306528

Goodwin, J. N., & Shoulders, C. W. (2013). The future of meat: A qualitative analysis of cultured meat media coverage. *Meat Science*, 95(3), 445–450. 10.1016/j.meatsci.2013.05.02723793078

Guan, X., Zhou, J., Du, G., & Chen, J. (2022). Bioprocessing technology of muscle stem cells: Implications for cultured meat. *Trends in Biotechnology*, 40(6), 721–734. 10.1016/j.tibtech.2021.11.00434887105

Gulzar, B., Mujib, A., Malik, M. Q., Mamgain, J., Syeed, R., & Zafar, N. (2020). *Plant tissue culture: agriculture and industrial applications*. Elsevier eBooks. 10.1016/B978-0-12-818632-9.00002-2

Haberlandt, G. (1902). Kulturversuche mit isolierten Pflanzenzellen. Sitzungser. Mat. Nat. K. *Kais. Akad. Wiss.(Wien)*, 111, 69–92.

Hajare, S. T., Chauhan, N. M., & Kassa, G. (2021). Effect of Growth Regulators on *In Vitro* Micropropagation of Potato (*Solanum tuberosum* L.) Gudiene and Belete Varieties from Ethiopia. *The Scientific World Journal*, 5928769, 1–8. Advance online publication. 10.1155/2021/592876933628138

Halder, M., Sarkar, S., & Jha, S. (2019). Elicitation: A biotechnological tool for enhanced production of secondary metabolites in hairy root cultures. *Engineering in Life Sciences*, 19(12), 880–895. 10.1002/elsc.20190005832624980

Hann, E. C., Harland-Dunaway, M., Garcia, A. J., Meuser, J. E., & Jinkerson, R. E. (2023). Alternative carbon sources for the production of plant cellular agriculture: A case study on acetate. *Frontiers in Plant Science*, 14, 1104751. 10.3389/fpls.2023.110475137954996

Haraguchi, Y., & Shimizu, T. (2021). Three-dimensional tissue fabrication system by co-culture of microalgae and animal cells for production of thicker and healthy cultured food. *Biotechnology Letters*, 43(6), 1117–1129. 10.1007/s10529-021-03106-033689062

Hasnain, A., Naqvi, S. A. H., Ayesha, S. I., Khalid, F., Ellahi, M., Iqbal, S., Hassan, M. Z., Abbas, A., Adamski, R., Markowska, D., Baazeem, A., Mustafa, G., Moustafa, M., Hasan, M. E., & Abdelhamid, M. M. A. (2022). Plants *in vitro* propagation with its applications in food, pharmaceuticals and cosmetic industries; current scenario and future approaches. *Frontiers in Plant Science*, 13, 1009395. 10.3389/fpls.2022.100939536311115

Hayek, M. N., Harwatt, H., Ripple, W. J., & Mueller, N. D. (2021). The carbon opportunity cost of animal-sourced food production on land. *Nature Sustainability*, 4(1), 21–24. 10.1038/s41893-020-00603-4

Helliwell, R., & Burton, R. J. (2021). The promised land? Exploring the future visions and narrative silences of cellular agriculture in news and industry media. *Journal of Rural Studies*, 84, 180–191. 10.1016/j.jrurstud.2021.04.002

Humbird, D. (2021). Scale-up economics for cultured meat. *Biotechnology and Bioengineering*, 118(8), 3239–3250. 10.1002/bit.2784834101164

Imseng, N., Schillberg, S., Schürch, C., Schmid, D., Schütte, K., Gorr, G., ... & Eibl, R. (2014). Suspension culture of plant cells under heterotrophic conditions. *Industrial scale suspension culture of living cells*, 224-258. 10.1002/9783527683321.ch07

Jankowski, M., Mozdziak, P., Petitte, J., Kulus, M., & Kempisty, B. (2020). Avian satellite cell plasticity. *Animals (Basel)*, 10(8), 1322. 10.3390/ani1008132232751789

Justine, A. K., Kaur, N., Savita, , & Pati, P. K. (2022). Biotechnological interventions in banana: Current knowledge and future prospects. *Heliyon*, 8(11), e11636. 10.1016/j.heliyon.2022.e1163636419664

Knežić, T., Janjušević, L., Djisalov, M., Yodmuang, S., & Gadjanski, I. (2022). Using vertebrate stem and progenitor cells for cellular agriculture, state-of-the-art, challenges, and future perspectives. *Biomolecules*, 12(5), 699. 10.3390/biom1205069935625626

Kollenda, E., Baldock, D., Hiller, N., & Lorant, A. (2020). Transitioning towards cage-free farming in the EU: Assessment of environmental and socio-economic impacts of increased animal welfare standards. *Policy report by the Institute for European Environmental Policy, Brussels & London*, 1-65. https://ieep.eu/wp-content/uploads/2022/12/Transitioning-towards-cage-free-farming-in-the-EU_Final-report_October_web.pdf

Kulus, M., Jankowski, M., Kranc, W., Golkar Narenji, A., Farzaneh, M., Dzięgiel, P., Zabel, M., Antosik, P., Bukowska, D., Mozdziak, P., & Kempisty, B. (2023). Bioreactors, scaffolds and microcarriers and in vitro meat production—Current obstacles and potential solutions. *Frontiers in Nutrition*, 10, 1225233. 10.3389/fnut.2023.122523337743926

Langer, R. (1997). Tissue engineering: A new field and its challenges. *Pharmaceutical Research*, 14(7), 840–841. 10.1023/A:10121313291489244137

Lee, S. Y., Jeong, J. W., Kim, J. H., Yun, S. H., Mariano, E.Jr, Lee, J., & Hur, S. J. (2023). Current technologies, regulation, and future perspective of animal product analogs—A review. *Animal Bioscience*, 36(10), 1465–1487. 10.5713/ab.23.002937170512

Lee, S. Y., Yun, S. H., Lee, J., Mariano, E.Jr, Park, J., Choi, Y., & Hur, S. J. (2024). Current technology and industrialization status of cell-cultivated meat. *Journal of Animal Science and Technology*, 66(1), 1–30. 10.5187/jast.2023.e10738618028

MacQueen, L. A., Alver, C. G., Chantre, C. O., Ahn, S., Cera, L., Gonzalez, G. M., O'Connor, B. B., Drennan, D. J., Peters, M. M., Motta, S. E., Zimmerman, J. F., & Parker, K. K. (2019). Muscle tissue engineering in fibrous gelatin: Implications for meat analogs. *NPJ Science of Food*, 3(1), 20. 10.1038/s41538-019-0054-831646181

Malekpour, A., & Chen, X. (2022). Printability and Cell Viability in Extrusion-Based Bioprinting from Experimental, Computational, and Machine Learning Views. *Journal of Functional Biomaterials*, 13(2), 40. 10.3390/jfb1302004035466222

Matheny, G., & Leahy, C. (2007). Farm-animal welfare, legislation, and trade. *Law & Contemp. Probs.*, 70, 325. https://scholarship.law.duke.edu/lcp/vol70/iss1/10

Mazzucato, M. (2023). Financing the Sustainable Development Goals through mission-oriented development banks. UN DESA Policy Brief Special issue. New York: UN Department of Economic and Social Affairs; UN High-level Advisory Board on Economic and Social Affairs; University College London Institute for Innovation and Public Purpose.

Mimuma, S., Kimura, N., Hirata, M., Tateyama, D., Hayashida, M., Umezawa, A., & Furue, M. K. (2011). Growth factor-defined culture medium for human mesenchymal stem cells. *The International Journal of Developmental Biology*, 55(2), 181–187. 10.1387/ijdb.103232sm21305471

Misawa, M. (1977). Production of natural substances by plant cell cultures described in Japanese patents. In *Plant Tissue Culture and Its Bio-technological Application: Proceedings of the First International Congress on Medicinal Plant Research, Section B, held at the University of Munich, Germany September 6–10, 1976* (pp. 17-26). Springer Berlin Heidelberg.

Morales-Rubio, M. E., Espinosa-Leal, C., & Garza-Padrón, R. A. (2016). Cultivation of plant tissues and their application in natural products. In Rivas-Morales, C., Oranday-Cardenas, M. A., & Verde-Star, M. J. (Eds.), *Investigación en plantas de importancia médica* (1st ed., pp. 351–410)., 10.3926/oms.315

Murthy, H. N., Georgiev, M. I., Kim, Y. S., Jeong, C. S., Kim, S. J., Park, S. Y., & Paek, K. Y. (2014). Ginsenosides: Prospective for sustainable biotechnological production. *Applied Microbiology and Biotechnology*, 98(14), 6243–6254. 10.1007/s00253-014-5801-924859520

Murthy, H. N., Georgiev, M. I., Park, S. Y., Dandin, V. S., & Paek, K. Y. (2015). The safety assessment of food ingredients derived from plant cell, tissue and organ cultures: A review. *Food Chemistry*, 176, 426–432. 10.1016/j.foodchem.2014.12.07525624252

Naing, A. H., Kim, S. H., Chung, M. Y., Park, S. K., & Kim, C. K. (2019). In vitro propagation method for production of morphologically and genetically stable plants of different strawberry cultivars. *Plant Methods*, 15(1), 36. 10.1186/s13007-019-0421-031011361

Navarro, M., Soto, D. A., Pinzon, C. A., Wu, J., & Ross, P. J. (2019). Livestock pluripotency is finally captured in vitro. *Reproduction, Fertility, and Development*, 32(2), 11–39. 10.1071/RD1927232188555

O'Neill, E. N., Ansel, J. C., Kwong, G. A., Plastino, M. E., Nelson, J., Baar, K., & Block, D. E. (2022). Spent media analysis suggests cultivated meat media will require species and cell type optimization. *NPJ Science of Food*, 6(1), 46. 10.1038/s41538-022-00157-z36175443

O'Neill, E. N., Cosenza, Z. A., Baar, K., & Block, D. E. (2021). Considerations for the development of cost-effective cell culture media for cultivated meat production. *Comprehensive Reviews in Food Science and Food Safety*, 20(1), 686–709. 10.1111/1541-4337.1267833325139

Odeleye, A. O. O., Baudequin, T., Chui, C. Y., Cui, Z., & Ye, H. (2020). An additive manufacturing approach to bioreactor design for mesenchymal stem cell culture. *Biochemical Engineering Journal*, 156, 107515. 10.1016/j.bej.2020.107515

Oecd, F. A. O. (2022). OECD-FAO agricultural outlook 2022-2031. Retrieved from https://policycommons.net/artifacts/2652558/oecd-fao-agricultural-outlook-2022-2031/3675435/

Okamoto, Y., Haraguchi, Y., Yoshida, A., Takahashi, H., Yamanaka, K., Sawamura, N., Asahi, T., & Shimizu, T. (2022). Proliferation and differentiation of primary bovine myoblasts using Chlorella vulgaris extract for sustainable production of cultured meat. *Biotechnology Progress*, 38(3), e3239. 10.1002/btpr.323935073462

Ong, S., Choudhury, D., & Naing, M. W. (2020). Cell-based meat: Current ambiguities with nomenclature. *Trends in Food Science & Technology*, 102, 223–231. 10.1016/j.tifs.2020.02.010

Orellana, N., Sánchez, E., Benavente, D., Prieto, P., Enrione, J., & Acevedo, C. A. (2020). A New Edible Film to Produce In Vitro Meat. *Foods*, 9(2), 185. 10.3390/foods902018532069986

Park, Y. H., Gong, S. P., Kim, H. Y., Kim, G. A., Choi, J. H., Ahn, J. Y., & Lim, J. M. (2013). Development of a serum-free defined system employing growth factors for preantral follicle culture. *Molecular Reproduction and Development*, 80(9), 725–733. 10.1002/mrd.2220423813589

Pathirana, R., & Carimi, F. (2024). Plant Biotechnology-An Indispensable Tool for Crop Improvement. *Plants*, 13(8), 1133. 10.3390/plants1308113338674542

Post, M. J. (2014). Cultured beef: Medical technology to produce food. *Journal of the Science of Food and Agriculture*, 94(6), 1039–1041. 10.1002/jsfa.647424214798

Purdy, C. (2019). The first cell-cultured meat will cost about $50. https://qz.com/1598076/the-first-cell-cultured-meat-will-cost-about-50

Reiss, J., Robertson, S., & Suzuki, M. (2021). Cell sources for cultivated meat: Applications and considerations throughout the production workflow. *International Journal of Molecular Sciences*, 22(14), 7513. 10.3390/ijms2214751334299132

Rubio, N. R., Fish, K. D., Trimmer, B. A., & Kaplan, D. L. (2019). In Vitro Insect Muscle for Tissue Engineering Applications. *ACS Biomaterials Science & Engineering*, 5(2), 1071–1082. 10.1021/acsbiomaterials.8b0126133405797

Santo, R. E., Kim, B. F., Goldman, S. E., Dutkiewicz, J., Biehl, E. M., Bloem, M. W., Neff, R. A., & Nachman, K. E. (2020). Considering plant-based meat substitutes and cell-based meats: A public health and food systems perspective. *Frontiers in Sustainable Food Systems*, 4, 134. 10.3389/fsufs.2020.00134

Sapontzis, S. (2004). Food for thought: The debate over eating meat. *Environmental Values*, 15(2), 264–267.

Saunders, R. E., & Derby, B. (2014). Inkjet printing biomaterials for tissue engineering: Bioprinting. *International Materials Reviews*, 59(8), 430–448. 10.1179/1743280414Y.0000000040

Schaefer, G. O., & Savulescu, J. (2014). The ethics of producing in vitro meat. *Journal of Applied Philosophy*, 31(2), 188–202. 10.1111/japp.1205625954058

Schumacher, H. M., Westphal, M., & Heine-Dobbernack, E. (2015). Cryopreservation of plant cell lines. *Cryopreservation and Freeze-Drying Protocols*, 423-429. https://doi.org/10.1007/978-1-4939-2193-5_21

Searchinger, T., Waite, R., Hanson, C., Ranganathan, J., Dumas, P., & Matthews, E. (2018). Creating a sustainable food future. https://research.wri.org/sites/default/files/2019-07/ WRR_Food_Full_Report_0.pdf

Sexton, A. E., Garnett, T., & Lorimer, J. (2019). Framing the future of food: The contested promises of alternative proteins. *Environment and Planning. E, Nature and Space*, 2(1), 47–72. 10.1177/2514848619827009 32039343

Spier, M. R., Vandenberghe, L. P. D. S., Medeiros, A. B. P., & Soccol, C. R. (2011). *Application of different types of bioreactors in bioprocesses. Bioreactors: Design, properties and applications*. Nova Science Publishers, Inc.

Suntornnond, R., Yap, W. S., Lim, P. Y., & Choudhury, D. (2024). Redefining the Plate: Biofabrication in Cultivated Meat for Sustainability, Cost-Efficiency, Nutrient Enrichment, and Enhanced Organoleptic Experiences. *Current Opinion in Food Science*, 101164, 101164. Advance online publication. 10.1016/j.cofs.2024.101164

The Good Food Institute. (2023). GOOD Meat and UPSIDE Foods approved to sell cultivated chicken following landmark USDA action - The Good Food Institute. Retrieved May 22, 2024, from https://gfi.org/press/good-meat-and-upside-foods-approved-to-sell-cultivated-chicken-following-landmark-usda action/#:~:text=WASHINGTON%20(June%2021%2C%202023),history%20of%20food%20and%20agriculture

Tuomisto, H. L. (2019). The eco-friendly burger: Could cultured meat improve the environmental sustainability of meat products? *EMBO Reports*, 20(1), e47395. 10.15252/embr.201847395 30552146

Tuomisto, H. L., & Teixeira de Mattos, M. J. (2011). Environmental impacts of cultured meat production. *Environmental Science & Technology*, 45(14), 6117–6123. 10.1021/es200130u 21682287

U.S. Food And Drug Administration. (2023, March 21). F. F. S. A. A. Human Food Made with Cultured Animal Cells. https://www.fda.gov/food/food-ingredients-packaging/human-food-made-cultured-animal-cells

Ushiyama, K. (1991). *Large scale culture of ginseng*. Plant Cell Culture in Japan. CMC.

Verbruggen, S., Luining, D., van Essen, A., & Post, M. J. (2018). Bovine myoblast cell production in a microcarriers-based system. *Cytotechnology*, 70(2), 503–512. 10.1007/s10616-017-0101-8 28470539

Vergeer, R. (2021). TEA of cultivated meat. Future projections of different scenarios.

Verzijden, K. (2019). *Regulatory pathways for clean meat in the EU and the US- differences & analogies*. Food Health Legal.

Wang, D., Liu, W., Han, B., & Xu, R. (2005). The bioreactor: A powerful tool for large-scale culture of animal cells. *Current Pharmaceutical Biotechnology*, 6(5), 397–403. 10.2174/138920105774370580016248813

Wang, S., Li, K., Gao, H., Liu, Z., Shi, S., Tan, Q., & Wang, Z. (2021). Ubiquitin-specific peptidase 8 regulates proliferation and early differentiation of sheep skeletal muscle satellite cells. *Czech Journal of Animal Science*, 66(3), 87–96. Advance online publication. 10.17221/105/2020-CJAS

Wang, Y., Xiao, X., & Wang, L. (2020). In vitro characterization of goat skeletal muscle satellite cells. *Animal Biotechnology*, 31(2), 115–121. 10.1080/10495398.2018.155123030602329

Chapter 3
The Environmental Impact of Cellular Agriculture

Gunavathy Selvarajh
https://orcid.org/0000-0002-6041-6660
Lincoln University College, Malaysia

Farzana Yasmin
Lincoln University College, Malaysia

Udugalage Isuru Harsha Kumara
Lincoln University College, Malaysia

Jayasree Kanathasan
https://orcid.org/0000-0001-7509-6746
Lincoln University College, Malaysia

Devi Nallapan
https://orcid.org/0000-0003-2673-5614
Lincoln University College, Malaysia

ABSTRACT

Cellular agriculture, known as cultured or lab-grown meat production, involves the cultivation of animal cells in vitro to generate meat products without the need for traditional animal rearing and slaughtering. There are numerous environmental issues stemming from traditional animal agriculture. These include land degradation, water scarcity, greenhouse gas emissions, and biodiversity loss, which collectively contribute to environmental degradation and climate change. Against this backdrop, cellular agriculture offers a promising solution by leveraging bio-

DOI: 10.4018/979-8-3693-4115-5.ch003

technological advancements to produce animal-derived products without the need for intensive livestock farming. This chapter evaluates the environmental impact of cellular agriculture across multiple dimensions. Further, this chapter also evaluates the water conservation, followed by biodiversity conservation. The final topic is on the reduced energy consumption due to the implementation of cellular agriculture.

INTRODUCTION TO CELLULAR AGRICULTURE

With the world's population estimated to reach 10 billion by 2050, there will be a significant increase in demand for food, particularly animal protein. Traditional animal agriculture, which provides a considerable amount of the world's protein, is increasingly seen as unsustainable due to its large environmental impact. According to the Food and Agriculture Organization (2017), livestock production accounts for roughly 14.5% of global greenhouse gas emissions, mostly from methane produced by enteric fermentation in ruminants and nitrous oxide from manure management (Gerber et al., 2013). Furthermore, the sector is a major cause of deforestation, water scarcity, and biodiversity loss (Steinfeld et al., 2006). These environmental challenges need a transition to more sustainable food production systems. Cellular agriculture, which involves the in vitro development of animal cells to produce meat products, is a viable alternative that has the potential to reduce many of the environmental problems associated with traditional livestock production practices. There are numerous benefits of cellular agriculture, and it is listed in Figure 1.

Figure 1. The benefits of cellular agriculture on environmental aspects

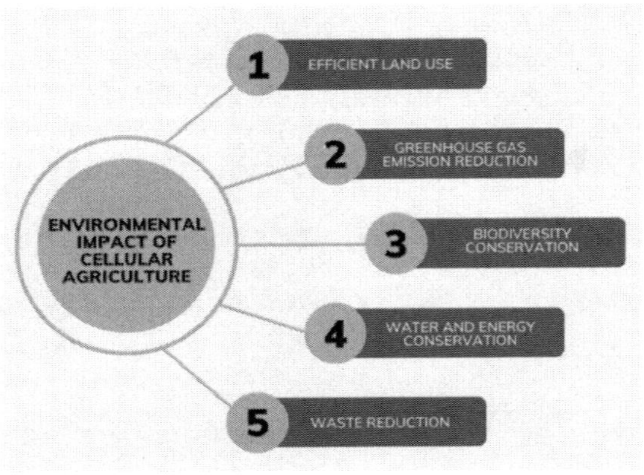

Traditional animal agriculture has a substantial impact on land and water resources. The growth of pastureland and the cultivation of feed crops like soy and maize have resulted in extensive deforestation, especially in tropical region (Nepstad et al., 2014). This deforestation not only causes habitat loss and biodiversity reduction, but it also exacerbates climate change by diminishing the planet's ability to retain carbon. Furthermore, cattle husbandry is extremely water intensive. According to Mekonnen and Hoekstra (2012), producing one kilogram of beef requires around 15,000 litres of water, which includes both direct consumption by the animals and water used to grow their feed. In water-stressed areas, the increased demand for water exacerbates the shortage of freshwater supplies, posing a serious hazard to both ecosystems and humans. It reduces the water consumption by 96%, reducing demand on freshwater resources and promoting water conservation (Tuomisto et al., 2015).

Cellular agriculture is a significant biotechnological innovation where the cultured meat production may require up to 99% less land than traditional cattle farming (Tuomisto and Teixeira de Mattos, 2011). The researchers estimated that the production of 1,000 kg of cultured meat would require 7–45% less energy, 99% less land, and 82–96% less water, depending on the species of meat. Furthermore, cellular agriculture can dramatically reduce animal waste production and greenhouse gas emissions by removing methane produced by enteric fermentation in livestock. This approach produces little waste and can be integrated into closed-loop systems that recycle nutrients, significantly reducing its environmental impact (Post, 2012). These reductions in emissions and waste make cellular agriculture a promising strategy for mitigating climate change and promoting sustainable food systems.

This showed that environmental benefits associated with the global replacement of livestock with the cellular agriculture products is promising and can be adopted worldwide. This can save the environment from degradation and pollution. The cellular agriculture needs to be developed further by implementing extensive research and collaboration, so that can further conserve the environment. Hence, discussion regarding the impact of cellular agriculture on the environment is well elaborated in each section for knowledge and further understanding on the topics.

GREENHOUSE GAS EMISSION REDUCTION

Cellular agriculture, which focuses on producing animal products from cell cultures rather than traditional cattle, has immense potential for reducing greenhouse gas (GHG) emissions. Traditional animal farming emits significant amounts of GHGs, including methane (CH_4), nitrous oxide (N_2O), and carbon dioxide (CO_2). The GHG emission is seen as a devastating form of gas that impacting environment.

The gas are being released from various sources such as farming activity, industry, and combustion.

Traditional livestock production accounts for a significant share of world GHG emissions. Livestock GHG emission is approximately 60% of all agricultural emissions (Ammar et al., 2020). Enteric fermentation accounts for approximately 63% of animal emissions, followed by manure management (12%) and the use of dung and urine on grassland (25%) (Tubeillo et al., 2015). It had been stated that dairy cows and granivorous animals produced the most GHG, accounting for 311,000 kg and 430,000 kg per farm, respectively (Golasa et al., 2021). Ruminant animals, such as cattle, sheep, and goats, emit a lot of methane (CH_4) as a byproduct of their digestion. CH_4 represents about 92% of the total GHGs and it has a 28-36 times higher global warming potential (GWP) than CO_2 during a 100-year time span. Cattle dairy and small ruminants (sheep and goats) are the first animal species producers of CH_4, participating with more than 52% and 40% of the total CH_4 emissions, respectively (Tubeillo et al., 2015). The author even stated that, farm animals also produce high amount of nitrous dioxide gas where, dairy cattle contribute with more than 50% of the total N_2O emissions, followed by small ruminants (>30%), poultry (>10%), and other species (<4%). The extensive GHG emission contributes to the world global warming. However, the rapid increment in cellular agriculture offers a great reduction in GHG emission.

Cellular agriculture, particularly the production of cultured meat, considerably reduces greenhouse gas (GHG) emissions through a variety of ways when compared to traditional livestock farming. First, cultured meat production removes enteric fermentation, a process in which ruminant animals release methane, a harmful greenhouse gas. Cellular agriculture avoids the digestive processes of ruminants by growing meat directly from animal cells in bioreactors, hence reducing methane emissions from enteric fermentation. Tuomisto and de Mattos (2011) found that cultured meat could cut GHG emissions by up to 96% compared to conventional beef production. This huge decrease is mostly due to the absence of methane production and the substantially lower energy. This is in accordance with El Wali et al., (2024) who stated that cultured meat production, require fewer resources and generate fewer greenhouse gases compared to conventional meat production, thus offering a more sustainable alternative.

Additionally, cellular agriculture reduces emissions from feed production and manure management. Traditional livestock farming produces nitrous oxide, a greenhouse gas with a global warming potential 298 times that of carbon dioxide, from manure and feed production. Decomposition of one metric ton of organic solid waste can potentially release 50 - 110 m^3 of carbon dioxide and 90 - 140 m^3 of methane into the atmosphere (Macias-Corral et al., 2008). On the other hand, producing meat through cultured meat uses no feed and less land, which means that nitrous oxide emissions are reduced. According to life cycle assessments by Mattick et al., (2015), cultured meat can lower carbon dioxide emissions by 78 - 96%. The production of cultured meat is more resource-efficient, therefore this reduction is made possible by using less land and energy. Thus, by eliminating methane emissions and significantly reducing carbon dioxide and nitrous oxide emissions, cellular agriculture offers a promising pathway to mitigate climate change impacts associated with food production.

WASTE MANAGEMENT

All of the cattle, poultry, and other farm animal-rearing sectors, are critical to the global food chain, but conventional animal rearing generates waste during the rearing process. Huge wastes are being generated from livestock farming (Table 1). The waste production is huge and not managed properly. The waste high in nitrogen and phosphorus run off and pollute water bodies, lead to eutrophication, promote toxic algal blooms, and deplete oxygen, all of which are fatal to aquatic life (Reiss et al., 2021). Furthermore, manure breakdown produces methane, a greenhouse gas that is twenty-eight times more powerful than carbon dioxide over a century, and nitrous oxide, a greenhouse gas that is two hundred and sixty-five times as efficient as carbon dioxide. Methane and nitrous oxide are greenhouse gases that have a warming potential of 28 and 265 times respectively, compared to carbon dioxide. Such emissions cause climate change through leading to high temperatures, storminess, and rise in sea levels (Voss et al., 2024).

Table 1. Animal waste generation

Animals	Quantity of waste production in a farm	References
Cattle cow	65,700 – 109,500 kg/year	Parihar et al. (2019)
Cow	130, 000 kg/year	Soyer and Yilmaz (2020)
Cow, fish, and goat	133, 225 kg/year	Jalil et al. (2017)
Livestock waste	6.2 million tonnes/year	Afazeli et al. (2014)
Livestock	3.2 million tonnes/year	Prasertsan and Sajjakulnukit (2006)

Wastewater produced through the animal production systems inclusive of animal housing, cleaning of the housing structures and systems used in management of animal waste run off to the soil or water bodies. The analysed samples of this wastewater have been characterized by high content of organic substances, suspensions, nutrients, and pathogenisms that threaten water quality and human health. Wastewater containing such substances can be treated energy–intensively, and the processes may result in producing one or more new waste streams which re-enters the cycle of polluting the environment (Reiss et al., 2021).

Furthermore, the animals supply other products such as manure, wastewater, and many by-products such as bones, hides, blood and inedible organs from the continuous slaughtering and processing of animals. The byproducts mentioned here cannot be handled, treated, and disposed of like the other kinds of wastes, further complicating waste management. Disposal of animal byproducts and wastes usually have adverse effects on the soil and water resources through polluting them hence endangering the lives of humans and animals through pathogenic means as well as through toxicity (Helliwell and Burton, 2021).

A waste by-product associated with the production of processed meat, dairy and eggs to different households and various market segments. Co-products like whey obtained while producing cheese, inedible offal from meat production, feathers produced from deboning poultry are difficult to dispose and would pollute the environment if no proper disposal system. The large quantities and variability of waste produced in conventional animal production makes it difficult and expensive to handle (Gasteratos, 2019).

However, cellular agriculture has the potential to eliminate the inefficiencies and environmental risks associated with traditional animal production. Because cellular agriculture involves the direct synthesis of animal products from cell cultures, it does not entail animal rearing, which eliminates several of the procedures that are prone to generate waste in agricultural production (Reiss et al., 2021). Thus, no animals would produce methane and manure to eventually become a threat to the water supply as well as contribute to the increased levels of greenhouse gases. This alone is a good improvement towards lessening the impact on environment due to the livestock production processes (Reiss et al., 2021).

In cellular agriculture, the amount of waste which is expected to be produced from cell culture is significantly low because most cell culture facilities are closed systems. These are designed for nutrient metered delivery and waste products removal to minimize the flow and load of wastewaters in reaching to those of conventional farming enterprises. It also serves the purpose of preserving water, lessening the energy required and hence the expenses for purification of the wastewater (Perreault et al., 2023). The waste produced in cellular agriculture is mainly volume-excessive spent cell culture media which is a lot more homogenous in comparison to the diverse wastes that are derived from conventional livestock farming. This homogeneity makes waste treatment a relatively easy task and there is considerable potential for valorisation to be achieved. For example, spent media can undergo further activities such as among them being separation processes with the aim of obtaining several product forms such as amino acids, lipids, as well as the vitamins, which in one way or the other can be used in food industries including the feed and the pharmaceutical industries. Not only does it decrease shrimps' wastage but it also optimises on the chances of making an extra sale hence improving the company's revenue streams (Haraguchi and Shimizu, 2021). Additionally, since cellular agriculture does not require animal slaughtering in supplying its products, issues concerning byproduct management, another point of focus on waste management, do not arise. It is clear that cellular agriculture reduces and prevents the production of waste.

WATER CONSERVATION

The rapidly emerging area of cellular agriculture aims to generate agricultural goods such as dairy, meat, and other animal-derived products directly from cell cultures rather than living animals. The meat grown in controlled environment is environmentally friendly and use lesser water for the meat production, which conserve the water (George, 2020).

Water conservation is needed for fostering efficient resource use, environmental sustainability, and long-term food security. According to the Mekonnen and Hoekstra (2012), agriculture accounts for around 70% of global freshwater withdrawals, with animal husbandry being particularly water-intensive. On the other hand, cellular agriculture uses significantly less water compared to traditional livestock farming (Jahir et al., 2023). This is because it does not involve raising entire animals, which consume vast amounts of water for feed production, drinking, and cleaning. In addition, cultured cells also require minimal water compared to maintaining live animals (Rischer et al., 2020). Installing closed-loop water systems in cellular agriculture enterprises significantly reduce water consumption. Treatment and recycling of the water used for cell culture and facility operations capable to reduce the amount of freshwater required and the amount of wastewater emitted. The cellular agriculture offers huge benefits in terms of water conservation in comparison to animal farming system.

Traditional livestock farming requires large amounts of water to grow feed crops such as alfalfa, soybeans, and corn. The irrigation of these crops' accounts for a major portion of global water consumption. For example, soybeans are known for having a high-water consumption because they require extensive irrigation to grow well. A kilogram of soybeans takes approximately 1800 litres of water. Cellular agriculture saves a lot of water by eliminating the need for huge quantities of cropland and associated irrigation by producing meat directly from cell cultures (George, 2020). Besides, in livestock farming, water also need to be provided for animals to drink, as well as other agricultural activities such as waste management and cleaning. This is opposite case in cellular agriculture because there are no live animals involved, water is either avoided entirely or used sparingly. Sharma et al., (2015) discovered that growing meat through culture media can use less water.

Additionally, water contamination due to livestock production could be prevented by growing meat cellularly. There is a frequents runoff of chemicals, animal excrement, which contains antibiotics, dangerous pathogens, and excess nutrients (such as nitrogen and phosphorus) into water bodies. Due to this, aquatic ecosystems, human health, rivers, lakes, and groundwater are in risk. Because cellular agriculture is a closed, regulated system, it reduces the possibility of water pollution by doing away with the need for extensive animal husbandry and waste management (Zhang et al., 2019). Due to the approach of cellular agriculture, drought season or water scarcity would not be a problem for meat production. There will be a continuous meat production despite of water scarcity because it is being produced in a controlled environments system which uses less water.

LAND USE EFFICIENCY

Cellular agriculture, which involves the production of lab-grown meat, promotes more efficient use of available land. Cellular agriculture uses substantially less area than conventional cattle farming, potentially reducing land use by up to 99% (Tuomisto & Teixeira de Mattos, 2011). This is because cellular agriculture requires little space since it involves cultivating animal cells directly in controlled conditions, such as bioreactors, rather than producing entire animals in farms.

This strategy eliminates the need for extensive pastures for grazing and large fields for cultivating feed crops. It had been stated that extensive grazing of beef cattle has led to conversion of forest to animal farms (Molossi et al., 2020). The process of growing meat in laboratories enables the generation of protein with a substantially smaller spatial footprint. This high land use efficiency is due to the elimination of the need for feed cultivation and the direct conversion of nutrients into animal tissue without the intermediate step of raising and sustaining whole animals. As a result, cellular agriculture can be performed in urban areas or locations that are unsuitable for conventional farming, reducing the demand for arable land and forests.

Unlike cellular agriculture, traditional livestock farming utilizes a huge land, and this leads to deforestation. Deforestation brings numerous adverse effects to both environment and living things because deforestation causes endangered species extinction, increase of temperature, and soil degradation (Lim et al., 2017). The removal of trees disrupts the soil structure, leading to increased erosion, nutrient loss, and a decline in soil fertility. Besides, it leads to many species' extinction. The nature is losing approximately 80,000 acres of tropical rainforest, and along with it around 50,00 species plant and insect species are being wiped out (George, 2020).

As forests are cleared, the natural places of numerous organisms are destroyed, disrupting the complex web of life that maintains ecological equilibrium. This habitat loss is especially severe for highly specialized or endemic species, which occur only in certain geographical locations and cannot easily relocate or adapt to new habitats (Laurance, 2010). Many tropical birds, amphibians, and large animals are particularly vulnerable. Tropical forests, which house more than half of the world's species, are extremely threatened, and their loss directly correlates with a substantial decrease in global biodiversity (Laurance and Peres, 2006). There are many drawbacks of traditional animal farming system. Hence, it can be stated that the rise of cellular agriculture offers numerous benefits in terms of land use efficiency by minimizing deforestation activity.

BIODIVERSITY ECOSYSTEM CONSERVATION AND HEALTH IMPROVEMENT

Biodiversity is the broad range of living things, including microbes, plants, and animals, as well as the complex webs of interactions that these species establish inside their respective ecosystems. Maintaining biodiversity is critical to maintaining ecosystem health and balance as well as the resilience of our food production systems (Chappell and LaValle, 2011). The preservation of the complex balance of ecosystems and the enhancement of human health and well-being depends heavily on the conservation of biodiversity through cellular agriculture. Studying the critical elements that affect cellular agriculture's ability to conserve biodiversity and the ways in which these methods can improve the health of ecosystems is imperative (Newman et al., 2023; Tuomisto, 2024).

Essential ecosystem services, or the many advantages that ecosystems provide to both humans and the environment, are largely provided by cellular agriculture. The protection of natural habitat, food production, and security stand out as being especially important among the many ecosystem services. By producing cultured meat, milk, and other animal products without the need for large-scale farming, cellular agriculture offers a ground-breaking possibility to revolutionize food production (Ercili-Cura and Barth, 2021; Mendly-Zambo et al., 2021). This innovative method minimizes the environmental impact of conventional agriculture while simultaneously conserving resources and reducing the amount of land used for farming.

Cellular agriculture also has the potential to improve soil quality. Traditional farming techniques, such as overgrazing, monoculture planting, and extensive use of chemical inputs, frequently cause soil deterioration, resulting in loss of fertility and erosion. The lab grown meat minimizes the need for large fields, encourage the rehabilitation of degraded lands and the reestablishment of natural plants. Eventually, this improves soil health and fostering higher biodiversity. It can be stated that cellular agriculture helps ecosystems recover and thrive by reducing the demand for arable land, resulting in a more sustainable and balanced environment.

Besides, the health of human also can be improved by consumption of cellular meat. The dietary composition of traditional animal-based items, such meat, is limited. The cultured meat composition and quality of the meat can be controlled, so cultured meat may be beneficial to health. According to Allan et al., (2019), it can change the flavour, fat content, and particularly the proportion of saturated to unsaturated fatty acids. Producing proteins with specific qualities is made possible by cellular agriculture (Webb et al., 2021). Furthermore, where protein synthesis is not naturally occurring, precise fermentation techniques allow for the synthesis of proteins in cellular factories. The use of cellular agriculture has the potential to provide protein products for safe human consumption in a way that is economical, environmentally responsible, and morally upright. This degree of personalization could perhaps provide healthier options.

Cellular meat is produced in a controlled and sterile technique that significantly reduces the danger of zoonotic illnesses. Since humans and animals are not in close proximity, zoonotic diseases like swine and bird flu are less likely to arise (Roy et al., 2021; Munteanu et al., 2021). Furthermore, the process of producing cultured meat in controlled laboratories or bioreactors guarantees a sterile environment, which effectively inhibits contamination and the formation of new diseases (Rubio, 2020). Also, cellular agriculture greatly lowers the possibility of disease transfer from animals to people and does away with the need for dioxins and antibiotics. This resolve concerns about antibiotic resistance in addition to improving public health (Munteanu et al., 2021).

DECREASED ENERGY CONSUMPTION

Cellular agriculture offers a viable way to reduce energy usage while minimizing the environmental implications of traditional animal rearing. Traditional cattle production is energy intensive, needing large amounts of feed, water, and land while emitting significant amounts of greenhouse gases. Other than that, overuse of pesticides and chemical pollution as well as efforts to support the biodiversity of livestock and fishery stock leads to immense energy use which in turn causes environmental damage (Kodaparthi et al., 2024). Cellular agriculture, on the other hand, produces animal products under regulated conditions using advanced biotechnologies such as tissue engineering, regenerative medicine, fermentation, and synthetic biology. This strategy considerably minimizes the requirement for large-scale agricultural inputs while also lowering total energy use. Cellular agriculture has the potential to reduce yearly agricultural greenhouse gas emissions by 52% and land utilization by 83% by 2050 (Wali et al., 2024).

One of the most significant benefits of cellular agriculture is its capacity to be fueled by renewable energy sources, which further reduces its carbon impact. Cultured beef and other cellular agriculture products can be produced in bioreactors using green energy technologies such as solar and wind power, resulting in low-carbon energy usage. Furthermore, using plant-based scaffolds in meat production eliminates the need for fetal bovine serum, which is commonly utilized in cell cultures, lowering the energy and environmental costs associated with animal agriculture (Messmer et al., 2022).

Another intriguing component of cellular agriculture is its ability to recycle waste resources, which increases energy efficiency. For example, seafood waste can be used to create cultured fisheries products, converting what would otherwise be garbage into important inputs for food production (Tsuruwaka et al., 2022). This strategy not only decreases waste but also the energy necessary for trash handling and disposal, helping to create a more sustainable food production system. Furthermore, the use of microbial and recombinant proteins in cellular agriculture provides an energy-efficient alternative to traditional animal farming, as these proteins may be generated with fewer resources and have a smaller environmental impact (Ge et al., 2023).

Despite the tremendous promise of cellular agriculture, obstacles persist, including the rising costs of raw materials used in renewable energy technologies, such as aluminium and nickel (IEA, 2023). However, with continued study, governmental assistance, and technological breakthroughs, cellular agriculture's energy usage can be reduced. Effective monitoring and the application of energy-efficient procedures are critical to maximizing the environmental benefits of cellular agriculture. Companies that create cultured meat, such as Shiok Meats, show how ecologically friendly procedures and reduced animal health resource utilization may attract significant investment and drive sustainable production. Furthermore, developing technologies such as 3-D printing, vertical farming, and digital agriculture can help to reduce the environmental impact of food production, making cellular agriculture important (Bapat et al., 2022). By integrating these technologies, cellular agriculture can become a cornerstone of sustainable food systems, providing a viable solution to meet the growing global demand for food while minimizing environmental impact.

CONCLUSION

The environmental impact of cellular agriculture is positive, providing solutions to some of the most serious issues linked with traditional animal agriculture. The transition to cultured meat production can reduce land use, conserve water, lower greenhouse gas emissions, and reduce waste, all of which contribute to a more sustainable and ecologically friendly food system. As global protein consumption rises, cellular agriculture offers a feasible and required alternative to traditional animal rearing. Continued research and development, together with supportive legislative frameworks, will be required to fully realize the environmental benefits of this novel method to food production.

REFERENCES

Afazeli, H., Jafari, A., Rafiee, S., & Nosrati, M. (2014). An investigation of biogas production potential from livestock and slaughterhouse wastes. *Renewable & Sustainable Energy Reviews*, 34, 380–386. 10.1016/j.rser.2014.03.016

Allan, S. J., De Bank, P. A., & Ellis, M. J. (2019). Bioprocess design considerations for cultured meat production with a focus on the expansion bioreactor. *Frontiers in Sustainable Food Systems*, 3, 44. 10.3389/fsufs.2019.00044

Ammar, H., Abidi, S., Ayed, M., Moujahed, N., deHaro Martí, M. E., Chahine, M., & Hechlef, H. (2020). Estimation of Tunisian greenhouse gas emissions from different livestock species. *Agriculture*, 10(11), 562. 10.3390/agriculture10110562

Bapat, S., Koranne, V., Shakelly, N., Huang, A., Sealy, M. P., Sutherland, J. W., Rajurkar, K. P., & Malshe, A. P. (2022). Cellular agriculture: An outlook on smart and resilient food agriculture manufacturing. *Smart and Sustainable Manufacturing Systems*, 6(1), 1–11. 10.1520/SSMS20210020

Bryant, C. J. (2020). Culture, meat, and cultured meat. *Journal of Animal Science*, 98(8), skaa172. 10.1093/jas/skaa17232745186

Chappell, M. J., & LaValle, L. A. (2011). Food security and biodiversity: Can we have both? An agroecological analysis. *Agriculture and Human Values*, 28(1), 3–26. 10.1007/s10460-009-9251-4

El Wali, M., Rahimpour Golroudbary, S., Kraslawski, A., & Tuomisto, H. L. (2024). Transition to cellular agriculture reduces agriculture land use and greenhouse gas emissions but increases demand for critical materials. *Communications Earth & Environment*, 5(1), 1–17. 10.1038/s43247-024-01227-8

Ercili-Cura, D., & Barth, D. (2021). *Cellular Agriculture: Lab Grown Foods* (Vol. 8). American Chemical Society.

Food and Agriculture Organization of the United Nations. (2017). *The future of food and agriculture: Trends and challenges*.

Gasteratos, K. (2019). 90 Reasons to consider cellular agriculture.

Ge, C., Selvaganapathy, P. R., & Geng, F. (2023). Advancing our understanding of bioreactors for industrial-sized cell culture: Health care and cellular agriculture implications. *American Journal of Physiology. Cell Physiology*, 325(3), C580–C591. 10.1152/ajpcell.00408.202237486066

George, A. S. (2020). The development of lab-grown meat which will lead to the next farming revolution. *Proteus Journal*, 11(7), 1–25.

Gerber, P. J., Steinfeld, H., Henderson, B., Mottet, A., Opio, C., Dijkman, J., . . . Tempio, G. (2013). *Tackling climate change through livestock: a global assessment of emissions and mitigation opportunities*. Food and agriculture Organization of the United Nations (FAO).

Gołasa, P., Wysokiński, M., Bieńkowska-Gołasa, W., Gradziuk, P., Golonko, M., Gradziuk, B., Siedlecka, A., & Gromada, A. (2021). Sources of greenhouse gas emissions in agriculture, with particular emphasis on emissions from energy used. *Energies*, 14(13), 3784. 10.3390/en14133784

Haraguchi, Y., & Shimizu, T. (2021). Microalgal culture in animal cell waste medium for sustainable 'cultured food' production. *Archives of Microbiology*, 203(9), 5525–5532. 10.1007/s00203-021-02509-x34426852

Helliwell, R., & Burton, R. J. (2021). The promised land? Exploring the future visions and narrative silences of cellular agriculture in news and industry media. *Journal of Rural Studies*, 84, 180–191. 10.1016/j.jrurstud.2021.04.002

Jahir, N. R., Ramakrishna, S., Abdullah, A. A. A., & Vigneswari, S. (2023). Cultured meat in cellular agriculture: Advantages, applications and challenges. *Food Bioscience*, 53, 102614. 10.1016/j.fbio.2023.102614

Jalil, A., Basar, S., Karmaker, S., Ali, A., Choudhury, M. R., & Hoque, S. (2017). Investigation of biogas Generation from the waste of a vegetable and cattle market of Bangladesh. *International Journal of Waste Resources*, 7(1). Advance online publication. 10.4172/2252-5211.1000283

Kodaparthi, A., Kondakindi, V. R., Kehkashaan, L., Belli, M. V., Chowdhury, H. N., Aleti, A., & Chepuri, K. (2024). Environmental Conservation for Sustainable Agriculture. In *Prospects for Soil Regeneration and Its Impact on Environmental Protection* (pp. 15–45). Springer Nature Switzerland. 10.1007/978-3-031-53270-2_2

Laurance, W. F. (2010). Habitat destruction: death by a thousand cuts. Conservation biology for all, 1(9), 73-88.

Laurance, W. F., & Peres, C. A. (Eds.). (2006). *Emerging threats to tropical forests*. University of Chicago Press.

Lim, C. H., Choi, Y., Kim, M., Jeon, S. W., & Lee, W. K. (2017). Impact of deforestation on agro-environmental variables in cropland, North Korea. *Sustainability (Basel)*, 9(8), 1354. 10.3390/su9081354

Macias-Corral, M., Samani, Z., Hanson, A., Smith, G., Funk, P., Yu, H., & Longworth, J. (2008). Anaerobic digestion of municipal solid waste and agricultural waste and the effect of co-digestion with dairy cow manure. *Bioresource Technology*, 99(17), 8288–8293. 10.1016/j.biortech.2008.03.05718482835

Mattick, C. S. (2018). Cellular agriculture: The coming revolution in food production. *Bulletin of the Atomic Scientists*, 74(1), 32–35. 10.1080/00963402.2017.1413059

Mekonnen, M. M., & Hoekstra, A. Y. (2012). A global assessment of the water footprint of farm animal products. *Ecosystems (New York, N.Y.)*, 15(3), 401–415. 10.1007/s10021-011-9517-8

Molossi, L., Hoshide, A. K., Pedrosa, L. M., Oliveira, A. S. D., & Abreu, D. C. D. (2020). Improve pasture or feed grain? Greenhouse gas emissions, profitability, and resource use for nelore beef cattle in Brazil's Cerrado and Amazon Biomes. *Animals (Basel)*, 10(8), 1386. 10.3390/ani1008138632785150

Munteanu, C., Mireşan, V., Răducu, C., Ihuţ, A., Uiuiu, P., Pop, D., Neacşu, A., Cenariu, M., & Groza, I. (2021). Can cultured meat be an alternative to farm animal production for a sustainable and healthier lifestyle? *Frontiers in Nutrition*, 8, 749298. 10.3389/fnut.2021.74929834671633

Nepstad, D., McGrath, D., Stickler, C., Alencar, A., Azevedo, A., Swette, B., & Hess, L. (2014). Slowing Amazon deforestation through public policy and interventions in beef and soy supply chains. *Science, 344*(6188), 1118-1123.

Newman, L., Fraser, E., Newell, R., Bowness, E., Newman, K., & Glaros, A. (2023). Cellular agriculture and the sustainable development goals. In *Genomics and the global bioeconomy* (pp. 3–23). Academic Press. 10.1016/B978-0-323-91601-1.00010-9

Parihar, S. S., Saini, K. P. S., Lakhani, G. P., Jain, A., Roy, B., Ghosh, S., & Aharwal, B. (2019). Livestock waste management: A review. *Journal of Entomology and Zoology Studies*, 7(3), 384–393.

Perreault, L. R., Thyden, R., Kloster, J., Jones, J. D., Nunes, J., Patmanidis, A. A., Reddig, D., Dominko, T., & Gaudette, G. R. (2023). Repurposing agricultural waste as low-cost cultured meat scaffolds. *Frontiers in Food Science and Technology*, 3, 1208298. 10.3389/frfst.2023.1208298

Post, M. J. (2012). Cultured meat from stem cells: Challenges and prospects. *Meat Science*, 92(3), 297–301. 10.1016/j.meatsci.2012.04.00822543115

Prasertsan, S., & Sajjakulnukit, B. (2006). Biomass and biogas energy in Thailand: Potential, opportunity and barriers. *Renewable Energy*, 31(5), 599–610. 10.1016/j.renene.2005.08.005

Reiss, J., Robertson, S., & Suzuki, M. (2021). Cell sources for cultivated meat: Applications and considerations throughout the production workflow. *International Journal of Molecular Sciences*, 22(14), 7513. 10.3390/ijms2214751334299132

Rischer, H., Szilvay, G. R., & Oksman-Caldentey, K. M. (2020). Cellular agriculture—Industrial biotechnology for food and materials. *Current Opinion in Biotechnology*, 61, 128–134. 10.1016/j.copbio.2019.12.00331926477

Roy, B., Hagappa, A., Ramalingam, Y. D., & Mahalingam, N. (2021). A review on lab-grown meat: Advantages and disadvantages. *Quest International Journal of Medical and Health Sciences*, 4(1), 19–24.

Rubio, N. R., Xiang, N., & Kaplan, D. L. (2020). Plant-based and cell-based approaches to meat production. *Nature Communications*, 11(1), 1–11. 10.1038/s41467-020-20061-y33293564

Sharma, S., Thind, S. S., & Kaur, A. (2015). In vitro meat production system: Why and how? *Journal of Food Science and Technology*, 52(12), 7599–7607. 10.1007/s13197-015-1972-326604337

Soyer, G., & Yilmaz, E. (2020). Waste management in dairy cattle farms in Aydın region. Potential of energy application. *Sustainability (Basel)*, 12(4), 1614. 10.3390/su12041614

Steinfeld, H., Gerber, P., Wassenaar, T. D., Castel, V., & De Haan, C. (2006). *Livestock's long shadow: environmental issues and options*. Food & Agriculture Org.

Tsuruwaka, Y., & Shimada, E. (2022). Reprocessing seafood waste: challenge to develop aquatic clean meat from fish cells. NPJ Science of Food, 6(1), 7.

Tubiello, F. N., Salvatore, M., Ferrara, A. F., House, J., Federici, S., Rossi, S., & Smith, P. (2015). The contribution of agriculture, forestry and other land use activities to global warming, 1990–2012. *Global Change Biology*, 21(7), 2655–2660. 10.1111/gcb.1286525580828

Tuomisto, H. L., Ellis, M. J., & Haastrup, P. (2015). *Environmental impacts of cultured meat: alternative production scenarios*. EU Sci. Hub-Eur. Comm.

Tuomisto, H. L., & Teixeira de Mattos, M. J. (2011). Environmental impacts of cultured meat production. *Environmental Science & Technology*, 45(14), 6117–6123. 10.1021/es200130u21682287

Voss, M., Valle, C., Calcio Gaudino, E., Tabasso, S., Forte, C., & Cravotto, G. (2024). Unlocking the Potential of Agrifood Waste for Sustainable Innovation in Agriculture. *Recycling*, 9(2), 25. 10.3390/recycling9020025

Webb, L., Fleming, A., Ma, L., & Lu, X. (2021). Uses of cellular agriculture in plant-based meat analogues for improved palatability. *ACS Food Science & Technology*, 1(10), 1740–1747. 10.1021/acsfoodscitech.1c00248

Chapter 4
Navigating Ethics and Animal Welfare

Devi Nallappan
https://orcid.org/0000-0003-2673-5614
Lincoln University College, Malaysia

Jayasree S. Kanathasan
https://orcid.org/0000-0001-7509-6746
Lincoln University College, Malaysia

ABSTRACT

Animal ethics encompasses moral, legal, and ethical standards governing human interactions with animals, including pre-clinical research and industrial settings. Animal welfare assesses animals' well-being and adaptability. Scientific evaluation of welfare has been significant, with scientists and philosophers working together to define the appropriate relationship between humans and animals. This chapter focuses on navigation ethics and animal welfare. It begins with an introduction to animal research ethics. Then, it delves into the general application, issues and development of animal ethics. The subsequent part highlights animal ethics in cellular agriculture and food production. It then discusses the connection between animal ethics and animal welfare. The chapter concludes with an overview of the challenges, limitations and prospects of integrating animal ethics and animal welfare. This review aims to provide new perspectives on ethics and animal welfare for researchers, policymakers, regulators, and others involved in animal research and welfare fields.

DOI: 10.4018/979-8-3693-4115-5.ch004

1. INTRODUCTION

Animal research has been a crucial component of numerous scientific advancements over the past century. It continues to help us understand various diseases and how they can be treated. The practice of using animals for experimental research dates back to ancient Greece, with Aristotle and Hippocrates being notable proponents (Franco, 2013). Although there have been documented reports of animal experiments since the 5th century B.C., the study has increased since the 19th century. Initially, animal experiments did not pose significant moral issues as they were in accordance with the Cartesian philosophy in the 17th century (Kiani et al., 2022). Today, most research institutions worldwide use non-human animals as experimental subjects to gain a better understanding of human diseases and explore potential treatments. Animal experimentation can advance medical science since animals like Zebrafish (*Danio rerio*), fruit fly (*Drosophila melanogaster*), mouse (*Mus musculus*), rat (*rattus*) and rabbit (*Oryctolagus cuniculus*) share genetic and physiological similarities with humans (Kari et al., 2007; Moraes & Montagne, 2021; Dutta & Sengupta, 2016; Shiomi, 2009; de Artinano & Castro, 2009).

It is pertinent to recognize that animal models are indispensable for researchers to evaluate the effectiveness and safety of potential medical drugs and treatments. These models are instrumental in detecting any harmful or unwanted side effects, such as toxicity, liver damage, infertility, birth defects and potential carcinogenic effects (Tobita et al., 2010; Fijak et al., 2018; Modarresi Chahardehi et al., 2020; Liu et al., 2013; Wanibuchi, 2004). Under current U.S. federal law, all new treatment options must be tested on non-human animals to ensure their effectiveness and safety before proceeding to human trials. In addition, animal testing serves a dual purpose by benefiting both human and animal populations. Many pharmaceuticals and medical treatments developed for human use also contribute to prolonging and improving the quality of life for animals (Kiani et al., 2022). Moreover, the Institutional Animal Care and Use Committees (IACUCs) is a well-established organization responsible for scrutinizing and supervising all facets of animal care and utilization within research institutions to guarantee adherence to ethical standards and regulatory mandates. Furthermore, national and international guidelines, including the U.S. Animal Welfare Act (AWA), the Guide for the Care and Use of Laboratory Animals and the European Union Directive 2010/63/EU set forth standards for the treatment of animals in research and govern the utilization of animals in scientific research (Beauchamp & DeGrazia, 2019; Palmer et al., 2020; National Research Council, 2011).

The ethical concerns surrounding the use of animals in research highlight the need to explore alternative methods that are both effective and compassionate. To address these issues, the 3Rs framework was proposed 60 years ago. The 3Rs principles: Replacement, Reduction, and Refinement are considered the gold standard for humane animal research, as they aim to balance animal welfare and research goals. The 3Rs principle intends to minimize animal use in research, but it can be challenging to apply (Graham & Prescott, 2015). The "Replacement" principle suggests that animals should be replaced with alternative methods or less sentient species wherever possible. However, in some cases, animals may benefit from participating in research and expressing their interests. Moreover, replacing more complex species with less complex ones may require using more individual animals, which goes against the "Reduction" principle. By reducing the number of animals used, fewer animals are subjected to more severe or multiple procedures, thereby minimizing harm and upholding the "Refinement" principle (Nannoni & Mancini, 2024).

Nevertheless, the use of animals for biomedical research and cellular agriculture raises ethical concerns in the field of animal ethics. Recently, animal research ethics has become a separate subfield within research ethics due to an increasing amount of literature on various topics related to animal research ethics. It is important to consider the growth of this field and the impact of placing the ethics of animal research within the broader context of research ethics.

2. GENERAL APPLICATION, ISSUES, AND DEVELOPMENT OF ANIMAL ETHICS

In general, animal ethics have been widely applied in several types of research including marine mammals, zoo animals and farm animals.

Ethics of Marine Mammals

Marine mammals are extensively studied in their natural habitats to enhance people's knowledge of their physiology, behaviour, ecology and life history (Parsons et al., 2015). The scientific interest in marine mammals has been increasing over time, driven by the need for conservation, economic, management and scientific research (Rilov et al., 2019; Allan et al., 2014; Rogers et al., 2021). This has resulted in the development of a wide range of tools and techniques that enable scientists to collect innovative data about marine mammals in the wild.

The ethical treatment of marine mammals, like other vertebrates, is an important concern that arises from research involving these animals. However, until recently, the scientific community studying marine mammals has not given sufficient attention to these concerns comprehensively. Research on these animals is conducted by scientists from various countries and cultures, who adhere to different ethical values, motivations and legislative controls. While this diversity has an impact on any global conversation regarding the ethics of marine mammals, there are still some essential ethical and scientific principles that are universally applicable, irrespective of cultural and legal differences (Ressurreição et al., 2012). Research on marine mammals raises ethical concerns that require careful consideration due to their advanced social structures and strong social bonds. Moreover, numerous marine mammal populations are highly threatened, which necessitates researchers to minimize any disturbance or negative impact, particularly when attempting to save a species from extinction (Papastavrou & Ryan, 2023). Studying marine mammals comes with its fair share of challenges, mainly because they inhabit remote ocean areas, making them hard to observe. Even those living in coastal regions or spending time on land are only visible for a portion of their lives. Additionally, it is challenging to handle marine mammals in order to attach tags, obtain blood samples or perform other invasive procedures due to their large size, which necessitates ethical considerations (Papastavrou & Ryan, 2023). The intricacies of specific physiological, anatomical, and behavioral adaptations including breath-hold diving, demand a meticulous approach to anesthesia. This necessitates a high level of experience, technical skill and robust infrastructure to ensure safe and effective management (Dold & Ridgway, 2014).

ETHICS OF FARM ANIMALS

Ethical animal farming places animal welfare as a top priority. This involves practices such as providing ample space for animals to move freely and express natural behaviors including dust-bathing and wallowing, avoiding practices like beak trimming, tail docking, or teeth pulling. Livestock raised on pasture in ethical farming systems tend to produce healthier meat, dairy and eggs (Alonso et al., 2020; Fraser & Weary, 2004). It is important to sustain the health of animals during the drought period. Moreover, the expansive nature of the farms presents a challenge in conducting regular inspections of all animals, potentially leading to instances of neglect and mortality due to injuries or dystocia (Mee, 2013). Furthermore, cattle reared in pastures bear a substantial burden of ticks and internal parasites, while the presence of predators poses a threat to the well-being of younger and older animals alike (Temple & Manteca, 2020).

ETHICS OF ZOO, WILD, AND LABORATORY ANIMALS

Interacting with animals can be beneficial for both humans and animals and can lead to pro-conservation and respect for nature behaviors. However, it can also lead to inappropriate wild-animal 'pet' ownership and exploitation of animals for "cheap titillation". Ethically run zoos should consider incorporating three ethical frameworks to ensure human-animal interactions are ethical: Conservation Welfare, Compassionate Conservation and Duty of Care. It is essential to closely monitor human-animal interactions in zoos using welfare tracking tools (Learmonth, 2020; Mancini, 2017).

In wild animal studies, applying the Reduction aspect is challenging due to several reasons. Wild animals display more genetic variation than captive populations, leading to diverse responses to conditions and requiring larger sample sizes. Additionally, the environmental variation of wild animals is much greater than in controlled lab conditions, necessitating larger sample sizes. Studies on wild animals also have a high probability of losing animals due to natural mortality or other unforeseen events. Therefore, it is essential to conduct a pre-study power analysis to ensure that the sample size is adequate (Steidl et al., 1997). Besides, the guidelines for refining research should be standardized to support researchers and ethical review bodies. Currently, there are no standardized approaches for planning or reporting on refinements. The current refinement guidelines only focus on welfare harms and do not take into account broader ethical harms (Nannoni & Mancini, 2024).

Besides, most laboratory research involves using rodents, which is not always ethical. Therefore, the 3Rs principles encourage scientists to use the lowest level of sentient life possible. Biomedical researchers focus on studying physiological, genetic, or biochemical responses through *in vitro* and *in silico* methods, which reduce the need for animal testing (Hirsch & Schildknecht, 2019; Viceconti et al., 2016). Sometimes, it is not feasible to find alternatives to using animals in research studies that involve observing individual animals and their interactions within the ecosystem. Very occasionally, less intelligent or less protected species may be used for testing hypotheses, but this requires approval at a higher or more protected level (Sneddon et al., 2017).

Various non-invasive methods have been developed for conducting research studies on wild animals. These methods differ from those used in laboratory-based studies because they aim to avoid harming the species or its population and to release animals back into the wild quickly. One well-established approach is DNA analysis from hair or faeces collection. However, it is important to collaborate with other disciplines to improve and refine these techniques (Cattet, 2013). Remote monitoring methods such as camera trapping or passive acoustic monitoring are increasingly used, as well as advanced analytical methods like machine learning

(Burton et al., 2015; Gibb et al., 2019; Tabak et al., 2019). Research involving wild animals encompasses a broad spectrum of species and techniques that can have effects on individual animals, groups and entire ecosystems. Fieldwork is often carried out under challenging conditions, including non-sterile environments, severe weather and regions with extreme climate conditions. The capturing and handling of animals during studies has the potential to cause harm to the subjects. Despite these challenges, it is crucial to prioritize animal welfare as a critical consideration in the design of studies involving wild animals (Soulsbury et al., 2020).

Researchers are required to define their scientific goals before using animals for research. Research should aim to have a reasonable expectation of contributing to the advancement of scientific knowledge in various aspects of biomedicine. It should also enhance understanding of the species under study or provide results that could improve the health or welfare of humans and other animals (Begley & Ioannidis, 2015; Broom, 2007). The research must have a significant scientific purpose to justify the use of animals. The species selected for the study should be the best suited to answer the research questions. It is important to note that good experimental design can help reduce the number of animals used in research. To prevent the repetition of experiments and the need for more animals, scientists should collect data using the minimum number of animals required. However, it is pertinent to use a sufficient number of animals to ensure precise statistical analysis and reliable results (Kilkenny et al., 2009). The ethical review board diligently assesses research proposals for ethical considerations to ensure compliance with ethical guidelines and minimize animal suffering (Schuppli, 2011).

Besides, scientists have an obligation to provide explicit instructions to all individuals who use animals under their supervision. This instruction manual must thoroughly cover experimental methods, as well as the maintenance, care and handling of the species being studied. It is imperative for researchers to refine the experimental procedures and management of pain. Additionally, they must rigorously evaluate the method of administration, the effects of the substance on the animal and the amount of handling and restraint required. (Percie du Sert et al., 2017). With regard to conducting experiments involving animals, it is essential for researchers to treat the animals with care and to ensure that appropriate anesthetics and analgesics are administered to minimize any discomfort or pain that the animals may experience. The establishment of a culture of care within the organization necessitates the implementation of stringent regulations pertaining to the treatment of animals. It is essential to ensure that all animal technicians and other personnel not only understand these regulations but also actively implement them in their daily work. This involves comprehensive training, regular monitoring, and strict enforcement to guarantee the well-being and ethical treatment of all animals under their care.

Therefore, it is essential to provide adequate training, which should be continually reviewed and improved to refine animal research (Auer et al., 2007).

3. ANIMAL ETHICS IN CELLULAR AGRICULTURE AND FOOD PRODUCTION

Cellular agriculture is an innovative method of food production that involves growing animal-based products such as meat and dairy directly from cells in a controlled environment, eliminating the need for traditional animal rearing and slaughtering (Jahir et al., 2023). Lab-grown meat, also known as cultivated meat, is produced by extracting animal cells and placing them in nutrient-rich bioreactors. These cells multiply and develop into muscle tissue that closely resembles conventional meat in taste and texture (Ercili-Cura et al., 2021). The approval of lab-grown meat for sale in Singapore signifies a significant milestone in the industry (Chodkowska et al., 2022). In addition, companies like Eat Just and Mzansi Meat have commercially introduced cultivated chicken and beef products (Tsvakirai et al., 2023; Guan et al., 2021).

Nevertheless, this innovative approach raises several ethical considerations. Cellular agriculture aligns with the non-harm principle, emphasizing the minimization of harm to animals. By enabling the production of animal products without causing harm to animals, this approach supports the ethical objective of reducing animal suffering. However, ethical queries emerge concerning the utilization of cell-donor animals, necessitating the assurance of their welfare and rights (Dutkiewicz & Abrell, 2021). Notably, cellular agriculture adheres to the principles of animal rights, seeking to mitigate the exploitation of animals for human consumption and advocating for the ethical treatment of animals as distinct from mere resources for human use (Poirier, 2022).

Besides, utilitarianism is important to be applied in cellular agriculture, as it aims to maximize overall happiness and minimize suffering. Cellular agriculture can be viewed as a utilitarian approach because it seeks to reduce animal suffering and environmental impact while fulfilling human demands for animal products (Moyano-Fernández, 2023). In line with the "respect for life" principle, cellular agriculture involves respecting the intrinsic value of animal life. By creating alternatives to traditional animal farming, cellular agriculture demonstrates respect for animal life and aims to treat animals with dignity (Moyano-Fernández, 2023). The solution for the problem arising in ethics requires a solution, which can be achieved through integration with animal welfare.

4. THE INTEGRATION OF ETHICS AND ANIMAL WELFARE

Animal welfare refers to the well-being of animals and how it is affected by human actions. It reflects how animals deal with the conditions in which they live (Hewson, 2003). Domesticated animals undoubtedly offer humans essential benefits, including milk, meat and labor power. Just as individuals expect personal benefits after a hard day's work, animals also merit unwavering support for their indisputable contributions. It is therefore imperative that humans uphold a moral obligation to ensure the well-being and quality of life of these animals. Prioritizing animal welfare in food production systems indisputably amplifies productivity, quality, food safety and economic returns, thereby unequivocally fortifying food security and economic prosperity (Madzingira, 2018).

Organizations like the World Organization for Animal Health and American Veterinary Medical Association believe that humans have a responsibility to ensure a certain level of welfare for animals (Bayvel & Cross, 2010). The field of animal welfare science has its roots in both scientific inquiry and public concern about the treatment and raising of animals, as expressed by ethicists and social critics. However, animal welfare scientists have largely overlooked the contributions of these ethicists and critics (Fraser et al., 1997). Animal welfare is a term used to describe a branch of science and a concept. It involves measuring an animal's state and its ability to cope with its environment. Research on animal welfare can present ethical challenges, as it may involve studying animals in compromised situations (Olsson et al., 2022).

The integration of ethics and animal welfare is essential to ensure the responsible and humane treatment of animals in scientific research. The incorporation of ethical considerations and animal welfare into the realm of genetic engineering, scientific research, cellular agriculture and animal agriculture demands a comprehensive approach (Kaiser, 2009; Poirier, 2022). This encompasses adherence to ethical principles, compliance with regulatory frameworks, implementation of best practices, thorough review and oversight, as well as a firm commitment to transparency and accountability. This all-encompassing approach ensures that the use of animals in the food industry and research is conducted responsibly and compassionately, with a primary focus on minimizing harm and promoting animal welfare (Hampton et al., 2021; Madzingira, 2018).

Typically, all procedures involving animals must undergo a review by a local animal care committee to ensure that they are suitable and humane. If establishing a local animal care committee is not possible, scientists are encouraged to seek advice from a similar committee at a cooperative institution (National Research Council, 2011). Researchers must ensure that animals are well cared for, whether they are being used in experiments or for teaching purposes. This responsibility is overseen by the institution's animal care committee and designated individuals (Monamy, 2017). Researchers are encouraged to provide enriching environments for laboratory animals, thanks to increased awareness and concern about the humane treatment of experimental animals. This has led to a reduction in the number of animals used in experiments, the elimination of unnecessary duplication, and minimized pain and distress. Although scientists are moving towards self-regulation, this approach will be grounded in the scientific method and may not completely bridge the gap between scientific and animal protection groups (Rollin, 2006).

Anesthesia plays a crucial role in ensuring the safety and well-being of both researchers and animals during the capture and handling process. It is vital for invasive procedures like surgery and blood collection, providing comfort and minimizing stress for the animals. Additionally, it is essential for non-invasive research activities such as taking measurements and fitting collars, allowing for accurate data collection while ensuring the welfare of the animals involved. However, anaesthesia usage can be challenging due to the limited information available on procedures, difficult environmental conditions and varied welfare outcomes. Anesthesia, despite having well-established protocols, still poses a heightened risk of mortality. The administration of anesthesia demands a high level of training and expertise, and may also be governed by national legislative regulations (Burton et al., 2015). Administering anaesthesia to smaller animals is particularly challenging due to the smaller margins of error with dosage. Continuous monitoring of stress levels and the depth of unconsciousness is essential to avoid over- or under-dosing and accurate record-keeping of anaesthetic events is important.

Furthermore, all animal-related treatments should be designed and carried out with a humane concern for the animal's well-being. Surgical techniques need strict observation and attention to humanitarian issues by the scientist. When feasible, laboratory animals and animals that require treatment should be handled using aseptic procedures (National Research Council, 2011). All surgical operations and anesthesia should be performed under the direct supervision of a skilled practitioner. If the surgical operation is anticipated to cause more discomfort than anesthesia, and there is no special reason to behave otherwise, animals should be kept under anesthesia until the treatment is completed. Repeated surgical procedures on animals should only be undertaken if they are necessary for research, surgical intervention, or the well-being of the animal. Special approval from the animal care committee is required for any repeated surgeries (Couto & Cates, 2019; Kiani et al., 2022).

Besides, blood sampling should always be justified in any study protocol because it is an invasive procedure. Several factors must be taken into consideration when it comes to blood sampling. These factors are specific to the species being studied. It is important to limit the amount of blood taken from an animal to ensure their well-being. Taking more than 10% of the total blood volume at once, which is approximately 1% of the body mass, can be harmful. When collecting multiple blood samples, it is essential to adhere to the guideline of taking no more than 1% of the blood volume within a 24-hour period. Researchers must also carefully consider animal welfare and plan for potential euthanasia if large or whole-body volumes of blood must be collected (Diehl et al., 2001). Animals can also be identified using various methods such as external marks like coloring, tattooing, branding, or clipping appendages (Silvy et al., 2005). Additionally, they can be identified by external tags or equipment like leg rings, ear tags, collars, harnesses and radio transmitters. It is also possible to utilize internal tags or markers, such as PIT tags and chemical markers. Each technique's efficacy varies according to the species and the study's objectives (Soulsbury et al., 2020).

There are three options when dealing with sick or injured animals: euthanasia, treatment, or no intervention (Kirkwood et al., 1994; Wolfensohn, 2010). There are situations where each of these is acceptable from a welfare standpoint. Treatment is justified if the animal is unlikely to recover on its own and can be managed and released after treatment with minimal stress on the animal, or if the animal is likely to recover without treatment but treatment will improve its welfare by shortening the time to recovery. Minimal treatments range from minimal stitches to washing wounds and giving antibiotics (Soulsbury et al., 2020). In the majority of nations, veterinary care must be administered, either directly or under supervision. Research on animals indicates that prompt *in situ* therapy is best. Thus, deciding how to handle animals is a crucial component of contingency planning in the design phase.

Apart from this, the transportation of animals requires major attention. Some animals will be transported from the outdoors to a captive habitation facility. Although greater trips need more preparation and attention, it is crucial to remember that any mode of transportation can be a severe stressor, affecting animal welfare and the research findings (Nielsen et al., 2011). The major goal should be to transport the animals in a way that does not endanger their well-being and guarantees that they arrive at their destination in excellent health and with little suffering. The route and journey plan, length of the trip, vehicle design, container design, type of food and water supplies, the skill of drivers and other transportation staff and plans for acclimatization following transportation are just a few of the many factors that need to be taken into account (Swallow et al., 2005). All applicable animal transport regulations must be observed; appointing someone in each business with responsibility for knowing and executing transport legislation will help to assure compliance (National Research Council, 2006).

ETHICS AND WELFARE IN CELLULAR AGRICULTURE AND FOOD PRODUCTION

The consideration of animal welfare in food production is of paramount importance due to its far-reaching impact on both animals and humans. Animals raised for food or fiber are frequently subjected to inhumane treatment and living conditions. Concentrated animal feeding operations (CAFOs), also known as factory farms, prioritize rapid growth and production over animal health and welfare (Moses & Tomaselli, 2017). Within these facilities, animals are confined to relatively small and densely populated spaces, resulting in issues such as overcrowding, stress and physical or mental ailments (Smith, 2005). The stress and discomfort experienced by these animals can potentially compromise the quality of meat, milk, or eggs. Stressed animals release hormones such as cortisol, which can detrimentally affect muscle texture and flavor. Moreover, stress can markedly impact milk production and its composition. Hens enduring pain or stress may lay fewer eggs or produce eggs of inferior quality (Mia et al., 2023; Hedlund & Jensen, 2022). It is imperative that proper handling and humane slaughter be enforced to minimize stress and ensure superior meat quality. Adhering to ethical treatment of animals not only guarantees healthier and more productive livestock but also serves the best interests of consumers and the animals themselves (Welty, 2007).

Figure 1. The animal ethics and welfare in cellular agriculture and food production

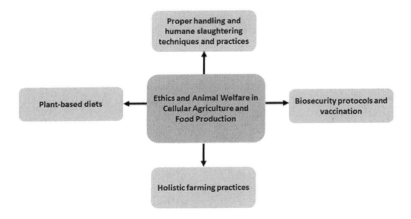

Aside from heat stress brought on by a lack of shade, other animal welfare issues on large beef cattle production farms include the mixing of incompatible animals, like bulls, which causes constant fights, inadequate calves monitoring, particularly for colostrum intake and painful husbandry procedures like castration, dehorning, animal identification and hot iron branding that are performed without analgesia (Madzingira, 2018). In addition, in order to protect dairy cattle from disease, vaccinations should be administered based on professional advice that takes local diseases into account (Chen et al., 2021).

In addition, it is crucial to support local, sustainable family farms that prioritize animal welfare. Holistic farming practices, including family farms, must prioritize the humane treatment of animals and provide them with better living conditions. Animals raised in such environments experience less stress and better overall well-being (Widdowson, 2013). These animal welfare practices not only reduce risks such as antimicrobial resistance but also contribute to more sustainable food production. By ensuring healthier living conditions and minimizing stress, the reliance on antibiotics can be significantly reduced (National Research Council, 1999).

Farm animals, such as broiler chickens, are reared in large groups, making it easy for diseases to spread and affect the entire flock. As a result, it is imperative to prioritize disease prevention and management on broiler farms to ensure the well-being of the animals. Strict adherence to health management and biosecurity protocols is essential to prevent the spread of diseases. It is also necessary to vaccinate birds against illnesses that are common in some regions to safeguard their health and stop outbreaks (Samanta et al., 2018). On the other hand, the ethical philosophy of caring significantly influences how animals are treated on farms. Recognizing the moral impact of this connection is crucial for improving welfare practices for farm animals. It is imperative to consider both the behavioral and cognitive needs of animals when evaluating animal farming to prevent subjecting them to poor welfare conditions (King, 2003).

Animal ethics in food production involves the crucial consideration of the moral principles guiding how animals are treated in the process of producing food for human consumption. This encompasses a thorough examination of the welfare of animals, the environmental impact of animal farming and the ethical implications of using animals for food (Fraser, 2012). It is imperative to ensure that animals are treated with utmost care and respect, and provided with unequivocal access to adequate food, water, shelter and medical care. The five commonly acknowledged freedoms for animal welfare, which include the freedom from pain, thirst, fear, hunger, injury, discomfort, sickness and are essential in averting animal suffering and subpar environmental conditions (McCulloch, 2013). The World Organisation for Animal Health (OIE), industries, regulatory authorities and animal welfare groups have all widely accepted these principles and they have been included in a number of standards of practice across the world (Bruckner, 2009). Despite the profit-oriented nature of commercial livestock farming, it is crucial to acknowledge and address the associated animal welfare infringements. Promoting animals' physical and mental health in a proactive manner is the main goal of animal welfare, since it contributes to their total well-being (McMillan, 2002).

Moreover, intensive farming practices invariably result in poor living conditions for animals, including overcrowding, lack of natural behaviors and painful physical alterations such as de-beaking and tail docking. It is essential to actively promote plant-based diets and products such as tofu, seitan, and plant-based milk to decrease the demand for animal products and address ethical concerns head-on (Alcorta et al., 2021). Opting for a vegetarian or vegan diet, or simply reducing meat consumption, is crucial in significantly reducing the ethical and environmental impact of food production (Collier et al., 2022). Additionally, actively supporting farms and companies that prioritize animal welfare and sustainable practices can be a driving force for positive change in the industry (Von Keyserlingk & Hötzel, 2015).

5. THE CHALLENGES AND LIMITATIONS OF THE INTEGRATION OF ETHICS AND ANIMAL WELFARE

Despite the ethical and welfare considerations on animals, several challenges and limitations raise concern for the researchers. The animal farming industry has been seen as a contribution to greenhouse gas emissions. This industry results in animal diseases which also affect humans upon consumption (Knowles et al., 2023). Knowing the fact that meat as a food increases the risk for many cancers, several concerns have been raised to implement animal ethics and protect animal welfare. Enforcing more rigid standard operating procedures could help with the enforcement of animal ethics. Apart from that, implementing a tax on the mean and animal products also raises concern over the usage of animal products. However, the implementation of tax is one of the major challenges considering the rising inflation and price hikes in food, especially in meat products (World Food Situation, 2022).

Besides that, monitoring of animal farming poses many limitations in their current practice. One of the major challenges faced by the farmers includes the general assessment of health, behavioural and psychological indicators which are conducted through visual inspections and observations which is impractical for large-scale operations (Neethirajan, 2024). Even though the implementation of the latest technologies in animal farming made it possible, one of the main challenges is the acceptance from the farmer community as they might feel threatened by their job security due to the technological advancement which could lead to staff reduction in farmer's community (Lioutas et al., 2021).

The reliability of digital technologies also poses a major challenge in the transformation of the new ethical considerations including animal privacy and secured data which raises new concerns about the maintenance of digital livestock (Neethirajan, 2024). This requires the cooperation of various personnel such as farmers, researchers, stakeholders and clients to welcome the technological advancements in digital livestock farming (Neethirajan, 2024). The ethical approach in livestock monitoring such as less stressful wearable devices for the livestock and a more organized monitoring system helps in creating more humane ethical procedures as well as improved animal welfare. Besides, antimicrobials are used in animal farming to combat the infection caused by pathogens (Cuong et al., 2018). Despite that, the use of antimicrobials was found to negatively affect animal welfare (Diana et al., 2020).

6. THE FUTURE PERSPECTIVES ON THE INTEGRATION OF ETHICS AND ANIMAL WELFARE

With the rising technological advancements, animal welfare has been one of the domains in Artificial Intelligence (AI) discussions. Animal welfare which involves the physical and mental well-being of livestock has always been an emphasis for improvement by farmers, industrialists and policymakers (Eastwood et al., 2021). Continuous monitoring and assessment methods are imperative in practising ethics in animals. In a study in Germany, the suggestion to impose a tax on meat to support animal welfare has received more support than to fight issues raised by climate change, while many countries still face challenges by policymakers for tax implementation on red meat (Perino & Schwickert, 2023). Besides, artificial intelligence has been one of the renowned applications used in all fields. This cutting-edge technology has transformed animal farming to the next level. For example, sensors including wearable devices with AI algorithms can monitor physiological changes as well as minor changes in behavior which could indicate psychological disturbances such as stress (Neerthirajan, 2023).

In addition, the occupancy of imaging technologies like thermal imaging and 3D imaging also enables the overall real-time assessment of animals (Fernandes et al., 2020). Integration of such AI-based systems helps in the development of more humane and efficient animal farming. Using the machine learning approach, the supervised learning algorithms can be utilized to differentiate any diseased animal from the group. While the predictive approach helps to forecast a potential disease condition (Franzo et al., 2023). For instance, sound from pigs enables to detection of signs of respiratory diseases, milk conductivity measurements can help in the detection of mastitis in cows and comfort or stress can be detected through the classification of animal behaviour from the videos (Neerthirajan, 2024). A recent study has shown that involving producers in animal welfare practices has increased the recognition of the company in complying with emotional components of animals through animal-based measures (Martinez et al., 2024). The digital livestock farming is expected to create a more holistic journey in livestock farming which ensures a symbiotic communication between technology, animals and humans.

7. CONCLUSION

The use of millions of animals each year in painful and distressing scientific procedures raises ethical concerns. In modern societies, regulation regarding animal testing and farm animals is predicated on the notion that it can be ethically acceptable when specified legal, technological, logistical and ethical requirements are satisfied. Animal research and animal farming are crucial for the well-being of both humans and animals. It is imperative to conduct these practices in a controlled and humane manner. While computer modeling and grown human cells are promising substitutes for animal research, their broad acceptance would depend on their accessibility to researchers and their ability to supplement rather than replace animal experiments. Animal ethics in food production involves a comprehensive consideration of the welfare of animals, the environmental impact of farming practices and the moral implications of using animals for food. By adopting humane treatment practices, promoting sustainable and ethical alternatives and making informed consumer choices, it is possible to create a more ethical and sustainable food production system. Regulatory measures and public awareness play crucial roles in driving this transformation. Stakeholders, including veterinarians, farmers, and the general public, play a role in shaping animal welfare standards.

REFERENCES

Alcorta, A., Porta, A., Tárrega, A., Alvarez, M. D., & Vaquero, M. P. (2021). Foods for plant-based diets: Challenges and innovations. *Foods*, 10(2), 293. 10.3390/foods1002029333535684

Allan, G. J., Lecca, P., McGregor, P. G., & Swales, J. K. (2014). The economic impacts of marine energy developments: A case study from Scotland. *Marine Policy*, 43, 122–131. 10.1016/j.marpol.2013.05.003

Alonso, M. E., González-Montaña, J. R., & Lomillos, J. M. (2020). Consumers' concerns and perceptions of farm animal welfare. *Animals (Basel)*, 10(3), 385. 10.3390/ani1003038532120935

Auer, J. A., Goodship, A., Arnoczky, S., Pearce, S., Price, J., Claes, L., von Rechenberg, B., Hofmann-Amtenbrinck, M., Schneider, E., Müller-Terpitz, R., Thiele, F., Rippe, K.-P., & Grainger, D. W. (2007). Refining animal models in fracture research: Seeking consensus in optimising both animal welfare and scientific validity for appropriate biomedical use. *BMC Musculoskeletal Disorders*, 8(1), 1–13. 10.1186/1471-2474-8-7217678534

Bayvel, A. D., & Cross, N. (2010). Animal welfare: A complex domestic and international public-policy issue—who are the key players? *Journal of Veterinary Medical Education*, 37(1), 3–12. 10.3138/jvme.37.1.320378871

Beauchamp, T. L., & DeGrazia, D. (2019). *Principles of animal research ethics*. Oxford University Press. 10.1017/S0962728600009052

Begley, C. G., & Ioannidis, J. P. (2015). Reproducibility in science: Improving the standard for basic and preclinical research. *Circulation Research*, 116(1), 116–126. 10.1161/CIRCRESAHA.114.30381925552691

Broom, D. M. (2007). Quality of life means welfare: How is it related to other concepts and assessed? *Animal Welfare (South Mimms, England)*, 16(S1), 45–53. 10.1017/S0962728600031729

Bruckner, G. K. (2009). The role of the World Organisation for Animal Health (OIE) to facilitate the international trade in animals and animal products: Policy and trade issues. *The Onderstepoort Journal of Veterinary Research*, 76(1), 141–146. 10.4102/ojvr.v76i1.7819967940

Burton, A. C., Neilson, E., Moreira, D., Ladle, A., Steenweg, R., Fisher, J. T., Bayne, E., & Boutin, S. (2015). Wildlife camera trapping: A review and recommendations for linking surveys to ecological processes. *Journal of Applied Ecology*, 52(3), 675–685. 10.1111/1365-2664.1243226211047

Cattet, M. R. (2013). Falling through the cracks: Shortcomings in the collaboration between biologists and veterinarians and their consequences for wildlife. *ILAR Journal*, 54(1), 33–40. 10.1093/ilar/ilt01023904530

Chen, Y., Wang, Y., Robertson, I. D., Hu, C., Chen, H., & Guo, A. (2021). Key issues affecting the current status of infectious diseases in Chinese cattle farms and their control through vaccination. *Vaccine*, 39(30), 4184–4189. 10.1016/j.vaccine.2021.05.07834127292

Chodkowska, K. A., Wódz, K., & Wojciechowski, J. (2022). Sustainable future protein foods: The challenges and the future of cultivated meat. *Foods*, 11(24), 4008. 10.3390/foods1124400836553750

Collier, E. S., Normann, A., Harris, K. L., Oberrauter, L. M., & Bergman, P. (2022). Making more sustainable food choices one meal at a time: Psychological and practical aspects of meat reduction and substitution. *Foods*, 11(9), 1182. 10.3390/foods1109118235563904

Couto, M., & Cates, C. (2019). Laboratory guidelines for animal care. *Vertebrate Embryogenesis: Embryological, Cellular, and Genetic Methods*, 407-430. 10.1007/978-1-4939-9009-2_25

Cuong, N. V., Padungtod, P., Thwaites, G., & Carrique-Mas, J. J. (2018). Antimicrobial Usage in Animal Production: A Review of the Literature with a Focus on Low- and Middle-Income Countries. *Antibiotics (Basel, Switzerland)*, 7(3), 75. 10.3390/antibiotics703007530111750

de Artinano, A. A., & Castro, M. M. (2009). Experimental rat models to study the metabolic syndrome. *British Journal of Nutrition*, 102(9), 1246–1253. 10.1017/S000711450999072919631025

Diana, A., Lorenzi, V., Penasa, M., Magni, E., Alborali, G. L., Bertocchi, L., & De Marchi, M. (2020). Effect of welfare standards and biosecurity practices on antimicrobial use in beef cattle. *Scientific Reports*, 10(1), 20939. 10.1038/s41598-020-77838-w33262402

Diehl, K. H., Hull, R., Morton, D., Pfister, R., Rabemampianina, Y., Smith, D., Vidal, J.-M., & Vorstenbosch, C. V. D. (2001). A good practice guide to the administration of substances and removal of blood, including routes and volumes. *Journal of Applied Toxicology*, 21(1), 15–23. 10.1002/jat.72711180276

Dold, C., & Ridgway, S. (2014). Cetaceans. *Zoo animal and wildlife immobilization and anesthesia*, 679-691. 10.1002/9781118792919.ch49

Dutkiewicz, J., & Abrell, E. (2021). Sanctuary to table dining: Cellular agriculture and the ethics of cell donor animals. *Politics and Animals*, 7, 1–15.

Dutta, S., & Sengupta, P. (2016). Men and mice: Relating their ages. *Life Sciences*, 152, 244–248. 10.1016/j.lfs.2015.10.02526596563

Eastwood, C. R., Edwards, J. P., & Turner, J. A. (2021). Review: Anticipating alternative trajectories for responsible Agriculture 4.0 innovation in livestock systems. *Animal: an international journal of animal bioscience*, 15, 100296. https://doi.org/10.1016/j.animal.2021.100296

Ercili-Cura, D., Barth, D., Pitkänen, J.-P., Tervasmäki, P., Paananen, A., Rischer, H., & Rønning, S. B. (2021). *Cellular Agriculture: Lab-Grown Foods* (Vol. 8). American Chemical Society ACS., 10.1021/acs.infocus.7e4007

Fernandes, A. F. A., Dórea, J. R. R., & Rosa, G. J. M. (2020). Image Analysis and Computer Vision Applications in Animal Sciences: An Overview. *Frontiers in Veterinary Science*, 7, 551269. 10.3389/fvets.2020.55126933195522

Fijak, M., Pilatz, A., Hedger, M. P., Nicolas, N., Bhushan, S., Michel, V., Tung, K. S. K., Schuppe, H.-C., & Meinhardt, A. (2018). Infectious, inflammatory and 'autoimmune' male factor infertility: How do rodent models inform clinical practice? *Human Reproduction Update*, 24(4), 416–441. 10.1093/humupd/dmy00929648649

Franco, N. H. (2013). Animal experiments in biomedical research: A historical perspective. *Animals (Basel)*, 3(1), 238–273. 10.3390/ani301023826487317

Franzo, G., Legnardi, M., Faustini, G., Tucciarone, C. M., & Cecchinato, M. (2023). When Everything Becomes Bigger: Big Data for Big Poultry Production. *Animals (Basel)*, 13(11), 1804. 10.3390/ani1311180437889739

Fraser, D. (2012). Animal ethics and food production in the twenty-first century. *The Philosophy of Food*, 39, 190–213. 10.1525/9780520951976-012

Fraser, D., & Weary, D. M. (2004). Quality of life for farm animals: linking science, ethics and animal welfare. *The well-being of farm animals: Challenges and solutions*, 39-60. 10.1002/9780470344859.ch3

Fraser, D., Weary, D. M., Pajor, E. A., & Milligan, B. N. (1997). A scientific conception of animal welfare that reflects ethical concerns. *Animal Welfare (South Mimms, England)*, 6(3), 187–205. 10.1017/S0962728600019795

Gibb, R., Browning, E., Glover-Kapfer, P., & Jones, K. E. (2019). Emerging opportunities and challenges for passive acoustics in ecological assessment and monitoring. *Methods in Ecology and Evolution*, 10(2), 169–185. 10.1111/2041-210X.13101

Graham, M. L., & Prescott, M. J. (2015). The multifactorial role of the 3Rs in shifting the harm-benefit analysis in animal models of disease. *European Journal of Pharmacology*, 759, 19–29. 10.1016/j.ejphar.2015.03.04025823812

Guan, X., Lei, Q., Yan, Q., Li, X., Zhou, J., Du, G., & Chen, J. (2021). Trends and ideas in technology, regulation and public acceptance of cultured meat. *Future Foods : a Dedicated Journal for Sustainability in Food Science*, 3, 100032. 10.1016/j.fufo.2021.100032

Hampton, J. O., Hyndman, T. H., Allen, B. L., & Fischer, B. (2021). Animal harms and food production: Informing ethical choices. *Animals (Basel)*, 11(5), 1225. 10.3390/ani1105122533922738

Hedlund, L., & Jensen, P. (2022). Effects of stress during commercial hatching on growth, egg production and feather pecking in laying hens. *PLoS One*, 17(1), e0262307. 10.1371/journal.pone.026230734982788

Hewson, C. J. (2003). What is animal welfare? Common definitions and their practical consequences. *The Canadian Veterinary Journal. La Revue Veterinaire Canadienne*, 44(6), 496.12839246

Hirsch, C., & Schildknecht, S. (2019). *In vitro* research reproducibility: Keeping up high standards. *Frontiers in Pharmacology*, 10, 492837. 10.3389/fphar.2019.0148431920667

Jahir, N. R., Ramakrishna, S., Abdullah, A. A. A., & Vigneswari, S. (2023). Cultured meat in cellular agriculture: Advantages, applications and challenges. *Food Bioscience*, 53, 102614. 10.1016/j.fbio.2023.102614

Kaiser, M. (2009). Ethical aspects of livestock genetic engineering. In *Genetic Engineering in Livestock: New Applications and Interdisciplinary Perspectives* (pp. 91-117). Springer Berlin Heidelberg. 10.1007/978-3-540-85843-0_5

Kari, G., Rodeck, U., & Dicker, A. P. (2007). Zebrafish: An emerging model system for human disease and drug discovery. *Clinical Pharmacology and Therapeutics*, 82(1), 70–80. 10.1038/sj.clpt.610022317495877

Kiani, A. K., Pheby, D., Henehan, G., Brown, R., Sieving, P., Sykora, P., ... International Bioethics Study Group. (2022). Ethical considerations regarding animal experimentation. *Journal of Preventive Medicine and Hygiene, 63*(2 Suppl 3), E255. 10.15167/2421-4248/jpmh2022.63.2S3.2768

Kilkenny, C., Parsons, N., Kadyszewski, E., Festing, M. F., Cuthill, I. C., Fry, D., Hutton, J., & Altman, D. G. (2009). Survey of the quality of experimental design, statistical analysis and reporting of research using animals. *PLoS One, 4*(11), e7824. 10.1371/journal.pone.000782419956596

King, L. A. (2003). Behavioral evaluation of the psychological welfare and environmental requirements of agricultural research animals: Theory, measurement, ethics, and practical implications. *ILAR Journal, 44*(3), 211–221. 10.1093/ilar.44.3.21112789022

Kirkwood, J. K., Sainsbury, A. W., & Bennett, P. M. (1994). The welfare of free-living wild animals: Methods of assessment. *Animal Welfare (South Mimms, England), 3*(4), 257–273. 10.1017/S0962728600017036

Knowles, T. G., Kestin, S. C., Haslam, S. M., Brown, S. N., Green, L. E., Butterworth, A., Pope, S. J., Pfeiffer, D., & Nicol, C. J. (2008). Leg disorders in broiler chickens: Prevalence, risk factors and prevention. *PLoS One, 3*(2), e1545. 10.1371/journal.pone.000154518253493

Learmonth, M. J. (2020). Human–animal interactions in zoos: What can compassionate conservation, conservation welfare and duty of care tell us about the ethics of interacting, and avoiding unintended consequences? *Animals (Basel), 10*(11), 2037. 10.3390/ani1011203733158270

Lioutas, E. D., Charatsari, C., & De Rosa, M. (2021). Digitalization of agriculture: A way to solve the food problem or a trolley dilemma? *Technology in Society, 67*, 101744. 10.1016/j.techsoc.2021.101744

Liu, Y., Meyer, C., Xu, C., Weng, H., Hellerbrand, C., ten Dijke, P., & Dooley, S. (2013). Animal models of chronic liver diseases. *American Journal of Physiology. Gastrointestinal and Liver Physiology, 304*(5), G449–G468. 10.1152/ajpgi.00199.201223275613

Madzingira, O. (2018). Animal welfare considerations in food-producing animals. *Animal Welfare (South Mimms, England), 99*, 171–179. 10.5772/intechopen.78223

Martinez, A., Donoso, E., Hernández, R. O., Sanchez, J. A., & Romero, M. H. (2024). Assessment of animal welfare in fattening pig farms certified in good livestock practices. *Journal of applied animal welfare science. Journal of Applied Animal Welfare Science*, 27(1), 33–45. 10.1080/10888705.2021.202153238314792

Mancini, C. (2017). Towards an animal-centred ethics for Animal–Computer Interaction. *International Journal of Human-Computer Studies*, 98, 221–233. 10.1016/j.ijhcs.2016.04.008

McCulloch, S. P. (2013). A critique of FAWC's five freedoms as a framework for the analysis of animal welfare. *Journal of Agricultural & Environmental Ethics*, 26(5), 959–975. 10.1007/s10806-012-9434-7

McMillan, F. D. (2002). Development of a mental wellness program for animals. *Journal of the American Veterinary Medical Association*, 220(7), 965–972. 10.2460/javma.2002.220.96512420769

Mee, J. F. (2013). Why do so many calves die on modern dairy farms and what can we do about calf welfare in the future? *Animals (Basel)*, 3(4), 1036–1057. 10.3390/ani304103626479751

Mia, N., Rahman, M. M., & Hashem, M. A. (2023). Effect of heat stress on meat quality: A review. *Meat Research*, 3(6). Advance online publication. 10.55002/mr.3.6.73

Modarresi Chahardehi, A., Arsad, H., & Lim, V. (2020). Zebrafish as a successful animal model for screening toxicity of medicinal plants. *Plants*, 9(10), 1345. 10.3390/plants910134533053800

Monamy, V. (2017). *Animal experimentation: A guide to the issues*. Cambridge University Press. 10.1017/9781316678329

Moraes, K. C., & Montagne, J. (2021). *Drosophila melanogaster*: A powerful tiny animal model for the study of metabolic hepatic diseases. *Frontiers in Physiology*, 12, 728407. 10.3389/fphys.2021.72840734603083

Moses, A., & Tomaselli, P. (2017). Industrial animal agriculture in the United States: Concentrated animal feeding operations (CAFOs). *International farm animal, wildlife and food safety law*, 185-214. 10.1007/978-3-319-18002-1_6

Moyano-Fernández, C. (2023). The moral pitfalls of cultivated meat: Complementing utilitarian perspective with eco-republican justice approach. *Journal of Agricultural & Environmental Ethics*, 36(1), 23. 10.1007/s10806-022-09896-136467858

Nannoni, E., & Mancini, C. (2024). Toward an integrated ethical review process: An animal-centered research framework for the refinement of research procedures. *Frontiers in Veterinary Science*, 11, 1343735. 10.3389/fvets.2024.134373538694478

National Research Council. (1999). *The Use of Drugs in Food Animals: Benefits and Risks*. The National Academies Press. 10.17226/5137

National Research Council. (2011). *Guide for the Care and Use of Laboratory Animals* (8th ed.). The National Academies Press. 10.17226/12910

National Research Council. (2006). *Guidelines for the Humane Transportation of Research Animals*. The National Academies Press. 10.17226/11557

Neethirajan, S. (2023). Artificial Intelligence and Sensor Technologies in Dairy Livestock Export: Charting a Digital Transformation. *Sensors (Basel)*, 23(16), 7045. 10.3390/s2316704537631580

Neethirajan, S. (2024). Artificial Intelligence and Sensor Innovations: Enhancing Livestock Welfare with a Human-Centric Approach. *Hum-Cent Intell Syst*, 4(1), 77–92. 10.1007/s44230-023-00050-2

Nielsen, B. L., Dybkjær, L., & Herskin, M. S. (2011). Road transport of farm animals: Effects of journey duration on animal welfare. *Animal*, 5(3), 415–427. 10.1017/S1751731111000198922445408

Olsson, I. A. S., Nielsen, B. L., Camerlink, I., Pongrácz, P., Golledge, H. D., Chou, J. Y., Ceballos, M. C., & Whittaker, A. L. (2022). An international perspective on ethics approval in animal behaviour and welfare research. *Applied Animal Behaviour Science*, 253, 105658. 10.1016/j.applanim.2022.105658

Palmer, A., Greenhough, B., Hobson-West, P., Message, R., Aegerter, J. N., Belshaw, Z., Dennison, N., Dickey, R., Lane, J., Lorimer, J., Millar, K., Newman, C., Pullen, K., Reynolds, S. J., Wells, D. J., Witt, M. J., & Wolfensohn, S. (2020). Animal research beyond the laboratory: Report from a workshop on places other than licensed establishments (POLEs) in the UK. *Animals (Basel)*, 10(10), 1868. 10.3390/ani1010186833066272

Papastavrou, V., & Ryan, C. (2023). Ethical standards for research on marine mammals. *Research Ethics*, 19(4), 390–408. 10.1177/17470161231182066

Parsons, E. C. M., Baulch, S., Bechshoft, T., Bellazzi, G., Bouchet, P., Cosentino, A. M., Godard-Codding, C. A. J., Gulland, F., Hoffmann-Kuhnt, M., Hoyt, E., Livermore, S., MacLeod, C. D., Matrai, E., Munger, L., Ochiai, M., Peyman, A., Recalde-Salas, A., Regnery, R., Rojas-Bracho, L., & Sutherland, W. J. (2015). Key research questions of global importance for cetacean conservation. *Endangered Species Research*, 27(2), 113–118. 10.3354/esr00655

Perino, G., & Schwickert, H. (2023). Animal welfare is a stronger determinant of public support for meat taxation than climate change mitigation in Germany. *Nature Food*, 4(2), 160–169. 10.1038/s43016-023-00696-y37117860

Percie du Sert, N., Alfieri, A., Allan, S. M., Carswell, H. V., Deuchar, G. A., Farr, T. D., Flecknell, P., Gallagher, L., Gibson, C. L., Haley, M. J., Macleod, M. R., McColl, B. W., McCabe, C., Morancho, A., Moon, L. D. F., O'Neill, M. J., Pérez de Puig, I., Planas, A., Ragan, C. I., & Macrae, I. M. (2017). The IMPROVE guidelines (ischaemia models: Procedural refinements of in vivo experiments). *Journal of Cerebral Blood Flow and Metabolism*, 37(11), 3488–3517. 10.1177/0271678X1770918528797196

Poirier, N. (2022). On the Intertwining of Cellular Agriculture and Animal Agriculture: History, Materiality, Ideology, and Collaboration. *Frontiers in Sustainable Food Systems*, 6, 907621. 10.3389/fsufs.2022.907621

Ressurreição, A., Gibbons, J., Kaiser, M., Dentinho, T. P., Zarzycki, T., Bentley, C., Austen, M., Burdon, D., Atkins, J., Santos, R. S., & Edwards-Jones, G. (2012). Different cultures, different values: The role of cultural variation in public's WTP for marine species conservation. *Biological Conservation*, 145(1), 148–159. 10.1016/j.biocon.2011.10.026

Rilov, G., Mazaris, A. D., Stelzenmüller, V., Helmuth, B., Wahl, M., Guy-Haim, T., Mieszkowska, N., Ledoux, J.-B., & Katsanevakis, S. (2019). Adaptive marine conservation planning in the face of climate change: What can we learn from physiological, ecological and genetic studies? *Global Ecology and Conservation*, 17, e00566. 10.1016/j.gecco.2019.e00566

Rogers, A. D., Baco, A., Escobar-Briones, E., Currie, D., Gjerde, K., Gobin, J., Jaspars, M., Levin, L., Linse, K., Rabone, M., Ramirez-Llodra, E., Sellanes, J., Shank, T. M., Sink, K., Snelgrove, P. V. R., Taylor, M. L., Wagner, D., & Harden-Davies, H. (2021). Marine genetic resources in areas beyond national jurisdiction: Promoting marine scientific research and enabling equitable benefit sharing. *Frontiers in Marine Science*, 8, 667274. 10.3389/fmars.2021.667274

Rollin, B. E. (2006). The regulation of animal research and the emergence of animal ethics: A conceptual history. *Theoretical Medicine and Bioethics*, 27(4), 285–304. 10.1007/s11017-006-9007-816937023

Samanta, I., Joardar, S. N., & Das, P. K. (2018). Biosecurity strategies for backyard poultry: a controlled way for safe food production. In *Food control and biosecurity* (pp. 481–517). Academic Press. 10.1016/B978-0-12-811445-2.00014-3

Schuppli, C. A. (2011). Decisions about the use of animals in research: Ethical reflection by Animal Ethics Committee members. *Anthrozoos*, 24(4), 409–425. 10.2752/175303711X13159027359980

Shiomi, M. (2009). Rabbit as a model for the study of human diseases. *Rabbit biotechnology*, 49-63. 10.1007/978-90-481-2227-1_7

Silvy, N. J., Lopez, R. R., & Peterson, M. J. (2005). Wildlife marking techniques. *Techniques for wildlife investigations and management*, 6, 339-376.

Smith, L. W. (2005). Helping industry ensure animal well-being. *Agricultural Research*, 53(3), 2–3.

Sneddon, L. U., Halsey, L. G., & Bury, N. R. (2017). Considering aspects of the 3Rs principles within experimental animal biology. *The Journal of Experimental Biology*, 220(17), 3007–3016. 10.1242/jeb.14705828855318

Steidl, R. J., Hayes, J. P., & Schauber, E. (1997). Statistical power analysis in wildlife research. *The Journal of Wildlife Management*, 61(2), 270–279. 10.2307/3802582

Soulsbury, C., Gray, H., Smith, L., Braithwaite, V., Cotter, S., Elwood, R. W., Wilkinson, A., & Collins, L. M. (2020). The welfare and ethics of research involving wild animals: A primer. *Methods in Ecology and Evolution*, 11(10), 1164–1181. 10.1111/2041-210X.13435

Swallow, J., Anderson, D., Buckwell, A. C., Harris, T., Hawkins, P., Kirkwood, J., Lomas, M., Meacham, S., Peters, A., Prescott, M., Owen, S., Quest, R., Sutcliffe, R., & Thompson, K. (2005). Guidance on the transport of laboratory animals. *Laboratory Animals*, 39(1), 1–39. 10.1258/0023677052886649315703122

Tabak, M. A., Norouzzadeh, M. S., Wolfson, D. W., Sweeney, S. J., VerCauteren, K. C., Snow, N. P., Halseth, J. M., Di Salvo, P. A., Lewis, J. S., White, M. D., Teton, B., Beasley, J. C., Schlichting, P. E., Boughton, R. K., Wight, B., Newkirk, E. S., Ivan, J. S., Odell, E. A., Brook, R. K., & Miller, R. S. (2019). Machine learning to classify animal species in camera trap images: Applications in ecology. *Methods in Ecology and Evolution*, 10(4), 585–590. 10.1111/2041-210X.13120

Temple, D., & Manteca, X. (2020). Animal welfare in extensive production systems is still an area of concern. *Frontiers in Sustainable Food Systems*, 4, 545902. 10.3389/fsufs.2020.545902

Tsvakirai, C., Nalley, L., Rider, S., Van Loo, E., & Tshehla, M. (2023). The Alternative Livestock Revolution: Prospects for Consumer Acceptance of Plant-based and Cultured Meat in South Africa. *Journal of Agricultural and Applied Economics*, 55(4), 710–729. 10.1017/aae.2023.36

Tobita, K., Liu, X., & Lo, C. W. (2010). Imaging modalities to assess structural birth defects in mutant mouse models. *Birth Defects Research. Part C, Embryo Today*, 90(3), 176–184. 10.1002/bdrc.2018720860057

Viceconti, M., Henney, A., & Morley-Fletcher, E. (2016). *In silico* clinical trials: How computer simulation will transform the biomedical industry. *International Journal of Clinical Trials*, 3(2), 37–46. 10.18203/2349-3259.ijct20161408

Von Keyserlingk, M. A., & Hötzel, M. J. (2015). The ticking clock: Addressing farm animal welfare in emerging countries. *Journal of Agricultural & Environmental Ethics*, 28(1), 179–195. 10.1007/s10806-014-9518-7

Wanibuchi, H., Salim, E. I., Kinoshita, A., Shen, J., Wei, M., Morimura, K., Yoshida, K., Kuroda, K., Endo, G., & Fukushima, S. (2004). Understanding arsenic carcinogenicity by the use of animal models. *Toxicology and Applied Pharmacology*, 198(3), 366–376. 10.1016/j.taap.2003.10.03215276416

Welty, J. (2007). Humane slaughter laws. *Law and Contemporary Problems*, 70(1), 175–206.

Widdowson, R. W. (2013). *Towards holistic agriculture: A scientific approach*. Elsevier.

Wolfensohn, S. (2010). Euthanasia and other fates for laboratory animals. *The UFAW handbook on the care and management of laboratory and other research animals*, 219-226. 10.1002/9781444318777.ch17

World Food Situation—FAO Food Price Index. (2022). https://www.fao.org/worldfoodsituation/foodpriceindex/en/

Chapter 5
The Lab Grown Plate:
A Deep Dive Into Cellular Agriculture

Rathimalar Ayakannu
Lincoln University College, Malaysia

Hemapriyaa Vijayan
Lincoln University College, Malaysia

Asita Elengoe
Lincoln University College, Malaysia

ABSTRACT

This chapter highlights the cutting-edge field of cellular agriculture, emphasizing the creation, principles, and applications of lab-grown food. The lab-grown plate refers to a concept representing food that is produced through cellular agriculture techniques in a laboratory setting. It covers the idea of creating food products, such as meat, dairy, and other animal products, through cell culture techniques, without the need for conventional animal husbandry. Instead, cells are cultured and grown in controlled settings to produce edible tissue, which is subsequently processed into a variety of food products. However, there are still challenges to overcome in scaling up cellular agriculture to commercial levels, including technological hurdles, regulatory considerations, and consumer acceptance. Nevertheless, continued research and investment in this field hold promise for transforming the way animal products are produced and consumed in the future.

DOI: 10.4018/979-8-3693-4115-5.ch005

THE LAB GROWN PLATE

Cell-based meat is edible muscle tissue produced by cultivating stem cells in a controlled culture and physiological environment in a laboratory using tissue engineering and computational simulation techniques (Mengistie, 2020). It is sometimes referred to as in vitro meat, lab-grown meat, clean meat, or synthetic meat (Jones, 2023). In this chapter, we have attempted to provide a summary of the procedures involved and current challenges in producing meat using cells. Due to high production costs and immature manufacturing technology, cell-based meat is still in its infancy. Consequently, researchers are working continuously to improve technical aspects of the product to hasten its industrialization and commercialization (Guan et al., 2021).

PRODUCTION METHODS OF LAB GROWN MEAT

The cell-based meat manufacturing technology combines tissue and food engineering techniques. Possible sources of stem cells include induced pluripotent stem cells, muscle stem cells, or myosatellite stem cells, and embryonic stem cells (Pandurangan & Kim, 2015). Since the number of satellite cells reportedly remains constant after multiple injuries, these cells are considered stem cells, which can certainly be maintained by self-renewal (Shi & Garry, 2006). These cells can be easily obtained from live animal tissue biopsies by enzymatic digestion and mechanical disruption, or by flow cytometry purification using specific surface markers (Ding et al., 2017). Although a multitude of cell sources can be used to make cell-based meat, each cell type requires unique in vitro expansion and differentiation techniques due to its growth and development characteristics (Fish et al., 2020).

After the stem cells are harvested, they must be cultured to obtain large numbers of cells. Once the necessary number of cells is obtained, the cells are stimulated to develop into myotubes, adipocytes, or other mature cell types in muscle tissue (Liu, 2019). Laboratory flask or petri dish cultures are not large enough to meet market demand, so large-scale bioreactor systems are needed (Post et al., 2020). The efficiency of scaling up cell-based meat production depends on adequate oxygen perfusion during cell seeding and scaffold development (Jahir et al., 2023). The cells are then grown in a carefully regulated environment with adequate nutrients, plenty of oxygen and the right temperature. The cells multiply and divide, starting to form tissue that resembles that of the animal (Mattick, 2018).

It is possible to produce meat with a soft texture or without bones by using scaffolds. As they allow vital nutrients to be transferred more easily, remove waste products, and encourage the development of functioning tissues and organs, scaffolds are essential in tissue engineering. One muscle tissue piece can be replicated into over a ton of components by arranging differentiated myotubes in a ring on a scaffold and letting them grow in size and protein content (Mengistie, 2023).

At the end of the cell-based meat manufacturing process, mature cells are gathered and treated, including moulding, dyeing, and seasoning, to generate the final product (Zhang et al., 2020). In addition, food processing techniques, including the addition of heme proteins and flavorings, can produce finished products, including sausages, burgers and other foods (Liu, 2019).

The production of lab meat requires large-scale, high-throughput technologies to produce high-quality meat in the laboratory that is suitable for the livestock and agricultural industry, the environment, animal welfare and human health (Ching et al., 2022). Recent scientific and technological developments in the field of cell-based meat production have led to the emergence of a number of novel techniques, including organ printing, nanotechnology, and biophotonics, which are considered potential approaches for the manufacture of cell-based meat (Benny et al., 2022).

Organ printing refers to the application of three-dimensional (3D) or four-dimensional (4D) bioprinting technologies in conjunction with tissue engineering principles to produce biological tissue constructs that mimic the anatomical, structural, and functional properties of original organs or tissues (Gillispie et al., 2019). 3D printing is a cutting-edge tissue engineering technology that uses sprayed cell-encapsulating hydrogels to generate 3D shapes and structures on printed scaffolds (Jung et al., 2016). An improvement over 3D printing, 4D printing allows for the gradual addition of a new dimension of transformation in situations where the target organs or tissues are susceptible to changes in temperature and humidity (Javaid & Haleem, 2019). 3D printing offers benefits such as fast manufacture, shape customization, consistent distribution of nutritional content, and easy preparation (Liu et al., 2017). Future developments in the 3D bioprinting of cultured meat will focus on cloud manufacturing, integrated fabrication, and multi-component products (Guo et al., 2023).

Since nanofibers are naturally present in meat and have an effect on the texture and colour of the meat after cooking, nanotechnology can be extremely important in the development and packaging of novel meat products (Picouet et al., 2014). The removal of preservatives and additives by nanotechnology is consistent with the growing consumer preference for natural and environmentally friendly products (Gupta et al., 2024). The application of nanotechnology in meat packaging offers a number of advantages, including as enhanced barrier properties, mechanical tolerance, heat resistance, and greater biodegradability. As packaging materials, they can also be employed with antibacterial boosters and spoilage detectors (Sharma et al., 2017). For example, Zinc oxide nanoparticles are effective in preserving meat quality by limiting microbial growth and preventing lipid/protein oxidation (Smaoui et al., 2023). Recently, Cherif et al. showed that ZnO-NPs are noteworthy antimicrobial agents in the food sector and exhibit high antimicrobial activity against a wide range of microorganisms (Chérif et al., 2023).

Additionally, another theoretical method for producing meat made of cells is biophotonics, a technology that holds matter particles together with light or laser (Bhat et al., 2019). It is a promising new technology that uses the action of lasers to move material particles into certain tissue structures to create in vitro meat (Gaydhane et al., 2018). Through the fusion of biophotonics and nanotechnology, growing cells can be better anchored, thereby enhancing coaxal chains in a systematic and consistent manner. This may be an alternative to using traditional scaffolding methods to immobilize cells (Ramachandraiah et al., 2015).

CONVENTIONAL AGRICULTURE AND CELLULAR AGRICULTURE

The cutting edge and quickly expanding field of cellular agriculture focuses on using the cells of animals to produce meat and other products directly from their cells, negating the need for traditional farming practices. There is currently a great deal of difficulty facing traditional agriculture. Food and other agricultural products are needed to feed a growing global population, which is expected to reach 9–11 billion people by 2050 (Röös et al., 2017).This demand is made worse by the scarcity of available land and the threat of climate change. In the near future, a sharp increase in traditional agricultural productivity is required to meet this goal. The adoption of cellular agriculture, a more ecologically friendly and sustainable form of agricultural production that uses less water and land and emits less greenhouse emissions than conventional farming techniques, is one possible answer to this issue. Contrary to popular assumption, human activity has actually lowered overall biomass on Earth by a factor of two rather than increased it. This includes domesticating livestock, adapting agriculture, and generating a massive increase in the human population through the industrial revolution (Bar-On et al., 2018).

The current livestock industry has been reported to contribute to approximately 18% of global greenhouse emission are caused by the livestock industry. An estimated 7516 million tonnes of carbon dioxide (CO_2) are released into the atmosphere annually (Ilea, 2009). According to Goodland and his co-workers, it is estimated that 32,564 million tonnes of carbon dioxide are released into the atmosphere each year, being accountable for at least 51% of these emissions by the global livestock industry (Goodland & Anhang, 2009). The majority of emissions associated with the livestock sector are methane (CH_4), ammonia (NH_3), nitrous oxide (N_2O), and carbon dioxide (CO_2). The emission of carbon dioxide by livestock has been reported to be a a major contributing factor for climate change (Leytem et al., 2011). It is reported that the emission of enterogenic nitrous oxide are caused by the livestock industry and this gas can remain in the atmosphere for a very long period and reported to be a contributor for ozone layer degradation which further causes global warming. Besides that, ammonia gas has been also reported as a major emission from the livestock industry which directly contributes to the acidification of ecosystem. Compared to carbon dioxide, methane has a higher potential to cause global warming. Methane emission from pigs and cattles have increased significantly by 37% and 50% respectively as reported by the U.S Environmental Protection Agency (Leytem et al., 2011).

Extensive amount of land is needed for the livestock industry to meet the increasing demand day by day. Approximately one third of world's land is used for cattle farm and while two thirds of land is used for agriculture due to the demand and this directly contributes deforestation and land clearing for this purpose (Ilea, 2009). Worldwide, over 40% of all crops are harvested are feed to animals. Therefore, we could feed every starving population on the planet and end world hunger if we could use the crops used for animal feed (Leitzmann, 2003). Animal extinction is one of the very devastating effects of massive deforestation and land clearing. Approximately, 140 species of flora and fauna are endangered due to this purpose. Nearly half (44%) of land degradation is caused only by meat and fish, mostly by the cultivation of terrestrial feed crops for large-scale livestock and aquaculture activities (Froehlich et al., 2018). According to Baden (2019), the high-intensity use of land for monoculture crop production results in a rapid rate of soil erosion and degradation, with about 90% of cropland losing soil at a rate 13 times above sustainable levels (Baden et al., 2019).

Water pollution in agriculture is mostly caused by the production of animal food items. Particularly in emerging nations, the trend of rising animal product consumption is detrimental to ecosystems and water sources. The animal excrement, antibiotics and hormones, fertilizer and insecticides in forage production is the source of the water contamination as they can leach through the soil and nearby water sources. Additionally, the livestock business wastes a lot of resources, especially water. For instance, in the United States, the amount of water used by individual homes accounts for around 5% of overall consumption, whereas the amount used by animal husbandry accounts for approximately 55% (Worm et al., 2006). In addition, the cattle sector generates enormous amounts of waste. Approximately 6,000 pounds of trash are produced by the US livestock industry every second. A farm with 2500 milking cows generates the same amount of garbage as a metropolis with 411,000 inhabitants, claim Haines and Staley (Pimentel & Pimentel, 2003). Critical water challenges in the United States caused by agricultural practices include evaporation and leakage from irrigation systems, over-drafting of aquifers, waterlogging and salinization of soils, loss and runoff from wetlands, and pollution of surface and groundwater sources (Marlow et al., 2009).

In current trend, lab grown meat has been an answer to the increasing demand of meat consumption worldwide. The cutting edge and quickly expanding field of cellular agriculture focuses on using the cells of animals to produce meat and other products directly from their cells, negating the need for traditional farming practices. It's essentially a whole new revolution in food production, doing away with conventional animal husbandry. For there to be global food security, food must be stable, usable, accessible, and available to everyone. With careful thought given to the establishment of the field, cellular agriculture may be able to support these requirements.

To start with, cellular agriculture production avoids some of the riskiest routes for illness in human history. COVID-19 is the most recent zoonotic disease pandemic to emerge from hunted or farmed animals, following the 1918 influenza and the H1N1 swine flu (Worobey et al., 2014). Meat processing facilities became key COVID-19 hotspots during the pandemic, prompting 57 plants to close three months into the outbreak (Marchant-Forde & Boyle, 2020). These animals also produce a lot of biowaste such as pus, mucus, blood and manure, which is a breeding ground for viruses, bacteria's such as E. Coli, and other infections that can contaminate meat and water sources (Sapkota et al., 2007). On the other hand, cellular agriculture generates isolated products from cultures in bioreactors, resulting in less biowaste and a quicker detection and removal of infections by producers.

On the other hand, cellular agriculture does not need the transportation of animals and will be more convenient and avoid interruptions in the supply chain sectors that rely on animal products. As an example, animals such as cattle, hog and chicken should be transported to the slaughterhouse in a given timeline as they have reached the ideal weight. Failure to do so will cause tremendous amount of loss in terms of production. Besides that, the bioreactor design used in cellular agriculture have been anticipated that each batch of culture will reach maturity in a few weeks' time, which enables prompt response in supply and demand (Allan et al., 2019). Cultures can also be cryopreserved and can be kept for a long term to provide food security.

Apart from that, cellular agriculture has the potential to provide nutrient-dense and culturally significant animal products. Many culturally significant and nutrient-dense meats, like salmon or octopus, are in high demand and hence expensive since producing them needs extra processing. Cellular agriculture have indicated that lean meat could be more accessible and affordable besides being a more healthy option for a healthy protein intake (Post et al., 2020). Comparing some cultured meats to their counterparts derived from animals, they may even offer superior nutrition (Simsa et al., 2019). In a nutshell, cellular agriculture has been demonstrated to produce more nutrient dense, safer and more resilient and makes culturally relevant food more accessible.

Lab-grown meat prevents animal killing and has fewer negative effects on the environment and public health (Post et al., 2020) . The emergences of cellular agriculture have been debated by many parties but the issues of food security associated with livestock industry can be solved by cellular agriculture. The problems with animal agriculture's current lack of food security may be solved by cellular agriculture. With cellular agriculture, zoonotic illnesses might not be as prevalent as they have been with centralised animal agriculture. Production of cellular agriculture could continue even in situations when infections necessitate social separation due to automation and reduced processing steps. Since cells grow in days or weeks rather than months or years, dealing with them will probably allow for more flexibility and manufacturing status. Cellular agriculture is an additional technique for producing food that would boost the availability of goods derived from animals without increasing the scope of animal agriculture.

CHALLENGES IN CELLULAR AGRICULTURE

The field of cellular agriculture is a recent development whereby agricultural products, usually sourced from animals, are created using cellular processes rather than whole-organism processes, which are commonly seen in farms. It includes using cell cultures to produce animal products like meat, milk, and eggs instead of rearing and killing animals, and it has many difficulties.

Cutting the production cost is one of the main obstacles. Currently, the cost of growing cells in a lab setting is high since facilities, growth media, and specialized equipment are required. In order for cellular agriculture to rival traditional animal agriculture, these costs must be reduced. The difficulty in making cultured meat lies in reproducing in a lab or factory the conditions necessary for muscle growth in a cow or other animal. Since muscle formation has developed over millions of years, it is an effective process that is well adapted to taking place in the body as part of a wide range of other processes. Muscle tissue engineering, like tissue engineering for other tissues, uses biochemical engineering concepts in conjunction with biological knowledge of tissue growth and development to simulate the in vivo environment. Up until now, the majority of tissue engineering research has been concentrated on medical applications including regenerative medicine and non-animal technologies for in vitro models utilized in toxicology and drug development. The production of cultured meat follows the same technical guidelines, but it requires a considerably greater scale and a commodity-level price for the product. Having said that, because cultured beef is a meal and not a medical product, it does not require the same strict regulations or the highest quality (purity) of raw ingredients as it does for biomedical uses (Stephens et al., 2018).

Although it is technically possible to generate animal products from cells, increasing output to satisfy customer demand is a difficult task. More research and development is needed to achieve large-scale production while upholding safety and quality standards. The process mode (batch, fed-batch, or continuous) dictates the size of the bioreactor and the amount of media that is used. For example, a continuous process mode can result in a ten-fold reduction in the size of the bioreactor (Eibl et al., 2009). Continuous or feed batch operations may be preferred when it comes to large-scale implementation. The amount, quality, and cost of the required cells should all be considered while choosing the process mode. It might also be looked into whether media components and microcarriers can be recycled.

Regulations governing products derived from cellular agriculture are still being developed by regulatory bodies worldwide. For these items to be accepted by consumers and gain access to markets, it is imperative that they adhere to safety and labelling regulations. A thorough safety monitoring system and an appropriate regulatory framework need to be designed in order to remove any hazards to the health and safety of customers. In order to guarantee that in vitro meat products are appropriately and safely labelled when they are introduced to the market in the next years, the US Department of Agriculture (USDA) and the US Food and Drug Administration (FDA) have been collaborating since 2018 (Crosser et al., 2020). The FDA and USDA signed a formal agreement in March 2019 to collaborate on the regulation of in vitro meat production from animals and poultry. The USDA is in charge of harvesting and postharvest facilities as well as product labelling, whereas the FDA is in charge of the early phases of in vitro meat production, such as cell collection and banks, cell proliferation, and differentiation. Furthermore, The Good Food Institute's research indicates that Singapore and India's governments are in favor of the introduction of the first in-vitro meat products with a progressive regulatory framework (Crosser et al., 2020). The Novel Food Regulation governs in vitro meat in the European Union, and the European Commission must approve it for sale (Ferranti et al., 2019). In an ideal scenario, the approval process might be finished in eighteen months.

Products derived from cellular agriculture must be accepted by consumers in order to succeed. It will be difficult to dispel skepticism and inform the public about the advantages of these products, such as their little impact on the environment and their consideration of animal welfare.

For cellular agriculture to be widely adopted, products must be produced that closely resemble the nutritional makeup, flavor, and texture of conventional animal products. To optimize the production process and reliably obtain these qualities, more investigation is required. Achieving equivalent structural and biochemical composition of in vitro meat in terms of flavor, color, and texture are the three primary sensory determinants that may provide another difficulty. Potential remedies to increase consumer acceptance of in vitro meat include the co-cultivation of muscle cells with adipose tissue, utilization of biotechnologically produced heme (legume hemoglobin), and additional process condition optimization (Bhat et al., 2019). Since myocytes are the main component of meat, tissue engineering for cultured meat typically focuses on producing these cells alone via the regenerative pathway. But it takes a combination of cell types to create muscle tissue that can mimic meat in its entirety. Most investigations of skeletal muscle have been done in two dimensions with cell lines (Burattini, 2004). Nevertheless, 3D structures, also known as "bioartificial muscle," are being researched for their potential to serve as a more accurate in-vitro model of natural skeletal muscle tissue and in regenerative medicine (Snyman et al., 2013). Whereas carcass meats require the development of thicker 3D structures with a food and oxygen supply and waste removal to sustain the inner core of cells, thin 3D cultures can be used to make processed meats (burgers, sausages). According to Lovett et al. (2009), the only viable muscle tissue structures to yet have been a few hundred microns thick, this is suitable for minced but not entire muscle slices. According to Hinds et al. (2013), cell sheets are being investigated for thicker tissue assembly; still, the creation of highly perfused scaffolds would be necessary for highly ordered and structured tissues. The focus of investigations has shifted to creating channels inside the tissue, with a particular focus on 3D structured tissue formation employing lithography (Muehleder et al., 2014), removable structures, and channeled networks made from sacrificial scaffolds (Mohanty et al., 2014). This allows for perfusion of flow throughout the tissue. Harvard researchers 3D-printed a perfusion network that was able to sustain a culture for six weeks, and cultured leather purveyors Modern Meadow patented a method and device for "scalable extrusion of cultured cells for use in forming three-dimensional tissue structures." These examples suggest that 3D printing is a promising concept for creating these channelled networks (Koleskyet al., 2016).

It is difficult to find and create scalable cell sources for many animal products. While some cell types such as muscle cells are ideal for producing meat, further research is needed to identify appropriate cell types for other goods, like milk and eggs. Most of the time, items derived from animals may be present in the culture media and synthetic scaffolds that are utilized. Naturally, substances found in the body, like collagen, which is frequently utilized as a substrate in cell culture systems, are the greatest places for cells to develop. Additional substances occurring in the body, such as growth factors and blood serum added to cell culture media, are frequently used in cell culture techniques. Fetal calf serum is commonly utilized in medical research, while it can also be obtained from other species including more mature mammals. Although growing cells without serum or with serum replacements is feasible, it is a little more challenging (Butler, 2015). Nevertheless, this field of study has not yet yielded a suitable and cost-effective substitute. Culture media contains components synthesized in crops, yeast, and bacteria, just like certain diets for people and animals do. Although the creation of ingredients by bacteria and yeast can be considered cellular agriculture and works in tandem with the development of cultured meat, the long-term sustainability of using crude oil derivatives to make components is questionable. Producing an economical, animal-free, and sustainable muscle cell culture medium is a significant task because these cultures are costly, sometimes even unaffordable when used on a big basis. The manufacturing of scaffolds has the same difficulties. Tissue engineering has made use of a variety of biomaterials generated from animals and non-animals. As would be expected, myogenic cells have a preference for living in materials derived from animals since these materials more closely resemble their native physiological niche. The bulk of bio-artificial muscle that has been successfully generated has been grown on collagen scaffolds (Snyman et al., 2013). Using synthetic biomaterials to achieve tissue contraction has proven difficult thus far (Bian & Bursac, 2009). For the development of meat's constituent parts (muscle, fat, and blood vessels), more studies must be done on biomaterials produced from non-animal sources or food-grade animal products.

Although traditional animal agriculture may have less of an environmental impact thanks to cellular agriculture, it is crucial to make sure that the production process is sustainable in and of itself. This entails managing trash and other environmental issues, as well as reducing the use of energy and resources.

We are still in the early stages of cellular agriculture, which uses animal cell and tissue cultures. It is evident that the production of meat including fish is the current priority. This is to be expected since by 2050, the world's meat consumption is predicted to double (Gaydhane et al., 2018). Nowadays, the scaffolding approach and satellite cells are frequently used in the in vitro meat production process. The first in vitro meat product prototypes have been shown thus far, and the first commercial items are anticipated to be released within the next two years. Many essential concerns remain unanswered if animal cell and tissue culture-based cellular agriculture is to compete with traditional meat production in the future. Chemically specified culture media, scaffolds and bioreactors, various process modes and controls (feed batch or continuous, expansion and differentiation inside the same bioreactor or independently, recycling of media components and microcarriers), and regulatory issues are the primary topics of concern. There could be valuable insights for the food industry from the biopharmaceutical sector. Even though the food business operates on far bigger scales and with fewer rigorous regulations than the pharmaceutical industry, it will be crucial for its operations to become more economically viable. It would be beneficial to create new study programs and training, produce cell culture technologists for agricultural products, and combine the essential knowledge from biochemical engineering, food technology, cell and tissue culture technology, and quality assurance and regulations with facility design. Beyond all technical constraints, consumer acceptance which may ultimately come down to taste is what determines the viability of in vitro meat and related goods.

REFERENCES

Allan, S., De Bank, P. A., & Ellis, M. J. (2019). Bioprocess Design Considerations for Cultured Meat Production With a Focus on the Expansion Bioreactor. *Frontiers in Sustainable Food Systems*, 3, 44. Advance online publication. 10.3389/fsufs.2019.00044

Baden, M. Y., Liu, G., Satija, A., Li, Y., Sun, Q., Fung, T. T., Rimm, E. B., Willett, W. C., Hu, F. B., & Bhupathiraju, S. N. (2019). Changes in Plant-Based Diet Quality and Total and Cause-Specific Mortality. *Circulation*, 140(12), 979–991. 10.1161/CIRCULATIONAHA.119.041014

Bar-On, Y. M., Phillips, R., & Milo, R. (2018). The biomass distribution on Earth. *Proceedings of the National Academy of Sciences of the United States of America*, 115(25), 6506–6511. 10.1073/pnas.1711842115

Benny, A., Pandi, K., & Upadhyay, R. (2022). Techniques, challenges and future prospects for cell-based meat. *Food Science and Biotechnology*, 31(10), 1225–1242. 10.1007/s10068-022-01136-6

Bhat, Z. F., Morton, J. D., Mason, S. L., Bekhit, A. E. A., & Bhat, H. F. (2019). Technological, Regulatory, and Ethical Aspects of In Vitro Meat: A Future Slaughter-Free Harvest. *Comprehensive Reviews in Food Science and Food Safety*, 18(4), 1192–1208. 10.1111/1541-4337.12473

Bian, W., & Bursac, N. (2009). Engineered skeletal muscle tissue networks with controllable architecture. *Biomaterials*, 30(7), 1401–1412. 10.1016/j.biomaterials.2008.11.015

Burattini, S., Ferri, P., Battistelli, M., Curci, R., Luchetti, F., & Falcieri, E. (2004). C2C12 murine myoblasts as a model of skeletal muscle development: Morphofunctional characterization. *European Journal of Histochemistry*, 48(3), 223–234.

Butler, M. (2015). Serum and protein free media. Animal Cell Culture, 223-236.

Chérif, I., Dkhil, Y. O., Smaoui, S., Elhadef, K., Ferhi, M., & Ammar, S. (2023). X-Ray Diffraction Analysis by Modified Scherrer, Williamson–Hall and Size–Strain Plot Methods of ZnO Nanocrystals Synthesized by Oxalate Route: A Potential Antimicrobial Candidate Against Foodborne Pathogens. *Journal of Cluster Science*, 34(1), 623–638. 10.1007/s10876-022-02248-z

Ching, X. L., Zainal, N. A. A. B., Luang-In, V., & Ma, N. L. (2022). Lab-based meat the future food. *Environmental Advances*, 10, 100315. 10.1016/j.envadv.2022.100315

Crosser, N., Bushnell, C., Derbes, E., Friedrich, B., & Lamy, J. (2020). 2019 state of the industry report: cultivated meat. Rep., Good Food Inst. https://www.gfi.org/files/soti/INN-CM-SOTIR2020_0512.pdf?utm_source=form&utm_medium=email&utm_campaign=SOTIR2019

Ding, S., Wang, F., Liu, Y., Li, S., Zhou, G., & Hu, P. (2017). Characterization and isolation of highly purified porcine satellite cells. *Cell Death Discovery*, 3(1), 17003. 10.1038/cddiscovery.2017.3

Eibl, R., Eibl, D., Pörtner, R., Catapano, G., Czermak, P., Eibl, R., & Eibl, D. (2009). Plant cell-based bioprocessing. Cell and Tissue Reaction Engineering: With a Contribution by Martin Fussenegger and Wilfried Weber, 315-356.

Ferranti, P., Berry, E., & Anderson, J. R. (2019). *Encyclopedia of Food Security and Sustainability* (1st ed.). Elsevier. 10.1038/cddiscovery.2017.3

Fish, K. D., Rubio, N. R., Stout, A. J., Yuen, J. S. K., & Kaplan, D. L. (2020). Prospects and challenges for cell-cultured fat as a novel food ingredient. *Trends in Food Science & Technology*, 98, 53–67. 10.1016/j.tifs.2020.02.005

Froehlich, H. E., Runge, C. A., Gentry, R. R., Gaines, S. D., & Halpern, B. S. (2018). Comparative terrestrial feed and land use of an aquaculture-dominant world. *Proceedings of the National Academy of Sciences of the United States of America*, 115(20), 5295–5300. 10.1073/pnas.1801692115

Gaydhane, M. K., Mahanta, U., Sharma, C. S., Khandelwal, M., & Ramakrishna, S. (2018). Cultured meat: State of the art and future. *Biomanufacturing Reviews*, 3(1), 1. 10.1007/s40898-018-0005-1

Gillispie, G. J., Park, J., Copus, J. S., Pallickaveedu Rajan Asari, A. K., Yoo, J. J., Atala, A., & Lee, S. J. (2019). Three-Dimensional Tissue and Organ Printing in Regenerative Medicine. In Atala, A., Lanza, R., Mikos, A. G., & Nerem, R. (Eds.), *Principles of Regenerative Medicine* (3rd ed., pp. 831–852). Academic Press. 10.1016/B978-0-12-809880-6.00047-3

Goodland, R., & Anhang, J. (2009). Livestock and climate change. *World Watch*, 22, 10–19.

Guan, X., Lei, Q., Yan, Q., Li, X., Zhou, J., Du, G., & Chen, J. (2021). Trends and ideas in technology, regulation and public acceptance of cultured meat. *Future Foods : a Dedicated Journal for Sustainability in Food Science*, 3, 100032. 10.1016/j.fufo.2021.100032

Guo, X., Wang, D., He, B., Hu, L., & Jiang, G. (2023). 3D Bioprinting of Cultured Meat: A Promising Avenue of Meat Production. *Food and Bioprocess Technology*. Advance online publication. 10.1007/s11947-023-03195-x

Gupta, R. K., Gawad, F. A. E., Ali, E. A. E., Karunanithi, S., Yugiani, P., & Srivastav, P. P. (2024). Nanotechnology: Current applications and future scope in food packaging systems. *Measurement Food*, 13, 100131. 10.1016/j.meafoo.2023.100131

Hinds, S., Tyhovych, N., Sistrunk, C., & Terracio, L. (2013). Improved tissue culture conditions for engineered skeletal muscle sheets. *TheScientificWorldJournal*, 2013(1), 2013. 10.1155/2013/370151

Ilea, R. C. (2009). Intensive Livestock Farming: Global Trends, Increased Environmental Concerns, and Ethical Solutions. *Journal of Agricultural & Environmental Ethics*, 22(2), 153–167. 10.1007/s10806-008-9136-3

Jahir, N. R., Ramakrishna, S., Abdullah, A. A. A., & Vigneswari, S. (2023). Cultured meat in cellular agriculture: Advantages, applications and challenges. *Food Bioscience*, 53, 102614. 10.1016/j.fbio.2023.102614

Javaid, M., & Haleem, A. (2019). 4D printing applications in medical field: A brief review. *Clinical Epidemiology and Global Health*, 7(3), 317–321. 10.1016/j.cegh.2018.09.007

Jones, N. (2023). Lab-grown meat: The science of turning cells into steaks and nuggets. *Nature*, 619(7968), 22–24. 10.1038/d41586-023-02095-6

Jung, J. W., Lee, J. S., & Cho, D. W. (2016). Computer-aided multiple-head 3D printing system for printing of heterogeneous organ/tissue constructs. *Scientific Reports*, 6(1), 21685. 10.1038/srep21685

Kolesky, D. B., Homan, K. A., Skylar-Scott, M. A., & Lewis, J. A. (2016). Three-dimensional bioprinting of thick vascularized tissues. *Proceedings of the National Academy of Sciences of the United States of America*, 113(12), 3179–3184. 10.1073/pnas.1521342113

Leitzmann, C. (2003). Nutrition ecology: The contribution of vegetarian diets. *The American Journal of Clinical Nutrition*, 78(3, Suppl), 657s–659s. 10.1093/ajcn/78.3.657S

Leytem, A. B., Dungan, R. S., Bjorneberg, D. L., & Koehn, A. C. (2011). Emissions of ammonia, methane, carbon dioxide, and nitrous oxide from dairy cattle housing and manure management systems. *Journal of Environmental Quality*, 40(5), 1383–1394. 10.2134/jeq2009.0515

Liu, W. (2019). A Review on the Genetic Regulation of Myogenesis and Muscle Development. *American Journal of Biochemistry and Biotechnology*, 15(1), 1–12. Advance online publication. 10.3844/ajbbsp.2019.1.12

Liu, Z., Zhang, M., Bhandari, B., & Wang, Y. (2017). 3D printing: Printing precision and application in food sector. *Trends in Food Science & Technology*, 69, 83–94. 10.1016/j.tifs.2017.08.018

Lovett, M., Lee, K., Edwards, A., & Kaplan, D. L. (2009). Vascularization strategies for tissue engineering. *Tissue Engineering. Part B, Reviews*, 15(3), 353–370. 10.1089/ten.teb.2009.0085

Marchant-Forde, J. N., & Boyle, L. A. (2020). COVID-19 Effects on Livestock Production: A One Welfare Issue. *Frontiers in Veterinary Science*, 7, 585787. 10.3389/fvets.2020.585787

Marlow, H. J., Hayes, W. K., Soret, S., Carter, R. L., Schwab, E. R., & Sabaté, J. (2009). Diet and the environment: Does what you eat matter? *The American Journal of Clinical Nutrition*, 89(5), 1699s–1703s. 10.3945/ajcn.2009.26736Z

Mattick, C. S. (2018). Cellular agriculture: The coming revolution in food production. *Bulletin of the Atomic Scientists*, 74(1), 32–35. 10.1080/00963402.2017.1413059

Mengistie, D. (2020). Lab-growing meat production from stem cell. *Journal of Nutrition & Food Sciences*, 3(1), 100015.

Mohanty, S., Larsen, L. B., Trifol, J., Szabo, P., Burri, H. V. R., Canali, C., Dufva, M., Emnéus, J., & Wolff, A. (2015). Fabrication of scalable and structured tissue engineering scaffolds using water dissolvable sacrificial 3D printed moulds. *Materials Science and Engineering C*, 55, 569–578. 10.1016/j.msec.2015.06.002

Muehleder, S., Ovsianikov, A., Zipperle, J., Redl, H., & Holnthoner, W. (2014). Connections matter: Channeled hydrogels to improve vascularization. *Frontiers in Bioengineering and Biotechnology*, 2, 52. 10.3389/fbioe.2014.00052

Pandurangan, M., & Kim, D. H. (2015). A novel approach for in vitro meat production. *Applied Microbiology and Biotechnology*, 99(13), 5391–5395. 10.1007/s00253-015-6671-5

Picouet, P. A., Fernandez, A., Realini, C. E., & Lloret, E. (2014). Influence of PA6 nanocomposite films on the stability of vacuum-aged beef loins during storage in modified atmospheres. *Meat Science*, 96(1), 574–580. 10.1016/j.meatsci.2013.07.020

Pimentel, D., & Pimentel, M. (2003). Sustainability of meat-based and plant-based diets and the environment1. *The American Journal of Clinical Nutrition*, 78(3), 660S–663S. 10.1093/ajcn/78.3.660S

Post, M., Levenberg, S., Kaplan, D., Genovese, N., Fu, J., Bryant, C., Negowetti, N., Verzijden, K., & Moutsatsou, P. (2020). Scientific, sustainability and regulatory challenges of cultured meat. *Nature Food*, 1(7), 403–415. 10.1038/s43016-020-0112-z

Post, M. J., Levenberg, S., Kaplan, D. L., Genovese, N., Fu, J., Bryant, C. J., Negowetti, N., Verzijden, K., & Moutsatsou, P. (2020). Scientific, sustainability and regulatory challenges of cultured meat. *Nature Food*, 1(7), 403–415. 10.1038/s43016-020-0112-z

Ramachandraiah, K., Han, S. G., & Chin, K. B. (2015). Nanotechnology in meat processing and packaging: Potential applications - a review. *Asian-Australasian Journal of Animal Sciences*, 28(2), 290–302. 10.5713/ajas.14.0607

Röös, E., Bajželj, B., Smith, P., Patel, M., Little, D., & Garnett, T. (2017). Greedy or needy? Land use and climate impacts of food in 2050 under different livestock futures. *Global Environmental Change*, 47, 1–12. 10.1016/j.gloenvcha.2017.09.001

Sapkota, A. R., Curriero, F. C., Gibson, K. E., & Schwab, K. J. (2007). Antibiotic-resistant enterococci and fecal indicators in surface water and groundwater impacted by a concentrated Swine feeding operation. *Environmental Health Perspectives*, 115(7), 1040–1045. 10.1289/ehp.9770

Sharma, C., Dhiman, R., Rokana, N., & Panwar, H. (2017). Nanotechnology: An Untapped Resource for Food Packaging. *Frontiers in Microbiology*, 8, 1735. 10.3389/fmicb.2017.01735

Shi, X., & Garry, D. J. (2006). Muscle stem cells in development, regeneration, and disease. *Genes & Development*, 20(13), 1692–1708. 10.1101/gad.1419406

Simsa, R., Yuen, J., Stout, A., Rubio, N., Fogelstrand, P., & Kaplan, D. L. (2019). Extracellular Heme Proteins Influence Bovine Myosatellite Cell Proliferation and the Color of Cell-Based Meat. *Foods*, 8(10), 521. Advance online publication. 10.3390/foods8100521

Smaoui, S., Chérif, I., Ben Hlima, H., Khan, M. U., Rebezov, M., Thiruvengadam, M., Sarkar, T., Shariati, M. A., & Lorenzo, J. M. (2023). Zinc oxide nanoparticles in meat packaging: A systematic review of recent literature. *Food Packaging and Shelf Life*, 36, 101045. 10.1016/j.fpsl.2023.101045

Snyman, C., Goetsch, K. P., Myburgh, K. H., & Niesler, C. U. (2013). Simple silicone chamber system for 3D skeletal muscle tissue formation. *Frontiers in Physiology*, 4, 65276. 10.3389/fphys.2013.00349

Stephens, N., Di Silvio, L., Dunsford, I., Ellis, M., Glencross, A., & Sexton, A. (2018). Bringing cultured meat to market: Technical, socio-political, and regulatory challenges in cellular agriculture. *Trends in Food Science & Technology*, 78, 155–166. 10.1016/j.tifs.2018.04.010

Worm, B., Barbier, E. B., Beaumont, N., Duffy, J. E., Folke, C., Halpern, B. S., Jackson, J. B., Lotze, H. K., Micheli, F., Palumbi, S. R., Sala, E., Selkoe, K. A., Stachowicz, J. J., & Watson, R. (2006). Impacts of biodiversity loss on ocean ecosystem services. *Science*, 314(5800), 787–790. 10.1126/science.1132294

Worobey, M., Han, G. Z., & Rambaut, A. (2014). A synchronized global sweep of the internal genes of modern avian influenza virus. *Nature*, 508(7495), 254–257. 10.1038/nature13016

Zhang, G., Zhao, X., Li, X., Du, G., Zhou, J., & Chen, J. (2020). Challenges and possibilities for bio-manufacturing cultured meat. *Trends in Food Science & Technology*, 97, 443–450. 10.1016/j.tifs.2020.01.026

Chapter 6
In Vitro Harvest

Zaliha Harun
Lincoln University College, Malaysia

Gunavathy Selvarajh
https://orcid.org/0000-0002-6041-6660
Lincoln University College, Malaysia

Norhashima Abd Rashid
Lincoln University College, Malaysia

Nik Nurul Najihah Nik Mat Daud
Lincoln University College, Malaysia

ABSTRACT

In vitro harvesting is a technique involving growing animal cells in controlled laboratories. This technology provides sustainable, ethical, and scalable options to meet the world's protein needs. This chapter addresses the environmental advantages and diminished impact of traditional farming methods. This chapter explains the cell culture techniques, different cell sources, nutritional media formulations, and bioreactor technologies in meat production. The ethical implications of the topic are discussed, with a specific focus on the well-being of animals and concerns related to the slaughter process. Successfully integrating in vitro harvests into mainstream markets depends on overcoming regulatory barriers and gaining customer acceptability. The chapter finishes by providing a concise overview of the significant impact that cellular agriculture can have on redefining sustainability, ethics, and nutritional diversity in the global food system. This transformation is made possible through the collaboration of many fields of study and the application of advanced technology.

DOI: 10.4018/979-8-3693-4115-5.ch006

INTRODUCTION

Technological advancements have resulted in substantial changes, particularly in cattle meat production. Scientists are investigating the potential of in vitro production or lab culturing to develop a meat substitute that can address consumer concerns, environmental impact, and animal welfare. The traditional method of meat production involves farming, which requires a substantial allocation of resources, including land, feed, and water (Bhat et al., 2015). Furthermore, it has a significant effect on the loss of biodiversity, the destruction of forests, and the emission of greenhouse gases. Scientists have been conducting research on in vitro meat since the 1900s. In 2013, researchers made a significant breakthrough by successfully producing the first laboratory-cultivated burger that closely resembled a conventional hamburger (Kulus et al., 2023). Presently, the primary objective of research and development in cultured meat is to obtain suitable cells and develop constituents that facilitate the construction of cultured meat structures. According to the research findings of Lee et al. (2023), these technologies are now in the developmental stage and do not possess the capability to be produced on a large scale. Only a few countries have authorized lab-grown meat consumption, including Israel, Singapore, and the United States. Other countries are expected to use this technology after the knowledge of the complex production process becomes available (Lee et al., 2020).

This chapter aims to elucidate the intricate scientific techniques and advancements driving the progress of in vitro meat technology. It involves utilizing cell culture and tissue engineering techniques to develop meat in laboratory environments. This chapter will highlight the significant environmental benefits of in vitro meat production. It offers a promising solution to pressing environmental concerns by reducing reliance on traditional animal agriculture. In addition, it provides a sustainable substitute for conventional meat production by decreasing the release of greenhouse gases, preserving land and water resources, and preventing deforestation. This technology can potentially reduce the environmental impact caused by conventional meat production. Furthermore, this chapter explores the regulatory elements that govern the progression of designing, manufacturing, and promoting in vitro meat, providing a comprehensive understanding of the complex landscape of policy-making and governance that encompasses this growing industry. Ultimately, this chapter encourages ethical discussion and critical study of the ethical concerns of laboratory meat production.

ENVIRONMENTAL BENEFITS OF IN VITRO HARVEST

In vitro harvest or cultured meat has often been seen as an alternative approach to solve the environmental problem caused by livestock production. Traditional livestock production increases greenhouse gas emissions and deforestation. Livestock production releases harmful contaminants, including nitrogen and phosphorus, into the water bodies, which causes eutrophication. Besides, livestock raising also consumes a large amount of water. However, in vitro harvest offers numerous advantages, especially regarding proper resource usage, increased land conservation, and reduced greenhouse gas emissions.

In vitro harvest, on the other hand, can utilize resources such as water and energy efficiently without a waste. In livestock rearing system, huge amounts of water, feeds, and energy are used. It had been stated by FAO (2013) that animal farming uses 70% of agricultural land and 8% of water to cater for the needs of the daily use of animals. This is the opposite of the in vitro meat harvest produced in the lab environment because less water is being used to develop cultured meat. According to Tuomisto and Teixeira de Mattos (2011), cultured meat could reduce water usage by up to 82-96% compared to traditional animal farming systems. The lesser water needs for drinking and feed production is mainly attributed to water usage reduction. Not only that, the in vitro cultured meat also uses minimum energy because feed production, transportation, and processing are unnecessary. It is agreed that initial energy demands for cultured meat are high, but technological advancements and adopting renewable energy could make it more energy-efficient than conventional meat in the long term (Mattick et al., 2015). The researcher found that cultured meat production consumes 7-45% less energy than conventional meat production (Mattick et al., 2015). Through this, it can be said that the in vitro harvest system utilizes a lesser water and energy to produce cultivated meat food.

Further, in vitro harvest system does not need vast land to produce cultured meat as livestock farming needs. Animal agriculture expansion had a tremendous impact on habitats, biodiversity, carbon storage and soil conditions. Traditional livestock farming needs a huge land for rearing, growing pastures, producing feeds, and slaughtering. Therefore, many forests are being cleared for animal farming to get meat and milk. However, due to excessive deforestation, many species are facing problems and in the brink of extinction. Excessive deforestation also contributes to the increase in world temperature due to the high release of C into the environment. It is stated that the global livestock farming system occupies around 30% of the terrestrial lands (FAO, 2013). Adopting the practice of cultivated meat could minimize the need for huge land. Tuomisto and Teixeira de Mattos (2011) estimated that cultured meat could reduce land use by up to 99%, eliminating the need for grazing lands and feed crop cultivation. The reduction of land necessity is attributed to the efficiency of cell culture systems, which produce cultured meat in controlled environments. This in vitro meat production process allows forests and ecosystems to be conserved and preserved (Nijdam et al., 2012).

Additionally, in vitro harvest reduces greenhouse gas emissions (GHG) compared to traditional animal farming systems. Livestock farming produces the highest percentage of greenhouse gases in the atmosphere, especially methane and nitrogen oxide, and consequently increases the carbon footprint (Tuomisto, 2019). This traditional rearing system contributes about 14.5% of GHG emissions (Gerber et al., 2013). This is because herbivore animals produce a lot of methane gas in their rumen due to their digestive process (Hocquette, 2016). Methane gas increases global warming 28 times more than the carbon dioxide itself. Hence, developing and producing in vitro meat is important and has a huge advantage in reducing GHG. This is in agreement with Tuomisto and Teixeira de Mattos (2011), who found that in vitro meat could reduce GHG emissions by up to 96% compared to conventional meat production systems. This reduction is due to the absence of enteric fermentation and excretion waste production like the conventional animal rearing system. Additionally, using renewable energy sources to produce cultured meat can further reduce the carbon footprint (Goodwin & Shoulders, 2013). Thus, it can be concluded that in vitro harvest meat production offers a wide variety of benefits in terms of environmental conservation.

PRINCIPLES OF IN VITRO HARVEST

The initial stage of producing lab-grown meat involves acquiring cells, which are crucial for obtaining the necessary raw ingredients for subsequent processing. Animals might be utilized for biopsy procedures or post-mortem to get microscopic cellular samples (Ben-Arye & Levenberg, 2019). These cell types, derived from various animal tissues such as muscle or fat, are the fundamental units for subsequent development. Fat tissues are selected for their contribution to flavour and texture, while muscle tissues are chosen for their high protein content and structural strength (Bhat & Fayaz, 2011).

After the cells have been gathered, the subsequent stage is cell culture, which involves cultivating the cells in a nutrient-rich medium. The medium contains essential nutrients, growth factors, and other crucial components necessary for cellular division and proliferation (Post, 2012). Typically, this medium consists of blood, vitamins, minerals, amino acids, and glucose. Scientists are actively researching serum-free alternatives to enhance scalability and tackle ethical concerns (Zhang et al., 2020). Cells must be kept in a regulated environment, such as a carbon dioxide (CO_2) chamber, to ensure their growth and survival. The controlled environment maintains optimal temperature, humidity, and gas levels, closely resembling the physiological parameters of living organisms.

Facilitating cellular differentiation is an integral aspect of the growth process (Stephens et al., 2018). These cells undergo differentiation into other types of cells present in meat, such as muscle cells or fat cells, in response to growth hormones, mechanical stimulation, and biochemical signals. Myogenic differentiation is initiated by employing myogenic regulatory elements such as MyoD and myogenin. Activation of peroxisome proliferator-activated receptor gamma (PPARγ) pathways initiates adipogenic differentiation (Ranga et al., 2014). To ensure the production of the appropriate type of tissue, differentiation must be meticulously strategized and coordinated. Cellular differentiation is the process responsible for forming three-dimensional structures that resemble the composition of meat tissue (Ben-Arye & Levenberg, 2019). Scaffold geometries are employed in tissue engineering to facilitate the organization and enhanced formation of tissues. This enables the production of tissue that closely resembles meat in terms of texture and composition. The scaffolds, composed of biomaterials such as collagen, fibrin, or synthetic polymers, offer mechanical support and facilitate the delivery of nutrients to the developing tissue (Huang et al., 2010).

Following that, cultured beef undergoes a maturing process that enhances its taste, texture, and nutritional value (Post, 2012). Currently, mechanical forces such as elongation or compression can be applied to improve the fibrousness and tenderness of meat by imitating the natural movement of muscles (Marga et al., 2007). Enhancing the physical attributes of a product can be achieved by optimizing the cultivation circumstances, the environment in which it is cultivated, and the post-production processes. This ensures that the outcome aligns with the desires of our customers. Comprehensive quality control measures are implemented at every stage of the cultivation process to ensure cultured beef products' safety, high quality, and uniformity (Stephens et al., 2018). It undergoes frequent assessments for cellular proliferation, viability, and potential contaminants to ensure the product's high quality and client satisfaction. High-performance liquid chromatography (HPLC) and mass spectrometry are sophisticated analytical techniques employed to examine the biological composition and detect impurities (van der Weele & Tramper, 2014).

CELL SOURCES

Myocytes, or muscle cells, are the essential units responsible for the synthesis of meat, as they form the central framework of meat. These cells are derived from muscle tissue and can proliferate and differentiate into mature muscular fibres under suitable conditions. Myoblasts, cells that come before muscle production, are frequently used because they may quickly divide and merge, creating myotubes. These myotubes eventually become muscle fibres (Bhat et al., 2019). During the process of contraction and organization, the myotubes create muscle tissue, which gives meat its distinct texture and protein makeup.

Adipose cells, or adipocytes, are essential for replicating the flavour and juiciness of meat. Adipose tissue significantly contributes to improving the taste and consistency of meat products. Adipose stem cells are used to produce adipocytes in the cultivation of lab-grown meat. These cells can accumulate lipids, replicating the marbling phenomenon observed in traditional pork slices (Specht et al., 2018). Adipocytes are necessary in lab-grown meat to reproduce the sensory experience of eating meat, as they directly impact both the taste and overall enjoyment of the experience.

Fibroblasts play a role in forming the extracellular matrix (ECM) and connective tissue, which helps provide structural support in meat production. The ECM is vital for the structural integrity and metabolic signalling necessary for tissue development and maintenance (Post, 2012). Fibroblasts secrete collagen and other ECM proteins, enabling the organization of muscle and fat cells into a cohesive structure. This arrangement is essential for developing a product that not only imitates the taste of meat but also accurately reproduces the ideal texture and sensation in the mouth.

Satellite cells are a distinct subset of stem cells found in muscle tissue, playing a vital role in the muscle's repair and regeneration. These cells are beneficial in meat production because they can differentiate into myoblasts and then develop into muscle fibres (Chargé & Rudnicki, 2004). Producers can enhance the efficiency and scalability of cultured meat production by harnessing the regenerative potential of satellite cells.

Induced pluripotent stem cells (iPSCs) offer a versatile approach to cultivate meat. These cells are generated by reprogramming adult cells into a pluripotent state, enabling them to differentiate into any cell (Takahashi & Yamanaka, 2006). iPSCs can be directed to differentiate into myocytes, adipocytes, and other particular cell types essential for meat production. Using iPSCs can streamline the manufacturing process and reduce reliance on animal-derived cells.

The effective advancement of cultured meat relies on integrating multiple cell types, each fulfilling a unique function in determining the final product's shape, flavour, and nutritional composition. Muscle cells play a significant role in the meat's overall protein content and texture, adipose cells enhance flavour and juiciness, fibroblasts provide structural support, satellite cells aid in regenerative capacity, and iPSCs serve as a versatile supply of many cell types. Continuing research and development in this field demonstrate the potential to convert cultivated meat into a viable and eco-friendly alternative to traditional meat production.

FACTOR INFLUENCE THE CULTIVATION

Meat cultivation in a laboratory setting is complex and influenced by various factors that might impact the efficiency, quality, and scalability of the finished product. Various factors, including cell source and type, culture media composition, bioreactor design, and ambient conditions, all impact laboratory meat cultivation. Each element influences cultured meat production's efficiency, quality, and scalability.

Properly selecting cell sources and kinds is critical to the success of cultured meat production. Myocytes (muscle cells), adipocytes (fat cells), and fibroblasts are frequently used, with each contributing to the final product's texture, flavour, and structure (Bhat et al., 2019). The selection between primary cells, satellite cells, and induced pluripotent stem cells (iPSCs) can substantially impact the cultivation process. Primary cells are obtained directly from animals and may have limited proliferative capacity, but satellite cells, which are muscle stem cells, and iPSCs have a higher potential for proliferation and differentiation (Chargé & Rudnicki, 2004; Takahashi & Yamanaka, 2006; Kadim et al., 2015).

Culture media composition is another crucial factor that influences cell proliferation and differentiation. Conventional culture media includes foetal bovine serum (FBS), which provides growth factors, hormones, and minerals necessary for cell growth and maintenance. FBS is commonly used due to its ability to maintain multiple cell types with diverse nutrients. Nevertheless, FBS gives rise to ethical and economic concerns. FBS extraction involves obtaining serum from bovine foetuses, which raises concerns regarding animal welfare (van der Valk et al., 2010). When used in large-scale cell culture applications, FBS is costly and can be affected by changes in the supply chain (Brunner et al., 2010).

Scientists are developing serum-free medium formulations using plant-based or synthetic alternatives to address these problems. These ideas aim to emulate the advantages of FBS while avoiding its ethical and economic disadvantages. Plant-based media formulations utilize soy or other plant extracts, resulting in sustainability and cost-effectiveness (Specht et al., 2018). Nevertheless, synthetic media formulations offer accurately measured quantities of essential nutrients, growth factors, and hormones tailored to specific cell types and applications (Humbird, 2020). These advancements remove components derived from animals and enhance the ability to replicate and scale up cell culture.

To produce superior, expandable cultured beef, it is necessary to optimize the culture medium to mimic the natural environment accurately. Cell proliferation, differentiation, and tissue development rely on intricate and ever-changing in vivo conditions. To enhance the efficiency and quality of cultured meat production, researchers can optimize the composition of the culture media based on specific conditions (Kolkmann et al., 2020). This involves altering the pH, oxygen, and growth hormones particular to the tissues.

Bioreactors are specialized tanks used to develop cells in controlled environments, and their design is critical in producing lab-grown meat. Bioreactors must create an environment that promotes cell attachment, development, and differentiation while assuring adequate nutrition, oxygen delivery, and waste disposal. Bioreactor design plays a pivotal role in the process. Different bioreactors are employed to meet specific requirements for cell growth, differentiation, and tissue formation, and they vary in their design, operational parameters, and scalability.

Stirred-tank bioreactors (STBs) are widely used due to their well-established design and scalability. These bioreactors are well-designed vessels with impellers that stir culture medium, ensuring uniform distribution of nutrients, oxygen, and cells. However, shear stress can harm some cell types. (Gstraunthaler, 2003; Martin et al., 2020). Another bioreactor design, which is perfusion bioreactors has a design that mimics the dynamic environment of living tissues and maintains optimal conditions for cell viability and productivity by ensuring a constant flow of fresh culture medium (de Zélicourt et al., 2010; Carpentier et al., 2021). These bioreactors are particularly useful for long-term cultures and growing tissues with higher cell densities. On the other hand, scaffold-based bioreactors employ three-dimensional scaffolds to provide structural support for cell adhesion, proliferation, and specialization, replicating the extracellular matrix seen in biological tissues. They enable the formation of tissue-like structures and generate a realistic texture and structure for meat. (Liu et al., 2007; Verbruggen et al., 2018). Besides, a bioreactor called rotary cell culture systems (RCCS) allows cells to aggregate and build 3D tissue constructions in low-shear microgravity. They're useful for sensitive cell culturing and decreased gravity cell behaviour research (Unsworth & Lelkes, 1998; Nickerson et al., 2004). Hollow fibre bioreactors can produce large-scale cells and tissues due to their high surface area-to-volume ratio, which supports significant cell growth and metabolic activity. These bioreactors use semi-permeable membranes to separate the cell culture chamber from the nutrient supply (Wurm, 2004; Malda et al., 2004).

Temperature, pH, oxygen concentration, and shear stress affect bioreactors' cell behaviour and tissue formation. These elements must be carefully managed for optimal cell and tissue growth. Temperature has a significant impact on cell metabolism and enzyme activity. Optimal temperatures are required for efficient cell metabolism. Any deviations outside of this ideal range might have detrimental effects on the viability and function of cells. It is necessary to maintain mammalian cell cultures at a temperature of 37°C (Sart et al., 2016). The pH level influences the solubility of nutrients and the activity of enzymes. The optimal pH for most mammalian cells is 7.4. Fluctuations in pH can disturb the electrical charge across the cell membrane and the balance of internal conditions. According to Kim et al. (2017), acidity and alkalinity have detrimental effects on cell growth.

Oxygen also has a crucial role in bioreactors. Aerobic respiration, a process that requires oxygen, supports cellular survival and the creation of ATP. Hypoxia can lead to necrosis or impaired cell proliferation in stem cell preparations. The exposure to excessive oxygen levels, known as hyperoxia, leads to oxidative stress, which causes damage to DNA, proteins, and lipids (Hussain et al., 2017). The amounts of oxygen influence the health and function of bioreactor cells. Shear stress is also crucial for cell behaviour. Mechanical forces alter the morphology, growth, and specialization of cells. Vascular endothelial cells are subjected to shear stress caused by blood movement. To generate muscle tissue that can endure significant stress, bioreactors need to replicate these specific mechanical conditions (Verbruggen et al., 2018). Bioreactors facilitate tissue growth by precisely controlling shear stress to align and develop muscle cells.

ETHICAL IMPLICATIONS

In traditional agricultural systems, the welfare of farm animals is a significant ethical concern. Animal welfare refers to the well-being of animals in various aspects, including their physical health, mental state, and ability to express natural behaviours. Traditional farming often involves confinement in small spaces, leading to physical and psychological stress. Overcrowding, lack of enrichment, and poor sanitation can result in diseases and injuries (Madzingira, 2018). Such conditions fail to meet the animals' needs for comfort and space, leading to compromised health and well-being. Besides that, adequate nutrition and timely veterinary care is essential for animal welfare. Inadequate feeding practices and insufficient medical attention also may lead to malnutrition, illness, and suffering (Harvatine, 2023).

Additionally, animals have intrinsic behavioural needs, such as rooting in pigs or dust bathing in chickens. Traditional farming practices often fail to accommodate these behaviours, leading to frustration and stress. When animals cannot engage in natural behaviours, it can result in behavioural issues and a lower quality of life (Madzingira, 2018).

Therefore, to overcome the animal welfare problem, in vitro meat has been introduced. Unexpectedly, one of the significant ethical concerns with current cultured meat production technologies is the suffering and death of animals (Treich, 2021). Recent production methods involve harvesting biopsies for stem cells from donor animals and using media based on fetal calf serum, which requires blood from fetuses obtained from slaughtered pregnant cows (Benny et al., 2022). Even though these animal biopsies are believed to be painless (or use tissues like feathers), and research on an animal-free growth medium is continuing, the current reality is that lab-produced meat and meat from start-up companies still depend on some aspects of animal involvement (Benny et al., 2022). The fact that it is unethical to promote unhealthy food, even if we think it will be produced ethically in the future. Another problem that is raising is the promotion of cultured meat. A significant campaign by health agencies is underway to stop the promotion and consumption of products that are known to be harmful to people's health, such as smoking and fast food for kids (Bala et al., 2017).

As we all know, eating and consuming processed red meat is related to an increased risk of heart disease and cancer (Wolk, 2017). The promotion of cultured meat would also encourage the use of processed meats because based on the current cultured meat technology, it can only be produced from processed meats (Siddiqui et al., 2022). It is said that it is ethical to use cultured meat as the primary source of food for the world's future population (Chriki & Hocquette, 2020). However, using plant sources directly is better and healthy than producing cultured meat based on plants or animals (Chriki & Hocquette, 2020; Treich, 2021). In addition, cultured meat is not guaranteed to be fresh because the tissues that are taken from the animal will be damaged or die (Izhar Ariff Mohd Kashim et al., 2023). Other problems in producing cultured meat are the possibility of cannibalism, the increased reliance on multinational food companies, and the decline in local self-sufficiency (Benny et al., 2022).

OVERVIEW OF THE REGULATORY FRAMEWORKS

In vitro harvesting is subject to complex regulatory environments that vary significantly across regions. These frameworks are designed to ensure lab-grown meat's safety, quality, and ethical standards, encompassing food safety, labelling, and environmental impacts. In the United States, the Food and Drug Administration (FDA) and the United States Department of Agriculture (USDA) jointly oversee the regulation of cellular agriculture. The FDA is responsible for pre-market consultation processes, ensuring the safety of cell cultures, while the USDA oversees the production and labelling of meat products derived from these cultures (FDA, 2019). This dual-agency approach aims to provide a comprehensive regulatory pathway but also introduces compliance complexity. However, the European Union (EU) follows a different regulatory path. Under the EU's Novel Food Regulation, lab-grown meat is classified as a novel food, requiring rigorous safety assessments and approval by the European Food Safety Authority (EFSA) before entering the market (EFSA, 2018). This process involves a detailed evaluation of the production process, potential allergens, and nutritional profile. In Asia, countries like Singapore have taken a proactive approach. Singapore became the first country to approve the sale of lab-grown meat in 2020, setting a precedent for regulatory frameworks prioritizing innovation while ensuring safety (Singapore Food Agency, 2020). This regulatory flexibility has positioned Singapore as a leader in the cellular agriculture sector.

Regulatory approval for lab-grown meat involves navigating scientific, legal, and ethical challenges. One of the primary scientific challenges is demonstrating the safety of the production process. The safety aspect of in vitro harvesting includes ensuring that cell lines are free from contaminants and that the culture media used do not pose any health risks (Post, 2012). Another significant challenge is the regulatory clarity and consistency. In the U.S., for example, the dual oversight by the FDA and USDA can lead to potential overlaps and gaps in regulation, creating uncertainty for producers. Consistency in regulatory frameworks across regions is also lacking, making it difficult for companies to scale their operations internationally (Stephens et al., 2018).

Opportunities lie in the potential for regulatory bodies to adopt more adaptive and forward-thinking approaches. A more streamlined and efficient regulatory pathway can be developed by fostering collaboration between scientists, industry stakeholders, and regulatory agencies. Singapore's regulatory framework serves as a model, emphasizing the importance of a risk-based approach that balances innovation with consumer safety. Moreover, regulatory approval can drive investment and consumer trust. Clear and rigorous regulatory processes can help build public confidence in lab-grown meat products, addressing concerns about safety and ethical production methods (Bryant & Barnett, 2018). Regulatory frameworks incorporating transparency and public engagement will likely be more successful in gaining consumer acceptance.

CONSUMER PERCEPTIONS AND ACCEPTANCE

Consumer acceptance is crucial for the success of lab-grown meat products. Awareness, perceived benefits, safety concerns, and ethical considerations shape public perception. Studies indicate that consumer awareness of lab-grown meat is still relatively low, though it is growing. Education and transparency about the production process can significantly influence acceptance. Consumers who are well-informed about the environmental and ethical benefits of lab-grown meat are likelier to try and purchase these products (Slade, 2018).

The perceived advantages, such as the decreased ecological footprint and improved treatment of animals, serve as powerful incentives for customer approval. Lab-grown meat has the potential to significantly reduce greenhouse gas emissions, land use, and water consumption compared to conventional meat production (Tuomisto &

Teixeira de Mattos, 2011). Highlighting these benefits can appeal to environmentally conscious consumers.

Safety concerns remain a significant barrier to acceptance. Consumers often question the long-term health implications of consuming lab-grown meat. Addressing these concerns through rigorous safety testing and transparent communication is essential. Providing clear information on nutritional content and safety standards can help alleviate consumer fears.

Ethical considerations also play a significant role. The idea of meat produced without animal slaughter appeals to many consumers, particularly those concerned about animal welfare. However, ethical debates about the naturalness of lab-grown meat and its potential impacts on traditional farming communities must be addressed thoughtfully (Hocquette, 2016).

Marketing and branding strategies can also influence consumer acceptance. Positioning lab-grown meat as a premium, sustainable, and ethical choice can attract early adopters and trendsetters. Collaborations with celebrity chefs and endorsements from influential figures can further enhance the product's appeal.

STRATEGIES TO ENHANCE CONSUMER ACCEPTANCE

Enhancing consumer acceptance of lab-grown meat requires multifaceted strategies. One practical approach is to increase consumer education through transparent communication about the production process, safety measures, and benefits of lab-grown meat. Educational campaigns can dispel myths and address common misconceptions, helping to build a positive perception. Engaging with consumers through interactive and participatory approaches can also foster acceptance. This approach could involve public tastings, open house events at production facilities, and interactive digital platforms where consumers can learn about cellular agriculture firsthand. Such initiatives can demystify the technology and create a sense of connection and trust.

In addition, developing clear and accurate labelling standards is another crucial aspect. Labels that indicate the origin, production process, and benefits of lab-grown meat can help consumers make informed choices. Transparency in labelling can also address ethical concerns and reinforce the message of sustainability and animal welfare.

Besides, collaborations with the culinary community can play a significant role in normalizing lab-grown meat. Chefs and food influencers can introduce lab-grown meat in innovative and appealing dishes, showcasing its versatility and taste. These endorsements can help shift public perception and increase acceptance among food enthusiasts.

CONCLUSION

Laboratory meat production, also known as in vitro meat production, shows enormous potential as an alternative to traditional animal husbandry, giving benefits regarding resource efficiency, ethics, and sustainability. This method involves cultivating cells collected through biopsy or post-mortem in nutrient-rich media under controlled circumstances to enhance proliferation and differentiation. Compared to traditional animal farming, this technology can dramatically reduce land, water, and greenhouse gas emissions linked with meat production.

Advances in cell culture and tissue engineering technology drive lab-grown meat growth. Key breakthroughs include optimizing the culture media, using scaffolds to encourage tissue creation, and purposely adding growth hormones and mechanical stimuli to mimic muscle and fat tissue development. The scalability and efficiency of the procedure are further improved by optimizing bioreactors explicitly made for cell growth. Despite these advances, significant obstacles remain. Ethical concerns about the use of animal-derived cells and medium remain, and there is an urgent need for robust regulatory frameworks to ensure product safety and quality. In the United States and the European Union, regulatory authorities such as the FDA, USDA, and EFSA conduct extensive safety evaluations and oversight. Precise and uniform regulations are critical to corporate performance and customer confidence.

Consumer acceptability is vital to the viability of cultured meat. While awareness and research are growing, honest and comprehensive communication and education are required to address safety issues and ethical debates over the product's naturalness. Clear labelling, public engagement through educational programmes, and forging culinary relationships with specialists to demonstrate the flexibility and taste of lab-grown meat are all strategies for increasing consumer acceptability.

In conclusion, in vitro meat production can potentially overhaul the meat business by offering a more sustainable, ethical, and ecologically friendly alternative to traditional meat. To fully realize the potential of cultured meat, continued study, technological developments, and regulatory coordination are required. Lab-grown meat production has the potential to contribute significantly to global food security and sustainability if implemented efficiently and with active public engagement.

REFERENCES

Alexander, P., Brown, C., Arneth, A., Dias, C., Finnigan, J., Moran, D., & Rounsevell, M. D. A. (2017). Could consumption of insects, cultured meat or imitation meat reduce global agricultural land use? *Global Food Security*, 15, 22–32. 10.1016/j.gfs.2017.04.001

Bala, M. M., Strzeszynski, L., & Topor-Madry, R. (2017). Mass media interventions for smoking cessation in adults. *Cochrane Database of Systematic Reviews*, 11(11), CD004704. Advance online publication. 10.1002/14651858.CD004704.pub429159862

Ben-Arye, T., & Levenberg, S. (2019). Tissue engineering for clean meat production. *Frontiers in Sustainable Food Systems*, 3, 46. 10.3389/fsufs.2019.00046

Benny, A., Pandi, K., & Upadhyay, R. (2022). Techniques, challenges and future prospects for cell-based meat. *Food Science and Biotechnology*, 31(10), 1225–1242. 10.1007/s10068-022-01136-635992324

Bhat, Z. F., & Fayaz, H. (2011). Prospectus of cultured meat—Advancing meat alternatives. *Journal of Food Science and Technology*, 48(2), 125–140. 10.1007/s13197-010-0198-7

Bhat, Z. F., Kumar, S., & Fayaz, H. (2019). In vitro meat production: Challenges and benefits over conventional meat production. *Journal of Integrative Agriculture*, 18(6), 221–235.

Brunner, D., Frank, J., Appl, H., Schöffl, H., Pfaller, W., & Gstraunthaler, G. (2010). Serum-free cell culture: The serum-free media interactive online database. ALTEX-. *Alternatives to Animal Experimentation*, 27(1), 53–62. 10.14573/altex.2010.1.5320390239

Bryant, C., & Barnett, J. (2018). Consumer acceptance of cultured meat: A systematic review. *Meat Science*, 143, 8–17. 10.1016/j.meatsci.2018.04.00829684844

Carpentier, A. V., Bonnet, M., Campard, D., & Moisan, A. (2021). Perfusion bioreactors for 3D skin cell culture: State of the art and perspectives. *Journal of Tissue Engineering and Regenerative Medicine*, 15(2), 109–123.

Chargé, S. B., & Rudnicki, M. A. (2004). Cellular and molecular regulation of muscle regeneration. *Physiological Reviews*, 84(1), 209–238. 10.1152/physrev.00019.200314715915

Chriki, S., & Hocquette, J. F. (2020). The Myth of Cultured Meat: A Review. *Frontiers in Nutrition*, 7(February), 1–9. 10.3389/fnut.2020.0000732118026

Datar, I., & Betti, M. (2010). Possibilities for an in vitro meat production system. *Innovative Food Science & Emerging Technologies*, 11(1), 13–22. 10.1016/j.ifset.2009.10.007

de Boer, J., & Aiking, H. (2011). On the merits of plant-based proteins for global food security: Marrying macro and micro perspectives. *Ecological Economics*, 70(7), 1259–1265. 10.1016/j.ecolecon.2011.03.001

de Zélicourt, D., Niklason, L. E., & Neidert, M. (2010). Perfusion bioreactors for tissue engineering. *Bioengineering & Translational Medicine*, 5(1), e10163.

European Food Safety Authority (EFSA). (2018). Novel foods: History and evolution. Retrieved from https://www.efsa.europa.eu/en/topics/topic/novel-foods

FAO. (2013). *Tackling Climate Change Through Livestock: A Global Assessment of Emissions and Mitigation Opportunities*. Food and Agriculture Organization of the United Nations.

Food and Drug Administration (FDA). (2019). FDA and USDA announce a formal agreement to regulate cell-cultured food products from cell lines of livestock and poultry. Retrieved from https://www.fda.gov/food/cfsan-constituent-updates/fda-and-usda-announce-formal-agreement-regulate-cell-cultured-food-products-cell-lines-livestock-and

Gerber, P. J., Steinfeld, H., & Henderson, B. (2013). *Tackling Climate Change Through Livestock: A Global Assessment of Emissions and Mitigation Opportunities*. FAO.

Goodwin, J. N., & Shoulders, C. W. (2013). The future of meat: A qualitative analysis of cultured meat media coverage. *Meat Science*, 95(3), 445–450. 10.1016/j.meatsci.2013.05.02723793078

Gstraunthaler, G. (2003). Alternatives to the use of fetal bovine serum: Serum-free cell culture. *ALTEX*, 20(4), 275–281. 10.14573/altex.2003.4.25714671707

Harvatine, G. (2023). Lack of nutrition in animals and its affects on animals health. *J Vet Med Allied Sci*, 7(1), 132. 10.35841/2591-7978-7.1.132

Hocquette, J. F. (2016). Is in vitro meat the solution for the future? *Meat Science*, 120, 167–176. 10.1016/j.meatsci.2016.04.03627211873

Huang, Y.. (2010). Biomaterials and scaffolds in tissue engineering. *Biotechnology Advances*, 28(7), 925–936. 10.1016/j.biotechadv.2010.08.011

Humbird, D. (2020). Scale-up economics for cultured meat. *Biotechnology and Bioengineering*, 118(8), 3239–3250. 10.1002/bit.2784834101164

Izhar Ariff Mohd Kashim, M., Abdul Haris, A. A., Abd. Mutalib, S., Anuar, N., & Shahimi, S. (2023). Scientific and Islamic perspectives in relation to the Halal status of cultured meat. *Saudi Journal of Biological Sciences*, 30(1), 103501. 10.1016/j.sjbs.2022.10350136466219

Kadim, I. T., Mahgoub, O., Baqir, S., Faye, B., & Purchas, R. W. (2015). Cultured meat from muscle stem cells: A review of challenges and prospects. *Journal of Integrative Agriculture*, 14(2), 222–234. 10.1016/S2095-3119(14)60881-9

Kolkmann, A. M., Post, M. J., & Rutjens, M. A. (2020). Cultured meat: The business of biotechnology. *Trends in Biotechnology*, 38(7), 683–689.

Kolkmann, A. M., Post, M. J., Rutjens, M. A., van Essen, A. L. M., & Moutsatsou, P. (2020). Serum-free media for the growth of primary bovine myoblasts. *Cytotechnology*, 72(1), 111–120. 10.1007/s10616-019-00361-y31884572

Liu, C., Xia, Z., & Czernuszka, J. T. (2007). Design and development of three-dimensional scaffolds for tissue engineering. *Chemical Engineering Research & Design*, 85(7), 1051–1064. 10.1205/cherd06196

Ma, X., Qu, J., Mei, H., Zhao, Y., & Jin, W. (2021). Effects of oxygen concentration on cell growth and differentiation in tissue engineering. *Journal of Biomedical Materials Research*, 109(5), 714–726.

Madzingira, O. (2018). Animal Welfare Considerations in Food-Producing Animals. IntechOpen. 10.5772/intechopen.78223

Malda, J., Woodfield, T. B. F., van der Vloodt, F., Wilson, C., Martens, D. E., Tramper, J., van Blitterswijk, C. A., & Riesle, J. (2004). The effect of PEGT/PBT scaffold architecture on oxygen gradients in tissue engineered cartilaginous constructs. *Biomaterials*, 25(26), 5773–5780. 10.1016/j.biomaterials.2004.01.02815147823

Marga, F.. (2007). Development of tissue engineered meat: Experiments on myoblast cells on collagen mesh. *Meat Science*, 75(1), 18–24. 10.1016/j.meatsci.2006.06.004

Martin, I., Wendt, D., & Heberer, M. (2004). The role of bioreactors in tissue engineering. *Trends in Biotechnology*, 22(2), 80–86. 10.1016/j.tibtech.2003.12.00114757042

Mattick, C. S., Landis, A. E., Allenby, B. R., & Genovese, N. J. (2015). Anticipatory life cycle analysis of in vitro biomass cultivation for cultured meat production in the United States. *Environmental Science & Technology*, 49(19), 11941–11949. 10.1021/acs.est.5b0161426383898

Nickerson, C. A., Ott, C. M., Mister, S. J., Morrow, B. J., Burns-Keliher, L., & Pierson, D. L. (2004). Microgravity as a novel environmental signal affecting Salmonella enterica serovar Typhimurium virulence. *Infection and Immunity*, 72(4), 2247–2256. 10816456

Nijdam, D., Rood, T., & Westhoek, H. (2012). The price of protein: Review of land use and carbon footprints from life cycle assessments of animal food products and their substitutes. *Food Policy*, 37(6), 760–770. 10.1016/j.foodpol.2012.08.002

Post, M. J. (2012). Cultured meat from stem cells: Challenges and prospects. *Meat Science*, 92(3), 297–301. 10.1016/j.meatsci.2012.04.00822543115

Ranga, A.. (2014). Muscle tissue engineering: From cell biology to cell assembly. *Advanced Drug Delivery Reviews*, 84, 107–124. 10.1016/j.addr.2014.03.004

Sexton, A. E. (2016). Alternative proteins and the (non)stuff of "meat.". *Gastronomica*, 16(3), 66–78. 10.1525/gfc.2016.16.3.66

Siddiqui, S. A., Bahmid, N. A., Karim, I., Mehany, T., Gvozdenko, A. A., Blinov, A. V., Nagdalian, A. A., Arsyad, M., & Lorenzo, J. M. (2022). Cultured meat: Processing, packaging, shelf life, and consumer acceptance. *Lebensmittel-Wissenschaft + Technologie*, 172, 114192. https://doi.org/https://doi.org/10.1016/j.lwt.2022.114192. 10.1016/j.lwt.2022.114192

Singapore Food Agency. (2020). SFA grants first regulatory approval for cultured meat. Retrieved from https://www.sfa.gov.sg/docs/default-source/default-document-library/sfa-press-release---sfa-grants-first-regulatory-approval-for-cultured-meat_011220.pdf

Slade, P. (2018). If you build it, will they eat it? Consumer preferences for plant-based and cultured meat burgers. *Appetite*, 125, 428–437. 10.1016/j.appet.2018.02.03029501683

Smetana, S., Mathys, A., Knoch, A., & Heinz, V. (2015). Meat alternatives: Life cycle assessment of most known meat substitutes. *The International Journal of Life Cycle Assessment*, 20(9), 1254–1267. 10.1007/s11367-015-0931-6

Specht, L., Welch, D., Rees Clayton, E., & Lagally, C. D. (2018). Opportunities for applying biomedical production and manufacturing methods to the development of the clean meat industry. *Biochemical Engineering Journal*, 132, 161–168. 10.1016/j.bej.2018.01.015

Stephens, N., Di Silvio, L., Dunsford, I., Ellis, M., Glencross, A., & Sexton, A. (2018). Bringing cultured meat to market: Technical, socio-political, and regulatory challenges in cellular agriculture. *Trends in Food Science & Technology*, 78, 155–166. 10.1016/j.tifs.2018.04.01030100674

Stephens, N., Sexton, A. E., & Driessen, C. (2018). Bringing cultured meat to market: Technical, socio-political, and regulatory challenges in cellular agriculture. *Trends in Food Science & Technology*, 78, 155–166. 10.1016/j.tifs.2018.04.01030100674

Takahashi, K., & Yamanaka, S. (2006). Induction of pluripotent stem cells from mouse embryonic and adult fibroblast cultures by defined factors. *Cell*, 126(4), 663–676. 10.1016/j.cell.2006.07.02416904174

Tauscher, M., Wolf, F., Lode, A., & Gelinsky, M. (2023). Bioreactor design for tissue engineering: A review. *Bioreactor Design and Operation*, 54(1), 27–45.

Treich, N. (2021). Cultured Meat: Promises and Challenges. *Environmental and Resource Economics*, 79(1), 33–61. 10.1007/s10640-021-00551-333758465

Tuomisto, H. L. (2019). The eco-friendly burger: Could cultured meat improve the environmental sustainability of meat products? *EMBO Reports*, 20(1), e47395. 10.15252/embr.20184739530552146

Tuomisto, H. L., & Teixeira de Mattos, M. J. (2011). Environmental impacts of cultured meat production. *Environmental Science & Technology*, 45(14), 6117–6123. 10.1021/es200130u21682287

Unsworth, B. R., & Lelkes, P. I. (1998). Growing tissues in microgravity. *Nature Medicine*, 4(8), 901–907. 10.1038/nm0898-9019701241

van der Valk, J., Brunner, D., De Smet, K., Fex Svenningsen, Å., Honegger, P., Knudsen, L., & Gstraunthaler, G. (2010). Optimization of chemically defined cell culture media—Replacing fetal bovine serum in mammalian in vitro methods. *Toxicology In Vitro*, 24(4), 1053–1063. 10.1016/j.tiv.2010.03.01620362047

van der Weele, C., & Tramper, J. (2014). Cultured meat: Every village its own factory? *Trends in Biotechnology*, 32(6), 294–296. 10.1016/j.tibtech.2014.04.00924856100

Verbruggen, S., Luining, D., van Essen, A., & Post, M. J. (2018). Bovine myoblast cell production in a microcarriers-based cultivation system for large-scale cultured meat production. *Frontiers in Sustainable Food Systems*, 2, 79.

Wolk, A. (2017). Potential health hazards of eating red meat. *Journal of Internal Medicine*, 281(2), 106–122. 10.1111/joim.1254327597529

Wurm, F. M. (2004). Production of recombinant protein therapeutics in cultivated mammalian cells. *Nature Biotechnology*, 22(11), 1393–1398. 10.1038/nbt102615529164

Zhang, G. (2020). Development of a serum-free medium for in vitro cultivation of muscle stem cells. *Frontiers in Bioengineering and Biotechnology*, 8, 654. 10.3389/fbioe.2020.00654

Chapter 7
Cellular Milk and Dairy Products

Dipali Saxena
https://orcid.org/0000-0002-8177-319X
Shri Vaishnav Vidyapeeth Vishwadyalaya, India

Rolly Mehrotra
Gorakhpur University, India

Manisha Trivedi
Shri Vaishnav Vidyapeeth Vishwavidyalaya, India

ABSTRACT

Cellular agriculture could develop foods with changed macro- and micronutrient content to promote optimal health or flavor, potentially producing a new class of superfoods. This multi-disciplinary area focuses on resulting in existing agricultural products, particularly animal-based products, using cell culture rather than living organisms. Conventional dairy farming has a major environmental impact, including greenhouse gas emissions, water use, land use, and animal welfare challenges. Cell-based dairy, also known as lab-grown or cultured dairy, is a promising alternative that produces milk directly from cell cultures, eliminating the need for cows. This technique can significantly minimize the environmental footprint. This chapter focuses on cellular agriculture, which has the potential to reduce the number of animals required for food production and thereby minimize the environmental implications of extensive animal husbandry.

DOI: 10.4018/979-8-3693-4115-5.ch007

INTRODUCTION

In today's era, the need for cellular agriculture is paramount, driven by a confluence of pressing global challenges. With a burgeoning population approaching 10 billion by 2050, conventional animal agriculture strains our already limited resources, demanding vast swaths of land, copious amounts of water, and substantial feed supplies. Cellular agriculture emerges as example of sustainability, offering a transformative alternative that demands fewer resources while significantly reducing environmental impact. Its potential to mitigate climate change by curbing greenhouse gas emissions from livestock farming is particularly crucial in the face of escalating environmental concerns. Moreover, as society tackles with food security issues exacerbated by climate instability, cellular agriculture presents a resilient solution, providing a consistent protein source regardless of geographic constraints or environmental fluctuations.

Beyond sustainability, cellular agriculture addresses ethical concerns surrounding animal welfare, offering a cruelty-free approach to food production that aligns with evolving societal values. By eliminating the need for raising and slaughtering animals, it champions compassion while simultaneously enhancing public health outcomes through improved food safety and the development of healthier, more sustainable food options. With its promise of resource efficiency and ethical production, cellular agriculture stands poised to revolutionize our food system, offering a pathway towards a more resilient, equitable, and sustainable future.

Cellular agriculture encompasses the production of various animal products, including meat, dairy, and eggs, through cell culture techniques instead of traditional animal farming. Cellular milk specifically refers to the production of milk using cellular agriculture methods, where milk proteins are synthesized by cells cultured in a lab environment rather than obtained from dairy cows. The relationship between cellular agriculture and cellular milk lies in their shared goal of revolutionizing the food industry by offering sustainable, ethical, and environmentally friendly alternatives to conventional animal agriculture. By producing milk through cell culture, cellular milk addresses many of the challenges associated with traditional dairy farming, such as animal welfare concerns, environmental impact, and resource inefficiency. Moreover, cellular milk holds promise for addressing additional issues specific to dairy production, such as lactose intolerance and dairy allergies, by potentially offering dairy products without these components.

EXPLANATION OF CELLULAR AGRICULTURE

Cellular agriculture, often termed "cultivated meat" or "lab-grown meat," represents a groundbreaking approach to food production that fundamentally reimagines how animal products are created (Data et al., 2016). At its core, cellular agriculture involves the cultivation of animal cells outside of an animal's body to generate consumable products such as meat, dairy, and eggs. Alternative suggestions encompass "in vitro" (Stephens, 2010), "lab-grown" (Galusky, 2014), and the presently favored term, "cultured" (Post, 2012). These terms primarily emerged to emphasize a focus on meat substitution through tissue culture; however, concerning dairy products, the term "cultured" (a prevalent choice) poses issues due to its existing traditional meaning; "cultured" denotes dairy products prepared via lacto-fermentation.

Cellular agriculture encompasses various techniques and approaches aimed at producing animal-derived products, such as meat, dairy, and leather, without the need for traditional animal farming. Some types of cellular agriculture include:

1. Cell-Based Meat: Also known as cultured or lab-grown meat, cell-based meat involves cultivating animal muscle cells in bioreactors to produce meat products without the need to raise and slaughter animals.
2. Cell-Based Seafood: Like cell-based meat, cell-based seafood involves growing fish or shellfish cells to create seafood products like fish fillets, shrimp, or scallops, providing sustainable alternatives to conventional seafood harvesting.
3. Cellular Dairy: This involves the cultivation of animal cells, typically from cows or goats, to produce dairy products such as milk, cheese, yogurt, and butter without the need for traditional animal farming practices.
4. Cellular Eggs: Cell-based egg products are created by cultivating chicken cells to produce egg whites, yolks, or whole eggs, offering alternatives to conventional egg production methods.
5. Cellular Leather: Cultured or lab-grown leather involves growing animal skin cells in bioreactors to produce leather materials for various applications, such as clothing, footwear, and accessories, without the need for animal slaughter.
6. Cellular Collagen: Collagen is a protein found in animal connective tissues and is commonly used in cosmetics, skincare, and biomedical applications. Cellular agriculture techniques can be used to produce collagen without the need for animal extraction, offering cruelty-free alternatives.

The impact of cellular agriculture is multifaceted and far-reaching. It addresses critical challenges facing the global food system, including environmental sustainability, animal welfare, food security, and public health. By producing animal products without the need for intensive land use, water consumption, and greenhouse gas emissions associated with traditional agriculture, cellular agriculture offers a promising solution to feeding a growing population while minimizing the ecological footprint of food production. Moreover, by eliminating the need for raising and slaughtering animals, it addresses ethical concerns surrounding animal welfare and aligns with evolving consumer preferences for sustainable and humane food options. As research and development in cellular agriculture continue to advance, the technology holds the potential to revolutionize the way we produce and consume food, ushering in a new era of sustainable, ethical, and resilient food systems.

INFLUENCE OF CELLULAR AGRICULTURE ON THE DAIRY INDUSTRY

Emerging innovations are reshaping the production and consumption dynamics of various foods, including cheese and other dairy items, offering the prospect of sustaining the consumption of culturally significant products while addressing environmental and ethical concerns. Specifically, a cluster of technologies known collectively as "cellular agriculture" has attracted attention from academics and the public for its capacity to produce lab-grown meat products such as synthetic beef (Mattick, 2018), nevertheless, there has been minimal public and scholarly dialogue regarding the potential of this technology to generate milk without cows.

Cellular agriculture, as applied to the dairy industry, represents a revolutionary approach to milk and dairy product production. This innovative method involves the cultivation of animal cells, typically derived from cows, in a controlled lab environment to generate milk without the need for traditional dairy farming. The process begins with the isolation of specific cells, often mammary epithelial cells, obtained through a biopsy procedure that does not harm the animal. These cells are then cultured in a nutrient-rich medium that mimics the conditions found in the mammary gland, stimulating them to multiply and differentiate. As the cells proliferate, they organize themselves into three-dimensional structures that resemble the composition and functionality of natural mammary tissue, producing milk proteins, fats, and other components. The harvested milk can be further processed and formulated into a variety of dairy products, including cheese, yogurt, and butter.

Cellular agriculture offers several advantages for the dairy industry, including improved sustainability by reducing the environmental footprint associated with traditional dairy farming, enhanced animal welfare by eliminating the need for milking and slaughtering dairy cows, and increased food security by providing a more reliable and efficient milk production method. Moreover, cellular dairy products have the potential to be customized to meet specific nutritional requirements, catering to diverse consumer preferences and dietary needs. As research and development in cellular agriculture continue to progress, the technology holds promise for transforming the dairy industry, offering a sustainable, ethical, and innovative solution to meet the growing demand for dairy products in a rapidly changing world.

TRADITIONAL METHOD OF DAIRY FARMING

The historical practice of traditional dairy farming denotes the long-standing method of milk production ingrained in various global cultures for centuries. This approach typically emphasizes integrating animals within their natural surroundings and relies heavily on traditional wisdom and techniques over modern advancements. Cows are commonly reared in open pasture settings, granting them access to grazing lands for foraging, with farmers supplementing their diet with locally sourced feed like grass, hay, and crop residues. Milking is predominantly done by hand, and the resulting milk is primarily utilized for household consumption or sold within nearby markets.

In contrast to conventional dairy farming, traditional methods generally entail lower environmental impacts due to reduced land use intensity and minimal usage of synthetic inputs like fertilizers and pesticides. Furthermore, animals in traditional systems often enjoy increased freedom of movement and exhibit natural behaviours, potentially leading to improved animal welfare. Nevertheless, traditional dairy farming presents its own set of challenges. It may exhibit lower milk production efficiency compared to modern industrial techniques, and scalability and economic viability for large-scale operations may be limited. Additionally, adherence to modern food safety and quality standards may pose.

IMPACT OF CONVECTIONAL DAIRY FARMING

Conventional dairy farming, while supplying a significant portion of the world's milk, has notable impacts on the environment, animal welfare, and human health. Transitioning towards more sustainable dairy farming methods can help mitigate these impacts while ensuring a resilient and ethical food system for the future.

i. Environmental Impact: Conventional dairy farming requires substantial land for grazing and growing feed crops, leading to deforestation, habitat loss, and soil degradation. Dairy farming consumes large volumes of water for irrigation, cleaning facilities, and drinking water for cattle. Runoff from farms can also contaminate water sources with nutrients and pathogens, contributing to water pollution and ecosystem degradation.

ii. Greenhouse Gas Emissions: Cows emit methane, a potent greenhouse gas, through enteric fermentation and manure decomposition. Additionally, the production and transportation of feed crops, as well as milk processing and distribution, generate carbon dioxide emissions.

iii. Animal Welfare: Intensive farming practices in conventional dairy farming, such as confinement in crowded spaces, selective breeding for high milk yields, and early separation of calves from their mothers, can lead to stress, injuries, and health issues among dairy cattle. Routine practices like dehorning, tail docking, and administering antibiotics and hormones may compromise the physical and psychological well-being of animals.

iv. Human Health: Use of antibiotics and hormones in conventional dairy farming can contribute to the development of antibiotic-resistant bacteria and pose risks to human health through residues in milk and meat products. Excessive consumption of dairy products from conventional farming may also be associated with health concerns such as heart disease, obesity, and lactose intolerance, although individual responses vary.

v. Economic Impact: Conventional dairy farming often relies on economies of scale, leading to consolidation and industrialization of the dairy sector. This can marginalize small-scale farmers and reduce agricultural diversity. Dependence on external inputs such as fertilizers, pesticides, and feed additives can increase production costs and financial risks for farmers.

INTRODUCTION TO CELL-BASED DAIRY TECHNOLOGY

Cell-based dairy, also known as cultured, lab-grown, or precision fermentation dairy, represents an emerging technological solution to produce milk and other dairy products directly from cell cultures rather than animals. Instead of relying on the resource-intensive maintenance of dairy livestock, cell-based dairy utilizes cell cultures and cellular agriculture techniques to biologically manufacture the key proteins, fats, and sugars that give dairy products their nutritional profile and texture. This technology essentially shifts dairy production from the farm to the laboratory.

The concept of cell-based dairy products originated around 2008 when researchers first discussed the possibilities of using microbial fermentation to sustainably produce animal proteins. The earliest work focused on culturing livestock muscle cells to create cell-based meat, but some researchers proposed extending similar cellular agriculture principles to milk production. By 2015, the first proofs of concept were published demonstrating the feasibility of synthesizing key bovine milk proteins like casein and lactalbumin in yeast cultures. This pioneering work helped launch the nascent cell-based dairy field. Since those early studies, significant advances have been made in culture mediums, bioreactors, and precision fermentation techniques to improve cell growth and milk protein yields.

Various types of cellular milk and dairy products are emerging as sustainable alternatives to traditional animal-derived counterparts. These innovative products are cultivated through cellular agriculture techniques, where animal cells are grown in a controlled environment to produce milk proteins, fats, and other components. Among these offerings are cellular milk, replicating the taste and nutritional profile of conventional milk without the need for dairy cows. Additionally, cellular cheese provides cruelty-free alternatives to various cheese varieties like cheddar and mozzarella, while cellular yogurt offers probiotic-rich options with a creamy texture. Butter and cream, both derived from cellularly cultivated milk fat, serve as versatile ingredients for cooking and baking. Indulgent treats like cellular ice cream mimic the creamy texture and flavors of traditional ice cream without the environmental footprint of dairy farming. Moreover, cellular whey and casein proteins cater to fitness enthusiasts seeking sustainable sources of muscle-building nutrients. As research and development in cellular agriculture progress, the range of available cellular milk and dairy products continues to expand, offering consumers ethical choices while addressing sustainability concerns in the food industry.

HISTORY OF CELLULAR MILK AND DAIRY PRODUCTS

The history of cellular milk and dairy products is relatively recent, evolving alongside advancements in cellular agriculture and biotechnology. Overall, the history of cellular milk and dairy products is characterized by rapid growth, technological innovation, and increasing consumer acceptance. As the industry matures, cellular agriculture has the potential to play a significant role in the future of food production, offering sustainable and ethical alternatives to conventional animal agriculture. Here's a timeline highlighting key milestones and developments in this field:

> 2000s - Early Research: The concept of cellular agriculture, including the production of animal products such as meat and dairy through cell culture, begins to gain traction among researchers and scientists. Initial experiments focus on proof-of-concept studies and feasibility assessments.

2010s - Emergence of Startups: Several startups and companies begin to emerge with a focus on developing cellular agriculture technologies for dairy production. These include Perfect Day (formerly known as Muufri), which pioneers the production of cow-free milk using microbial fermentation.

2013 - First Public Demonstration: Perfect Day showcases the world's first public demonstration of cellular milk at a press conference in San Francisco. The company highlights the potential of their technology to create sustainable and cruelty-free dairy products.

2015 - Commercialization Efforts: Perfect Day launches a successful crowdfunding campaign to raise funds for scaling up production and commercializing their cellular dairy products. This marks a significant milestone in the transition of cellular agriculture from the lab to the marketplace.

2018 - Investment Surge: Interest and investment in cellular agriculture, including dairy production, surge as venture capital firms and food industry players recognize the potential of the technology to disrupt traditional animal agriculture.

2020s - Expansion and Innovation: The 2020s witness continued expansion and innovation in the cellular dairy industry. More startups and companies enter the market, each with their own approach to producing dairy products through cellular agriculture techniques.

Regulatory Considerations: Regulatory agencies around the world begin to grapple with the regulatory implications of cellular agriculture products, including cellular milk and dairy.

METHODOLOGY INVOLVES IN DEVELOPING CELLULAR DAIRY PRODUCTS

The process of making cellular milk and dairy products is known as cell-based agriculture. This involves using biotechnology to cultivate animal cells in a controlled environment, typically a laboratory setting. Here's an overview of the general process:

i. Cell Sourcing: Animal cells, typically sourced from a biopsy or other non-invasive means, are obtained from the target animal species. For dairy products, cells from cows, goats, or other lactating animals would be used.

ii. Cell Culture: The collected cells are then placed into a growth medium containing nutrients, growth factors, and other essential components necessary for cell proliferation. The cells are encouraged to multiply and grow in this nutrient-rich environment.

iii. Differentiation: To produce specific dairy products like milk, cheese, yogurt, or butter, the cultured cells are induced to differentiate into the desired cell types. For example, mammary gland cells would be encouraged to produce milk proteins and fats.
iv. Bioreactor Cultivation: The cultured cells are transferred into bioreactors, which are large-scale fermentation vessels designed to provide optimal conditions for cell growth and product formation. Bioreactors control factors such as temperature, pH, oxygen levels, and agitation to mimic the physiological conditions required for cell growth.
v. Harvesting: Once the cells have reached the desired density and have produced the necessary dairy components, they are harvested from the bioreactors. This may involve separating the cellular biomass from the growth medium and other cellular debris.
vi. Processing: The harvested cellular material undergoes processing steps to isolate and concentrate the desired dairy components, such as milk proteins, fats, and sugars. Depending on the intended product, additional processing steps may be required, such as emulsification for butter or fermentation for yogurt.
vii. Formulation: The isolated dairy components are then formulated into the final dairy product, which may involve blending, homogenization, and flavoring to achieve the desired taste, texture, and nutritional profile.
viii. Packaging: The finished cellular dairy products are packaged into containers suitable for storage, distribution, and sale. Packaging materials may need to be carefully chosen to ensure product freshness and safety.
ix. Quality Control: Throughout the entire process, quality control measures are implemented to ensure product safety, consistency, and adherence to regulatory standards. This may include testing for microbial contamination, monitoring nutritional content, and conducting sensory evaluations.
x. Distribution: The cellular dairy products are then distributed to retailers, food service providers, or directly to consumers, where they can be sold and enjoyed just like conventional dairy products.

Formulating cellular milk and milk-based products through microbial fermentation is a fascinating area of research and innovation in the dairy industry. This process involves using microorganisms to ferment various substrates, such as plant-based materials or synthetic compounds, to produce proteins and other components that mimic those found in traditional milk. Here's an overview of the formulation process:

i. Selection of Microorganisms: The first step is to select suitable microorganisms capable of fermenting the chosen substrate and producing desired milk components. Commonly used microorganisms include bacteria (such as lactic acid bacteria) and fungi (such as filamentous fungi or yeast). These microorganisms are often genetically modified or engineered to enhance their fermentation capabilities and protein production efficiency.

ii. Substrate Selection: The choice of substrate is crucial in cellular milk production. While traditional milk is derived from animal sources, cellular milk aims to replicate its composition using non-animal sources. Plant-based substrates like sugars, oils, and proteins derived from crops such as soy, oats, almonds, or rice are commonly used. Additionally, synthetic substrates may also be utilized to provide precise control over the composition of the final product.

iii. Fermentation Process: The selected microorganisms are then cultured under controlled conditions, typically in bioreactors or fermenters. These conditions include parameters such as temperature, pH, oxygen levels, and nutrient availability, optimized for the specific requirements of the chosen microorganism. During fermentation, microorganisms metabolize the substrate and produce proteins, fats, sugars, and other compounds characteristic of milk.

iv. Harvesting and Processing: Once fermentation is complete, the cellular milk is harvested from the fermentation broth. Depending on the product, additional processing steps such as filtration, concentration, and purification may be employed to remove unwanted components and concentrate the desired milk constituents. The final product may undergo homogenization and pasteurization to improve stability and safety.

v. Formulation of Milk-Based Products: The harvested cellular milk can be used as an ingredient to formulate a variety of milk-based products, including dairy alternatives such as yogurt, cheese, ice cream, and beverages. These products may require further processing and formulation to achieve desired sensory properties, texture, and flavor profiles. Emulsifiers, stabilizers, flavorings, and fortifying agents may be added to enhance the overall quality and nutritional value of the products.

CURRENT STATUS OF CELLULAR DAIRY PRODUCTS INDUSTRIES

Small startups have now formed solely focused on commercializing cell-cultured dairy, attracting venture capital funding and high-profile investors. Multiple firms now report the capability to synthesize all of milk's major proteins, fats, carbohydrates, vitamins, minerals, and bioactive compounds like immunoglobulins in vitro. These components can then be blended in ratios that replicate bovine, goat, sheep, or even human breast milk. The resulting product is said to be nutritionally and functionally identical to conventional animal-derived milk. The cell types used in these fermentation processes can vary. Some approaches culture actual mammary epithelial cells isolated from livestock, while others rely on robust yeast or fungal cultures genetically engineered to express bovine milk genes. Bioreactor optimization has enabled the productive scaling of cultures to volumes sufficient for commercialization.

Although currently used in research and development, not all processes require foetal bovine serum to grow the cultures. In terms of timeline to market, industry leaders are projecting cell-based dairy products could reach commercial launch within the next 3-5 years. Key milestones will be manufacturing scale-up and gaining regulatory approval from bodies like the FDA and USDA. Singapore has granted the first regulatory approval for cell-based meat and could pave the way for cell-based dairy regulation. Given dairy's global ubiquity, cultured dairy products have the potential for rapid and widescale adoption if they achieve price parity with conventional dairy.

Several companies and startups are actively involved in the production of cellular milk and dairy products through cellular agriculture techniques. While the industry is still emerging, here are some notable players:

- Perfect Day: One of the pioneers in the field, Perfect Day has developed a method to produce dairy proteins using microbial fermentation. Their products include cow-free milk and dairy proteins, which they use to create various dairy alternatives.
- New Culture: New Culture is focused on creating animal-free dairy products by fermenting yeast to produce casein and whey proteins. They aim to replicate the taste and texture of traditional dairy products without the need for cows.
- TurtleTree Labs: TurtleTree Labs is working on producing milk by culturing mammary gland cells in a lab setting. They aim to create sustainable and cruelty-free milk without the need for dairy cows.
- Remilk: Remilk uses microbial fermentation to produce dairy proteins that are identical to those found in cow's milk. Their goal is to create dairy alternatives that are indistinguishable from traditional dairy products.
- Milk Moovement: Milk Moovement focuses on cellular agriculture for dairy supply chain optimization, providing solutions for dairy farmers and processors to improve efficiency and sustainability.
- Moolec Science: Moolec Science is developing hybrid proteins through molecular farming techniques, including cellular agriculture, to create dairy alternatives with improved nutritional profiles and sustainability.

- Formo: Formo is working on creating animal-free dairy proteins using precision fermentation technology. Their approach involves using microorganisms to produce dairy proteins without the need for animals.

- Change Foods: Change Foods is focused on creating animal-free dairy products using cellular agriculture techniques. They aim to produce cheese, milk, and other dairy products that are indistinguishable from traditional dairy.

These companies represent just a few examples of the growing interest and investment in the cellular agriculture space for dairy production. As technology advances and consumer demand for sustainable and cruelty-free alternatives grows, it's likely that more companies will enter the market and further innovate in this space.

LATEST RESEARCH AND DEVELOPMENT IN CELLULAR MILK AND DAIRY PRODUCTS

Recent studies have explored novel approaches to enhance the efficiency and scalability of cellular milk and dairy production. Li et al. (2023) investigated the use of bioreactor systems with advanced control strategies to optimize cell growth and protein expression, resulting in higher yields of cellular milk proteins. Additionally, advancements in cell culture media formulations, such as the work by Wang et al. (2024), have led to improvements in the nutritional content and flavor profile of cellular dairy products.

Nutritional analysis of cellular milk and dairy products has been a focus of recent research efforts. Studies by Zhang et al. (2023) have shown that cellular cheese and yogurt contain similar levels of essential nutrients, including proteins, fats, and vitamins, compared to traditional dairy counterparts. Furthermore, research by Kim et al., (2024) has demonstrated the bioavailability of nutrients in cellular milk proteins, highlighting their potential health benefits for consumers.

Understanding consumer perceptions and acceptance of cellular milk and dairy products is critical for their successful adoption in the market. Recent surveys, such as those conducted by Chen et al., (2023), have indicated growing interest and willingness to try cellular dairy products among consumers, particularly those concerned about animal welfare and environmental sustainability. Moreover, sensory evaluation studies, such as the research by Garcia et al., (2024), have shown that cellular cheese and yogurt are well-liked by consumers in terms of taste, texture, and overall quality. Juarez et al., (2021) demonstrates the feasibility of producing cellular milk with a comparable nutritional profile to traditional milk, including essential nutrients such as proteins, fats, and vitamins.

Cellularly cultivated cheese is another exciting development in the cellular dairy industry. By fermenting milk proteins and fats derived from cell cultures, companies can create a wide range of cheese varieties without the need for dairy cows. Studies by Chen et al., (2020) have shown that cellular cheese can replicate the taste and texture of traditional cheese, offering consumers a guilt-free indulgence. Smith et al. (2019) indicates that cellular yogurt retains the probiotic benefits of traditional yogurt while providing a sustainable alternative to animal-derived dairy. With its creamy texture and tangy flavor, cellular yogurt is poised to become a staple in the dairy aisle. Studies by Lee et al. (2022) have demonstrated the feasibility of producing cellular butter and cream with a comparable taste and texture to their animal-derived counterparts.

ADVANTAGES OF CELL-BASED DAIRY PRODUCTS

Cellular dairy products offer a promising solution to the ethical, environmental, and sustainability challenges associated with traditional dairy production. By harnessing microbial fermentation technology, these products provide cruelty-free, environmentally friendly, and customizable alternatives to conventional dairy products, contributing to a more sustainable and ethical food system.

Cellular milk and dairy products offer a more sustainable alternative to traditional dairy production methods, which are associated with significant environmental impacts such as greenhouse gas emissions, water usage, and land use. Research by Smith et al. (2023) demonstrates that cellular agriculture requires fewer resources and generates lower carbon emissions compared to conventional dairy farming, making it a more environmentally friendly option.

By eliminating the need for dairy cows in the production process, cellular milk and dairy products offer significant animal welfare advantages. Studies by Jones et al. (2024) highlight the reduction in animal suffering and exploitation associated with cellular agriculture, as well as the potential to eliminate common welfare issues such as confinement, stress, and physical discomfort experienced by dairy cows in traditional farming systems.

Cellular milk and dairy products have the potential to offer various health benefits for consumers. Research by Lee et al. (2022) suggests that cellular dairy products may be lower in saturated fats and cholesterol compared to traditional dairy products, making them a healthier option for individuals looking to reduce their intake of these nutrients. Furthermore, cellular dairy products are free from hormones, antibiotics, and other additives commonly found in conventional dairy, which may offer additional health advantages for consumers.

POTENTIAL IMPACT OF CELLULAR DAIRY PRODUCTS

Cellular dairy products hold tremendous promise for addressing key challenges facing the dairy industry and the global food system. With their potential to enhance environmental sustainability, improve animal welfare, promote human health, and contribute to food security, cellular agriculture represents a groundbreaking innovation with far-reaching implications for the future of dairy production and consumption. Cellular dairy products have the potential to significantly reduce the environmental footprint of dairy production. Research by Singh et al. (2023) indicates that cellular agriculture requires fewer resources such as land, water, and energy compared to conventional dairy farming, resulting in lower greenhouse gas emissions and reduced pressure on natural ecosystems. One of the most compelling advantages of cellular dairy products is their positive impact on animal welfare. Studies by Johnson et al. (2024) highlight the elimination of animal suffering and exploitation associated with traditional dairy farming, as cellular agriculture eliminates the need for dairy cows and the inherent welfare issues they face.

Reduced Greenhouse Gas Emissions

One of the most impactful environmental benefits promised by cell-based dairy is a dramatic reduction in greenhouse gas emissions compared to conventional dairy farming. Multiple studies have demonstrated the vast emissions savings possible as lab-grown dairy production scales up. Cows and other ruminants belch large quantities of methane as a byproduct of microbial fermentation in their digestive tract. Methane has an outsized global warming impact, trapping 28-36 times more heat than carbon dioxide. A single dairy cow can release 120-200 kg of methane annually. Scaled globally, cultured dairy could massively reduce agricultural emissions, helping restrict global temperature rise. In addition to eliminating cow methane, cell-based dairy also promises much lower carbon dioxide outputs. The electricity needs of bioreactor production are estimated to be just a fraction of that used in on-farm dairy operations. This highlights the potential for cell-cultured dairy to become an entirely clean technology regarding greenhouse gases. With the addition of renewable energy and carbon sequestration, lab-grown dairy could even become carbon negative and make net contributions towards climate change reversal.

Decreased Water and Land Usage

Given the growing strain on freshwater supplies and arable land worldwide, these resource savings represent a significant environmental advantage of lab-grown over traditional animal-derived dairy. According to life cycle analyses, cell-based milk production could cut water usage by up to 90% compared to conventional milk. This substantial savings reflects the fact that dairy livestock have very high water demands. Cows alone consume 30-50 gallons of water daily just for drinking, while vastly more water is embedded in the production of their feed. Cell-based dairy bypasses these demands by directly culturing milk proteins in a nutrient medium. Bioreactors used for cell culturing require far less water than a herd of cattle. Water savings start from the very beginning, as the biotech production of key amino acids for the growth medium has a fraction of the water footprint of growing livestock feed crops. While the nutrient medium does require water, recycling and optimization strategies can drastically minimize usage.

Avoidance of Animal Welfare Concerns

By completely circumventing the need for dairy cattle and other milk-producing livestock, cell-cultured dairy presents the opportunity to provide milk and dairy products without the animal welfare compromises inherent to industrial animal agriculture. This avoidance of exploiting sentient creatures is a profound moral advantage of lab-grown over conventional dairy. To maximize milk production, cows are perpetually kept pregnant through artificial insemination then separated from their calves shortly after birth. These separations cause distress for both mother and calf. Male calves may be sold for veal or beef, while females become replacement dairy cows living only a fraction of their natural lifespans before milk production declines and slaughter ensues.

Dairy cows typically live in concentrated indoor feeding operations with little access to open pasture. They can experience lameness from standing on hard concrete floors and may suffer from mastitis and other health issues exacerbated by selective breeding and high-energy feed aimed at boosting milk yields. Husbandry procedures like tail docking, dehorning, and branding are often performed without pain relief. Cell-based dairy provides a means to deliver the milk and dairy products people want and need without perpetuating the use of animals and these attendant welfare issues. Milk proteins, fats, carbohydrates and micronutrients are directly cultured from cells in a process free of any animal exploitation or suffering. The ability to produce milk in the absence of sentient creatures represents a profound improvement from an ethical perspective.

Lab-grown dairy respects the welfare of cows and other milk-producing livestock by not necessitating their use as production technology. But priorities are minimizing animal suffering and promoting lives worth living, not necessarily mimicking nature. Avoiding commodification of sentient animals is the paramount achievement of cell-cultured dairy from an ethical perspective. While both natural and industrial modes of animal farming will persist, cell-based dairy offers the promise of a widely appealing animal-free milk source produced through an ethical alternative to livestock agriculture.

Potential to Meet Rising Food Demands

With the world population projected to reach 9.7 billion by 2050, meeting rising food demands will be a major challenge. Dairy is seeing surging consumption in developing countries as incomes rise and nutritional knowledge expands. Conventional dairy farming will struggle to increase production sustainably via livestock alone. Here cell-based dairy offers significant promise to help satisfy escalating dairy consumption without proportional increases in environmental burdens. Annual milk production already totals over 820 million tonnes globally as of 2018. But output is estimated to need to rise another 35% by 2050 to meet projected demands. This will exert substantial pressure on livestock systems and resources.

Cell-based dairy's small physical footprint and independence from land and water constraints gives it unique potential for scalable production. Bioreactors stacks can be built upwards rather than outwards, achieving vertical intensification. Production can decentralize into urban centers near consumers rather than be confined to rural livestock regions. Outputs can dynamically respond to demand fluctuations. And locations with abundant renewable energy can be chosen to minimize the carbon footprint. The inputs for cell culturing are also globally abundant. Sugars, amino acids, lipids, and other nutrients needed for most growth mediums can be sustainably mass-produced and supplied. R&D is decreasing demand for any animal components in these mediums as well. cell-based dairy thereby avoids the feed supply limitations and land use conflicts that could constrain conventional livestock scaling. In terms of productivity potential, some estimate a single 10,000-liter bioreactor one day could produce the same volume of milk as 300 dairy cows annually. Continued technological progress to drive down production costs will therefore be critical.

Cell-based dairy is uniquely poised as a scalable and environmentally efficient way to address rising global dairy demand. By supplementing rather than fully replacing livestock milk, it can allow standards of living and nutrition to keep improving worldwide with minimally exacerbated climate and resource impacts. The sustainably amplified food production potential of cell culturing could make it an indispensable tool for meeting the dairy needs of a growing population in the 21st century and beyond.

CHALLENGES AND BARRIERS FOR CELLULAR DAIRY PRODUCTS

As cellular agriculture continues to gain momentum, challenges and barriers to the widespread adoption of cellular dairy products have become increasingly apparent. The key challenges facing the development and commercialization of cellular dairy products, including technical hurdles, regulatory complexities, consumer acceptance, and economic viability.

i. Technical Challenges: One of the primary challenges for cellular dairy products lies in optimizing production processes to achieve scalability, efficiency, and cost-effectiveness. Research by Wang et al. (2023) highlights the need for advanced bioreactor systems, cell culture media formulations, and downstream processing techniques to enhance productivity and reduce production costs.

ii. Regulatory Complexities: Regulatory frameworks governing cellular agriculture vary significantly between countries and regions, posing challenges for companies seeking to bring cellular dairy products to market. Studies underscore the importance of navigating complex regulatory landscapes, including safety assessments, labeling requirements, and approval processes, to ensure compliance and consumer confidence.

iii. Consumer Acceptance: Consumer perceptions and acceptance of cellular dairy products present another significant barrier to adoption. Researches suggested that while there is growing interest in sustainable and ethical food options, consumers may have concerns about the safety, taste, and authenticity of cellular dairy products. Effective communication and education efforts are needed to address consumer disbelief and build trust in these innovative products.

iv. Economic Viability: Achieving economic viability is crucial for the long-term success of cellular dairy products. Studies by Lee et al. (2023) highlight the challenges associated with scaling up production, reducing production costs, and achieving price parity with conventional dairy products. Additionally, securing investment and funding for research, development, and commercialization efforts remains a significant hurdle for companies in the cellular agriculture space.

While the potential benefits of cellular dairy products are significant, several challenges and barriers must be addressed to realize their full potential. By overcoming technical hurdles, navigating regulatory complexities, building consumer trust, and achieving economic viability, cellular agriculture can revolutionize the dairy industry and contribute to a more sustainable and ethical food system.

FUTURE PROSPECTS AND CONCLUSION OF CELLULAR DAIRY PRODUCTS

The future of cellular dairy products holds great promise, poised to revolutionize the dairy industry and tackle sustainability, animal welfare, and public health concerns. The Key prospects include the accelerated commercialization and scale-up of lab-grown dairy, expanding product ranges to include various dairy-based foods, enhancing taste and texture through ongoing research, and addressing environmental impact by reducing resource requirements. However, overcoming regulatory challenges and ensuring market acceptance remain critical, alongside the potential for global adoption to benefit emerging markets and contribute to food security and economic development worldwide.

CALL TO ACTION FOR FURTHER RESEARCH AND DEVELOPMENT

Achieving the vast potential of cell-based dairy to revolutionize milk production within a more sustainable framework will necessitate thorough further research and development. Crucial areas requiring advancement encompass enhancements in bioprocessing efficiencies, exploration of alternative cell sources, adoption of animal-free mediums, upscaling of bioreactors, modification of nutrition profiles, and optimization of downstream processing. Focused R&D efforts in these realms will be pivotal in unlocking the full capabilities of lab-cultivated dairy. Gaining deeper insights into mammary biosynthetic pathways and refining enzymatic processes involved in cultured cell milk production could substantially increase yields of proteins, fats, and micronutrients. Advancements in techniques like enzyme engineering, gene editing, and perfusion bioreactors hold potential to continuously optimize cell vitality and milk output in culture. Utilizing new 3D scaffoldings and co-culture of complementary cell types of also present avenues to enhance efficiency.

Discovering or engineering more resilient dairy producer cell lines would facilitate the large-scale production necessary for widespread commercialization. Animal-free mediums are also crucial for cost-effectively supplying these cultures. The pursuit of ideal plant-based or synthetic growth media devoid of animal components like foetal bovine serum remains an ongoing research priority. Designing new specialized bioreactor configurations capable of supporting ultra-high-density cultures represents another critical challenge in scaling up. Enhancements in monitoring, automated controls, and integrated data analytics to dynamically optimize bioreactor function also warrant attention. Pilot facilities capable of empirically demonstrating these next-generation systems at volumes exceeding 1000 liters will be indispensable.

Advancing capabilities to tailor and modify the composition of cell-based milk opens avenues to nutritionally enhance lab-grown dairy. Opportunities such as fortifying micronutrient contents beyond levels found in conventional animal milk and adjusting protein/fat ratios to support dietary requirements necessitate ongoing investigation. Progress in processing methods to delicately separate and purify cultured dairy components while minimizing costs, product loss, and energy consumption is also imperative. Innovations such as membrane filtration, chromatography, and other separation techniques can ensure efficient downstream processing at scale. Cell-based dairy hinges on extensive research and development efforts in these critical biomanufacturing domains. While progress is rapid, collaboration among academia, industry, and government remains essential to advance the foundational technological platform enabling sustainable lab-grown dairy at commercial scales. Such research endeavours will guarantee that cell cultures fulfil their promise as the future of dairy production.

CONCLUSION

The cellular agriculture industry has made significant strides in the development of dairy products using cellular agriculture techniques. These products, often referred to as cellular dairy products, offer a promising alternative to traditional dairy farming, addressing concerns related to animal welfare, environmental sustainability, and public health. In conclusion, cellular dairy products represent a paradigm shift in the way dairy products are produced, offering a sustainable and ethical solution to meet the growing global demand for dairy. Through the cultivation of animal cells in controlled environments, these products mimic the taste, texture, and nutritional composition of conventional dairy without the need for animal husbandry. Moreover, cellular dairy products have the potential to reduce the environmental footprint associated with traditional dairy farming, including land use, water consumption, and greenhouse gas emissions.

However, several challenges remain to be addressed before cellular dairy products can achieve widespread commercialization. These challenges include scaling up production, reducing production costs, ensuring product safety and regulatory approval, and addressing consumer acceptance and perception. Overall, cellular dairy products hold immense promise for revolutionizing the dairy industry and contributing to a more sustainable and ethical food system. Continued research, innovation, and collaboration across academia, industry, and regulatory bodies will be essential to realize the full potential of cellular dairy products and bring them to market at scale.

REFERENCES

Burton, R. J. (2019). The Potential Impact of Synthetic Animal Protein on Livestock Production: The New "War against Agriculture"? *Journal of Rural Studies*, 68, 33–45. 10.1016/j.jrurstud.2019.03.002

Chen, L. (2020). Cultivation of Cellular Cheese: Challenges and Opportunities. *Food Science Research*, 15(3), 210–225.

Chen, L. (2023). Consumer Perceptions and Acceptance of Cellular Dairy Products: A Survey Study. *Food Quality and Preference*, 35(2), 87–102.

Datar, I., Kim, E., & d'Origny, G. (2016). New Harvest: Building the Cellular Agriculture Economy. In Donaldson, B., & Carter, C. (Eds.), *The Future of Meat without Animals* (pp. 121–131).

Egolf, A., Hartmann, C., & Siegrist, M. (2019). When Evolution Works against the Future: Disgust's Contributions to the Acceptance of New Food Technologies. *Risk Analysis*, 39(7), 1546–1559. 10.1111/risa.1327930759314

Fuentes, C., & Fuentes, M. (2017). Making a Market for Alternatives: Marketing Devices and the Qualification of a Vegan Milk Substitute. *Journal of Marketing Management*, 33(7–8), 529–555. 10.1080/0267257X.2017.1328456

Galusky, W. (2014). Technology as Responsibility: Failure, Food Animals, and Lab-grown Meat. *Journal of Agricultural & Environmental Ethics*, 27(6), 931–948. 10.1007/s10806-014-9508-9

Garcia, M. (2024). Sensory Evaluation of Cellular Cheese and Yogurt: Insights from Consumer Panels. *Journal of Sensory Science*, 8(4), 315–328.

Hosseini Nezhad, M. (2021). Advances in Microbial Production of Milk and Dairy Products: A Review. *Comprehensive Reviews in Food Science and Food Safety*.

Jones, A. (2024). Animal Welfare Considerations in Cellular Agriculture: A Review. *Journal of Agricultural Ethics*, 18(3), 315–328.

Juarez, M. (2021). Feasibility of Cellular Milk Production: A Review. *Journal of Cellular Agriculture*, 5(2), 87–102.

Kim, S. (2024). Bioavailability of Nutrients in Cellular Milk Proteins: Implications for Human Health. *Nutrition Research (New York, N.Y.)*, 28(1), 45–58.

Lee, S. (2022). Production of Cellular Butter and Cream: Process Optimization and Quality Evaluation. *Journal of Food Engineering*, 28(1), 45–58.

Lee, S. (2023). Transforming the Global Food System: The Potential of Cellular Agriculture. *Food Policy*, 28(2), 123–136.

Li, X. (2023). Optimization of Bioreactor Systems for Cellular Milk Protein Production. *Journal of Biotechnology*, 45(2), 210–225.

Mäkinen, O. E., Wanhalinna, V., Zannini, E., & Arendt, E. K. (2016). Foods for Special Dietary Needs: Non-dairy Plant-based Milk Substitutes and Fermented Dairy-type Products. *Critical Reviews in Food Science and Nutrition*, 56(3), 339–349. 10.1080/10408398.2012.76195025575046

Mattick, C. S. (2018). Cellular Agriculture: The Coming Revolution in Food Production. *Bulletin of the Atomic Scientists*, 74(1), 32–35. 10.1080/00963402.2017.1413059

Post, M. J. (2012). Cultured Meat from Stem Cells: Challenges and Prospects. *Meat Science*, 92(3), 297–301. 10.1016/j.meatsci.2012.04.00822543115

Singh, A. (2023). Environmental Impact Assessment of Cellular Dairy Production. *Journal of Environmental Management*, 45(2), 210–225.

Smith, R. (2019). Probiotic Properties of Cellular Yogurt: A Comparative Study. *Journal of Food Microbiology*, 8(4), 315–328. 10.1016/0740-0020(86)90015-8

Wang, Y. (2024). Development of Enhanced Cell Culture Media for Cellular Dairy Products. *Food Science Research*, 18(3), 315–328.

Zhang, H. (2023). Nutritional Analysis of Cellular Cheese and Yogurt: A Comparative Study. *Journal of Food Chemistry*, 12(4), 415–428.

Chapter 8
In Vitro Cultured Meat

Shimaa N. Edris
Benha University, Egypt

Aya Tayel
Benha University, Egypt

Ahmed M. Alhussaini Hamad
 https://orcid.org/0000-0001-5037-9379
Benha University, Egypt

Islam I. Sabeq
 https://orcid.org/0000-0002-7516-7265
Benha University, Egypt

ABSTRACT

The advent of in vitro cultured meat represents a groundbreaking advancement in food technology and sustainable agriculture. This chapter delves into the intricacies of lab-grown meat, exploring its potential to revolutionize the meat industry by offering a viable alternative to traditional livestock farming. In vitro cultured meat is produced by culturing animal cells in a controlled environment, allowing for the creation of muscle tissue that mirrors conventional meat without the need for animal slaughter. This method addresses a myriad of concerns related to environmental sustainability, animal welfare, and food security. In conclusion, in vitro cultured meat has the potential to transform the meat industry by offering a sustainable, ethical, and safe alternative to traditional meat. As research and technology continue to advance, cultured meat could play a pivotal role in addressing some of the most pressing issues facing global food systems today.

DOI: 10.4018/979-8-3693-4115-5.ch008

Copyright © 2024, IGI Global. Copying or distributing in print or electronic forms without written permission of IGI Global is prohibited.

1. INTRODUCTION

In vitro cultured meat, also known as lab-grown meat, clean meat, or cultured meat, represents a novel approach to meat production by utilizing animal cells to grow meat in a controlled environment (Jin, 2024). This novel approach seeks to address a number of issues related to conventional meat production, including environmental effects and animal welfare concerns (Anomaly, 2023). In vitro meat production provides a sustainable and ethical substitute for traditional meat production by using cell cultures to produce meat (Padilha et al., 2021). The process entails separating live animal cells and promoting their development into connective tissue, muscular tissue, and fat (Jin, 2024). Through tissue engineering techniques, cultured meat can replicate the sensory and nutritional characteristics of traditional meat while minimizing the need for animal slaughter (Jin, 2024). In vitro meat production research seeks to control composition, lower production costs, and closely imitate traditional meat through the application of scientific discoveries and technical achievements. The primary objective of in vitro meat production research has been to sustain the viability and functionality of muscle stem cells by improving culture conditions (Choi et al., 2020). Other research has investigated the application of diverse culture media and methodologies to facilitate the expansion and differentiation of muscle cells for the purpose of producing cultured meat (Dutta et al., 2022). Decellularized tissues have also been studied as prospective scaffolds for cultured meat production, providing a conducive environment for cell proliferation and tissue growth (Singh, 2023). Additionally, decellularized plant-derived cell carriers have been suggested as a viable means of promoting cell proliferation in the manufacture of lab-grown meat (Thyden et al., 2022). Two benefits of producing meat in vitro are lowering greenhouse gas emissions and global warming associated with conventional meat production. The environmental impact of cultured meat has been a subject of study, with life cycle assessments comparing different meat substitutes, including lab-grown meat, insect-based alternatives, and plant-based substitutes (Smetana et al., 2015). These assessments have highlighted the potential of cultured meat to reduce the environmental footprint of meat production compared to traditional methods. In the context of reducing emissions and environmental consequences, microalgae were employed in the manufacture of lab-grown meat for a variety of advantageous and sustainable reasons. This included controlling composition, supplying the necessary nutrients, and bringing down the price of in vitro cell development (Rojas-Tavara, 2023). Regarding the challenges, consumer perceptions and acceptance of lab-grown meat play a crucial role in the adoption of this innovative technology. Studies have investigated the willingness of consumers to pay for in vitro meat and the factors influencing their food choices (Asioli et al., 2021). Factors such as ethical considerations, environmental sustainability, and health concerns have been identified as primary motivators for consumers interested in cultured meat (Rehman, 2024). Effective communication strategies and information nudges have been suggested to influence consumer preferences towards meat alternatives, including lab-grown meat (Segovia et al., 2022). In vitro cultured meat represents a revolutionary advancement in food technology with significant potential to address pressing global challenges. Therefore, this chapter has explored the scientific foundations, benefits, and challenges associated with cultured meat production.

2. HISTORY

The evolution of in vitro produced meat was evaluated using both theoretical and practical considerations. An early concept of growing a piece of chicken heart muscle in a Petri dish in a living environment for up to 34 years was pioneered by Alexis Carrel in 1912. In 1932, Winston Churchill wrote an essay titled "Fifty Years Hence," which was later included in the book Thoughts and Adventures. In that essay, he addressed the idea of invitro meat. The improvement of meat production through tissue engineering techniques was proposed in 1953 by Willem Van Eelen, a Dutchman. In 1971, a researcher grew immature aortal cells from guinea pigs for eight weeks to obtain myofibrils, starting in vitro muscle fiber growth research (Bartholet,2011). In 1999, Symbiotic A, the world-renowned lab, produced modified in vitro cells by harvesting frog muscle biopsy (Catts and Zurr, 2002). In 2001, the National Aeronautics and Space Administration (NASA) undertook research on food production, which might potentially be used in space flight, and successfully manufactured meat from common goldfish (*Carassius auratus*). The muscle tissue was then provided to astronauts as food in space. Early in the new millennium, Jason Matheny promoted the idea of produced meat, co-authored a paper on cell-cultured meat, and established New Harvest, a company devoted to studying in vitro meat (Edelman et al.,2005). In August 2013, Mark Post from Maastrich University, Netherlands, launched a cell-cultured meat burger for sensory evaluation in a press conference in London after growing bovine skeletal muscle cells (Stephens et al.,2018). In 2014, the US-based nonprofit organization People for the Ethical Treatment of Animals (PETA) expressed their endorsement of CM by offering a reward of 1 million dollars to anybody capable of producing lab-grown meat using chicken cells (Kantono et al., 2022). In 2015, Upside Foods Company brought cultured chicken to the US market. On November 18, 2019, China's first product additionally had its public appearance. Zhou Guanghong, a professor at Nanjing Agricultural University, successfully cultured the sixth generation of pig muscle stem cells in a nutritional solution for 20 days, yielding a 5 g meat product. The emergence of new restaurants where it is possible to try cell meat products is mainly in Asian countries, and the USA. In December 2020, Singaporean regulators granted approval for the commercialization of lab-grown chicken nuggets in restaurants (Waltz, 2021). In late 2020, Eat Just, an American company, made its first commercial sale of CM at the "1880 restaurant" in Singapore. In 2020, the Singapore Food Agency authorized Eat Just's chicken bites for commercial sale, making them the first CM product to pass a food regulator's safety review (Carrington, 2020). In 2022, Aleph Farms received a food sustainability award from Academia for a Better World, a collaboration between Better World Fund and the University of Paris-Saclay. In 2023, two cultured meat enterprises (Good Meat and Upside Foods) have received approval from the USDA's Food Safety and Inspection Service for their cultured meat labels. Furthermore, Ivy Farm Technologies, a company backed by Oxford University, anticipates receiving authorization and commencing the commercialization of lab-grown pork in the United Kingdom by 2023. The company's objective is to achieve an annual production of 12,000 tonnes of pork, which is equivalent to the meat gained from the slaughter of 170,000 pigs (Mridul,2023). Moreover in 2024, in Alephs Farm has been granted approval to commercialize its cultured meat. Nevertheless, twelve European countries, including Italy, France, and Australia, as well as certain American states, including Alabama and Florida, have prohibited the consumption of cultured meat in 2024. Figure 1 presents a concise overview of the progression flow diagram for cultivated meat.

Figure 1. Historical aspect of cultured meat

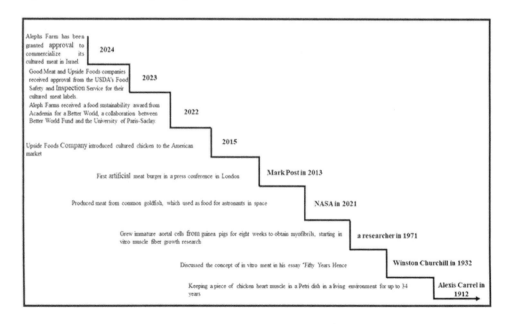

3. FUNDAMENTAL

Myogenesis, the process by which muscle tissue is formed, is the first step in the production of tissue cultured meat (CM). Multipotent myoblasts from the mesoderm layer undergo a number of processes during embryonic development, including fusion, proliferation, and differentiation, to form myotubes, which eventually unite to become muscle fibers (Kantono et al., 2022). Making CM is therefore very similar to growing skeletal muscles.

4. PROCESSING AND TECHNOLOGY

4.1 Cellular Tissue Origin (Cell Sourcing)

Cultured meat processes rely on the extensive proliferation of cells to generate sufficient biomass. Primary cell sources for CM biomanufacturing, including adult stem cells (ASCs), are obtained via biopsy or postmortem tissue from the designated location of the animal species of interest. The second alternative is to use pluripotent stem cells. ASCs have a limited replicative capacity of 50–60 divisions. The ASCs can be categorised into three types of satellite stem cells: myosatellite cells (MCs), adipose-derived stem cells (ADSCs), and mesenchymal stem cells (MSCs). Myosatellite cells have the capacity for self-renewal and proliferation, but their differentiation potential is limited to muscle cells. The second one is adipose tissue-derived stem cells, which develop from subcutaneous fat in adipose tissues and may develop into adipogenic, myogenic, chondrogenic, or osteogenic. The last one of ASCs, namely mesenchymal stem cells (MSC), are multipotent cells, which can be derived from non-muscle tissues such as bone marrow, adipose tissues, and placental tissues (Jervis et al., 2019), which can differentiate into different kinds of cells such as myocytes, adipocytes, and fibroblasts (Okamura et al., 2018). However, the regenerative properties of MSCs declined over time. Additionally, the myogenic differentiation of MSCs alone is inadequate. Therefore, MSCs are employed for co-cultivation with myoblasts, which secrete many growth factors that play a role in muscle regeneration and stimulate myoblast migration, proliferation, and differentiation. Pluripotent stem cells are categorised into embryonic stem cells (ESCs) and induced pluripotent stem cells (iPSCs). ESCs are obtained from the inner mass of blastocysts in the early stages of embryonic development. They possess pluripotent characteristics (Williams et al., 2012), meaning they can differentiate into the three primary germ layers: ectoderm, mesoderm, and endoderm. These ESCs are the best source since they can differentiate and proliferate into any cell type without restriction (Reiss et al., 2021). The challenges are somewhat determined by the cell type selected, including primary stem cells, which invariably lose proliferation and differentiation capacity during long-term culture (loss of stemness), whereas the stable maintenance of pluripotent cells can necessitate complex, expensive medium formulations (Bar-Nur et al., 2018). Invariably, primary cells undergo a senescent state during long-term culture, which is defined by permanent cell-cycle exit, widespread gene expression changes, and remarkable cellular flattening and enlargement (Di Micco et al., 2021). Decreases in proliferation are observed even before complete senescence. Proliferation rates in bovine SCs decrease to approximately 0.6–0.8 population doublings (PDs) per day until cells enter senescence at 20 to 30 PDs (Stout et al., 2022). To produce industrially cultured meat, the proliferative capacity must be significantly increased beyond these constraints (Melzener et al., 2021). Understanding and overcoming this obstacle require an understanding of the causal link between cellular senescence or death and decreased proliferation. The aging process involves a variety of cellular changes, including the accumulation of mutations at the genetic and epigenetic levels, changes in metabolism and morphology, and changed signaling system activity (Ogrodnik, 2021).

Table 1. Different cell types from different animal species for cultured meat production

1	Types of cells	Characters of cells	Disadvantages	Site of cells	References
Cattle	1. Adult stem cells (AS) (progenitor cells)	These cells may develop into adipogenic, myogenic, chondrogenic, or osteogenic.			Okamura et al., 2018 Witt et al., 2017
	a. Adipose tissue-derived stem cells	Have ability to self-renew and proliferate.	they can only differentiate into muscle cells.	subcutaneous fat in adipose tissues	
	b. Satellite stem cells (Myosatellite cells, myoblast cells)	Multipotent cells, have the ability to differentiate into different kinds of cells such as myocytes, adipocytes, fibroblasts and chondrocytes.		Skeletal muscle cells	

continued on following page

Table 1. Continued

1	Types of cells	Characters of cells	Disadvantages	Site of cells	References
	c. mesenchymal stem cells (MSCs)		The myogenic differentiation of MSCs alone is inadequate. Therefore, MSCs are employed for co-cultivation with myoblasts, which secrete many growth factors that play a role in muscle regeneration, stimulate myoblast migration, proliferation and differentiation.	Bone marrow, adipose tissue and placental tissues	
	2. Pluripotent stem cells	It is the best origin because they can differentiate and proliferate to all cell kinds without limits.			**Reiss et al., 2021**
	a. Embryonic stem cells (ES)			Inner cell mass of the blastocyst stage of embryo	
	b. Induced pluripotent stem cells (iPSCs)			iPSCs are obtained by inducing adult somatic cells using a set of identified pluripotency factors	
Chicken	Chicken muscle satellite cell				**Siddiqui et al., 2022**
	1-Satellite cells from the slow muscle	differentiate into slow and fast muscle fibers.			
	2--Satellite cells from the fast muscle	only differentiate into fast muscle fibers.			
Fish	1. Continuous fish cell lines (CAM) derived from *Carassius auratus* and *cromileptes altivelis*				**Li, Guo, & Guo, 2021;**

continued on following page

Table 1. Continued

1	Types of cells	Characters of cells	Disadvantages	Site of cells	References
	2. Fathead minnow (FHM) cell line from *pimephales promelaus*				Chen et al., 2020
	3. Cell lines developed from muscle cells of: a) *Danio rerio* b) *Paralichthys olivaceus* c) *Lates calcarifer*				Vishnolia et al.,2020 Peng et al., 2016 Lai et al., 2008

4.2 Seeding of Cells on Scaffold (Scaffolding of Cells, Cell Adsorption on Scaffold)

Stem cells (progenitor cells) are seeds on a scaffold; a biomaterial substrate used in the preparation of cultured meat. Scaffolding biomaterials, complex frameworks, are added to the culture media to provide structural support to the cells, enhance nutrient transfer, and facilitate cellular respiration (Schätzlein & Blaeser, 2022). Furthermore, scaffolding is required to replicate traditional meat tissue's 3D configuration by forming dense, integrated saturated meat with medium perfusion and a vascular system (Seah et al., 2022). Furthermore, scaffolding can promote cell cultivation without the use of serum, which can be achieved by incorporating bioactive chemicals into an edible scaffold rather than introducing them into the culture medium (Chen et al., 2024). Bioactive, flexible, having a large surface area, allowing growth media diffusion, and being edible, non-toxic, and allergen free are the fundamental features of scaffolding employed in cultured meat technology (Alam et al., 2024). The natural extracellular matrix (ECM) is composed of proteoglycans, collagen, and glycoproteins. Consequently, proteins and polysaccharides are expected to be the primary components of scaffold biomaterials (Bomkamp et al., 2022). For CM production, the ideal porosity range for scaffolds is 30% to 90%, with hole sizes ranging from 50 to 150 μm or larger. The scaffold thickness typically relies on processing methods (Bomkamp et al., 2022b). These scaffold biomaterials are animal- and plant-derived, as well as synthetic polymer biomaterials. ECM-enriched biomaterials such as elastin, gelatin, collagen, and fibronectin are all animal-derived biomaterials (Reddy et al., 2021), which are distinguished by their high extracellular matrix (ECM) content, ability to enhance cellular proliferation, and complete absorption by the human body. The second type of animal derived biomaterial, decellularized animal tissues, is mainly used in the biomedical field. Contrarily, plant-derived biomaterials are the best choice for developing meat biomaterials due to their nutritional content, low cost, good cellular compatibility, and perfect consumer acceptance (Ben-Arye, and Levenberg, 2019). For CM production, plant proteins, including soy, pea, zein, and glutenin, are abundant, competitive, and capable of being converted into scaffolding films with appropriate mechanical properties (Dong et al., 2004). Scaffold films, mainly produced from zein and glutenin, are used to stimulate the proliferation of stem cells and facilitate their differentiation into myotubes (Xiang et al., 2022). For the production of fibrous scaffolds, texture vegetable proteins (TVP) can be used (Bakhsh et al., 2022). Alginate, on the other hand, can be utilized to make self-assembling hydrogels with good mechanical characteristics, that disintegrate as cells move and secrete their own ECM (Sahoo and Biswal, 2021). For biomaterials that do not readily allow cell attachment, functionalization with short peptide sequences is a promising alternative. RGD-alginate is a well-studied tissue engineering system (Sandvig et al., 2015), and it could be a potential technique for cultured meat, however, additional peptide sequences that more closely approximate native ECM may be useful for attachment, migration, and maturation.

4.2.1 Scaffolding Types

1. Microcarriers

Microcarrier (MC) scaffolds are used for large-scale cell proliferation and are composed mainly of polystyrene, cross-linked dextran, cellulose, gelatin, or polygalacturonic acid (PGA) and coated with collagen, peptides containing adhesion motifs, or positive charges to promote cell adhesion. Their diameters are typically between 100 and 200 µm (Bodiou et al., 2020). According to Norris et al. (2022), larger MC enhances cell adhesion, while smaller MC results in higher growth rates because of increased shear stress. According to Bomkamp et al. (2022b), there are three possible scenarios for the use of microcarriers in cultured meat processing: first, cells are temporarily transported to MC to promote cell growth, and then they are withdrawn and processed. Second, a temporary carrier that dissolves or decomposes into free cells. Finally, the MC is an edible carrier that is incorporated into the finished product, which eliminates the costly cell harvesting steps and their corresponding yield losses, thereby reducing the cost of cultured meat. Hence, edible MC is a promising approach for cultured meat scaffolding that generates industry-scale cell mass while maintaining reduced costs (Levi et al., 2022). Edible biopolymers, including chitosan and alginate (Chui et al., 2019), starch (Zhang et al., 2017), zein (Li et al., 2016), and gelatin (Radaei et al., 2017), have been used to create MCs for various biomedical purposes. These MCs have the potential to be utilized in cultured meat production. For MC production, various technologies have been employed, primarily electrospray, spray drying, air jet milling, micro-grinding, dispersion polymerization, emulsion polymerization, photopolymerization, solvent evaporation, microfluidics, spherulitic crystallization, and air spraying (Morais et al., 2020). The production of microcarriers for cultured meat production is hampered by two challenges. The first is compliance with safety regulations regarding the use of food-grade and non-toxic crosslinkers, solvents, and surfactants in the formation of edible MC polymers. For instance, chitosan can be crosslinked using sodium tripolyphosphate (TPP) or genipin instead of toxic glutaraldehyde, and alginate can be crosslinked using $CaCl_2$ instead of $BaCl_2$. The final one is inadequate cell adhesion. To improve adherence, biopolymers can be modified with functional domains like RGD or integrin-recognized sequences (Yang et al., 2014), bioactive polymers (Chui et al., 2019), or crosslinked (Chui et al., 2019; Radaei et al., 2017; Zhang et al., 2017). Pig skeletal muscle, mouse muscle, and mouse adipose cells were grown on a 3D porous gelatin microcarrier (PoGelat-MC). Using 3D-printed molds and glutamine transaminase, minced pig muscle tissue was formed into centimetre-scale meatballs with excellent mechanical qualities and protein content (Liu et al., 2022). Additionally, MCs can be implemented in the production of cultivated flesh by facilitating cell proliferation or incorporating them into the final product. Matrix Meats (Brennan et al., 2021) and OMeat are two examples of commercial enterprises that have recently begun to develop edible MCs for use in cultivated meat.

2. Porous Scaffolds

Scaffolds with apertures that encompass a range of 10-100 μm (Zeltinger et al., 2001) are sponge-like structures that offer the mechanical stability necessary for seeded cells to establish tissues and deposit extracellular matrix. Moreover, porous scaffolds are particularly appealing due to their low cost and simple construction procedures. The scaffolds provide a three-dimensional platform for cell survival, proliferation, and maturation (Zhou et al., 2021). The size and distribution of scaffold pores play a vital role in cell culture during CM production. Larger pores are desirable for media perfusion as they facilitate the efficient transfer of nutrients and oxygen, mimicking the function of blood vessels (pseudo-vascularization) in CM (Singh et al., 2023). Additionally, a high surface-area-to-volume ratio and appropriate pore size contribute to achieving high cell density and help regulate cellular behavior (Carletti et al. 2011). However, the present porous scaffolds for cell-grown meat have challenges in matching food-grade material requirements with cell adhesion and proliferation capacities. Their cell differentiation efficiencies are also limited, which results in poor retention of typical meat properties like texture and nutrients (Chen et al., 2024). The selection of components and nutritional value is crucial for this scaffold, as it is designed to remain within the final product. Plant proteins, including soy (Ben-Arye et al., 2020), and plant polysaccharides, including cellulose (Abitbol et al., 2016), are frequently employed as scaffolding materials. The extrusion of soy protein powder results in the production of textured vegetable protein (TVP), a vegetable protein that is used as scaffolds for cultured bovine skeletal muscle. Coating the extruded TVP with fibrinogen enables effective adherence of bovine satellite cells, resulting in a cell seeding efficiency exceeding 80% (Ben-Arye et al., 2020). Moreover, the rapid freezing of a solution of hyaluronic acid and gelatin results in the formation of ice crystals concurrently with the formation of cross-links, which in turn leads to the formation of a porous scaffold that exhibited more than 90% porosity and was able to support the attachment, proliferation, and differentiation of porcine adipose-derived stem cells (Chang et al., 2013).

3. Hydrogel Scaffolds

Hydrogels are 3-D crosslinked hydrophilic polymer matrixes characterized by high water-absorption capacity. In this network, water is the dispersion phase and makes up at least 70% of the gel weight (Tan and Joyner, 2020). Hydrogels are created by the process of physically or chemically crosslinking synthetic, natural, or copolymers. Cell proliferation, motility, and differentiation are significantly influenced by hydrogel stiffness and diffusion kinetics. This is because an optimal rate of diffusion of micronutrients and signaling molecules is necessary to penetrate the hydrogel's thickness and reach the growing cells, which supports cell proliferation and growth. Conversely, high hydrogel stiffness restricts cell proliferation and migration (Freeman and Kelly, 2017). These hydrogels possess impressive mechanical capabilities, but they lack biocompatibility and adaptability. Additionally, they exhibit some level of cytotoxicity and may pose a risk to food safety (Ye et al., 2023). As a result, the focus of current research has shifted to natural polymer hydrogel. Natural polymer materials are typically derived from polysaccharides or proteins to form hydrogels (Ghanbari et al., 2021a). Proteins possess inherent advantages over polysaccharides in the development of hydrogels (Ghanbari et al., 2022). Proteins are composed of numerous amino acids, and numerous reactive groups can be employed as locations for chemical modification and crosslinking to generate polymer structures (Cuadri et al., 2016). Protein-based hydrogels have been widely developed and studied by researchers due to their excellent properties, which include high nutritional value, biocompatibility, biodegradability, adjustable mechanical properties, and low toxicity when compared to synthetic polymers (Farwa et al., 2022). Collagen, silk fibroin, and gelatin are among the most used protein hydrogel materials. However, most of these proteins are animal proteins with large application costs, and because of their complicated structure, structural alterations are frequently limited (Ghanbari et al., 2021b). Furthermore, plant-derived proteins may be safer than animal-derived proteins since they are less likely to transmit zoonotic infections (Surya et al., 2023). Soybean protein, being one of the most prevalent plant protein sources, is high in nutritional content, environmentally friendly, and available from a variety of sources, making it widely employed in the food business (Liu et al., 2023). Additionally, carrageenan, extracted from seaweed, can be utilized to produce food-grade hydrogels, which are extensively employed in meat processing (Yegappan et al. 2018).

4. Fibrous Scaffolds

Nanofibers are produced using spinning techniques such as electrospinning and have features that promote cell functions, including adhesion, penetrability, and a three-dimensional structure. Spinning processes can be utilized on various materials, such as soy (Phelan et al., 2020), gelatin (MacQueen et al., 2019), and polystyrene (Lerman et al., 2018). The fibres that are produced through spinning techniques are analogous to natural substances. The fibrous structure's advantage is its resemblance to the texture of a piece of meat, which can be advantageous for the 3D assembly of the meat and its flavor (Kolodkin-Gal et al., 2023).

4.2.2. Scaffolding Biomaterials

The structure and properties of scaffolds are influenced by the biomaterials used for scaffolding. Typically, scaffolding biomaterials have great porosity, biocompatibility, and ECM mimicking, as well as mechanical strength to direct cell adhesion, proliferation, and morphological changes (Sharma et al., 2015). Polysaccharides and proteins are two of the most frequent biopolymers with extracellular matrix-like properties. Edible biomaterials, derived from natural sources, have garnered considerable attention and are widely utilized in the field of cultured meat. This is mainly due to their abundant availability and the fact that they closely mimic real tissue in terms of chemical and biological properties (Su et al., 2021). Proteins and polysaccharides, which are polymers, are considered the fundamental components of scaffold biomaterials. These polymers may originate from natural or synthetic sources. Thus, polymer-based biomaterials are regarded as a promising option for scaffold fabrication (Khan and Tanaka, 2018). The FDA has certified many polymers for food use, confirming their safety. The edible polymers that have been approved by the FDA include pectin, chitosan (CS), gluten, gellan gum (GG), cellulose, gelatin (GL), collagen (COL), soy protein isolate (SPI), starch, glucomannan, and alginate (Alg) (Ali & Ahmed, 2018).

1. Natural Polymer (Biopolymer)

Natural polymers are materials derived from natural sources. They are divided into protein-based and polysaccharide-based biomaterials. Both types are naturally occurring polymers of animal and plant origin. Natural polymers are biocompatible, toxic-free, cell-adherent, and promote proliferation and differentiation. Despite this, their mechanical strength is low, and they are vulnerable to high temperatures (Del Bakhshayesh et al., 2018).

a. Protein-Based Biopolymer (Animal or Plant-Derived Protein Biopolymer)

Animal protein, plant protein, and fungal protein are the sources of protein-based scaffold biomaterials. Soy, pea, zein, and glutenin are plant proteins that are abundant, competitive, and capable of being converted into films with the necessary mechanical properties for the development of CMs (Dong et al., 2004). The growth of aligned cells and the subsequent development of aligned myotubes were effectively stimulated by the protein films composed of zein and glutenin (Xiang et al., 2022). So zein and glutenin may be promising candidates for future research in the production of CM. Textured vegetable proteins (TVP) are currently in high demand in a variety of culinary goods, including frozen dumplings, ham, sausages, and fish balls (Zhang et al. 2017). This suggests that global acceptance and satisfaction with TVP products are gradually increasing (Jones, 2016). Moreover, TVP may be implemented to generate fibrous scaffolds (Bakhsh et al., 2022). The absence of the arginine-glycine-aspartic acid (RGD) sequence in the plant protein scaffold hinders its ability to adhere to cells. To address this, Lee et al. (2022) applied a coating of fish gelatin/agar matrix onto textured vegetable protein, creating a more favorable environment for cell adhesion. Despite its high biodegradability and low cost, plant material has the potential to induce allergic reactions (Post, 2014). Hence, decellularized plants have lately been the focus of researchers interested in an alternative plant source (Thyden et al., 2022). Decellularizing tissues create an extracellular matrix with a vascular network that transports nutrients and oxygen (Contessi et al., 2020). Decellularized plant-based tissues show a natural fluidic transport system with plant arteries diverging from big, major veins into small capillaries. This system resembles mammalian tissue's branching vascular network (Harris et al., 2021). As a result, the unique structural properties of decellularized plant tissues were identified as promising scaffolding for cultured meat. To circumvent biopolymer-based edible scaffold problems, researchers have produced edible scaffolds from decellularized apple hypanthium and spinach leaves (Modulevsky et al., 2016; Jones et al., 2021). For cultured meat production, decellularized plant tissue scaffolds are a good choice because they do not contain any animal components, are cost-effective, ecologically friendly, easily scalable, and provide the necessary morphological and biochemical microenvironment for growing muscle cells (Jones et al., 2021). Broccoli, sweet pepper, spinach leaves, and green onion were among the plant tissues that were used for decellularization. Decellularized amenity grass was employed by Allan et al., 2021 for in vitro myoblast culture. The grass scaffold's striated topography facilitated the alignment and differentiation of myoblasts, which were preserved by their natural long, narrow structure and parallel vasculature system. Moreover, broccoli florets were chosen to serve as decellularized microcarriers in bioreactors to facilitate the scalability of cell proliferation (Thyden et al., 2022). Another protein biopolymer source is of animal origin. Animal-derived biomaterials, including elastin, gelatin, collagen, and fibronectin, have a high ECM content and promote cellular development. They are also fully absorbed by the human body. Nevertheless, scaffolds of a single material have inferior mechanical properties and certain collagen, produced from fish skin, and gelatin, derived from pigs and cowhide, components have challenges in terms of sustainability because of their high cost and vulnerability to ethical and environmental problems (Li et al., 2022b). A collagen gel-based meat model including smooth muscle cells (SMCs) was recently published by Zheng et al., 2021. In this model, SMCs reduced pressure loss, increased collagen, and made meat firmer, springier, and chewier than controls. These findings show that SMCs improve cultured meat texture by generating ECM proteins. Thus, another scaffolding biomaterial, such as a polysaccharide biomaterial, including alginate, is employed in conjunction with natural animal protein scaffolding, like Enrione et al., 2017 constructed an edible porosity scaffold utilizing freeze-drying technology that incorporated salmon gelatin, agar, and sodium alginate. This scaffold allowed muscle stem cells to adhere and grow, resulting in the necessary myogenic responses. Furthermore, by employing electrospinning, porcine gelatin, TG enzyme, and chemical crosslinking, the resulting microgelatin fibres facilitated the growth and alignment of muscle cells in a single orientation (Mendes et al., 2017). Moreover, Park et al. have recently devised a technique for producing enhanced cultured meat by utilizing fish gelatin's MAGIC powder and myoblast sheets. The powder, characterized by its edible gelatin microsphere (GMS) structure, displayed changes in shape and connection depending on the process of crosslinking. The researchers discovered that GMSs greatly improved the cultivation of myoblast sheets, resulting in more efficient cell sheets with meat-like properties compared to conventional methods. Due to the varied surface qualities resulting from crosslinking, the production of GMSs on a large scale was simply achieved. This research also determined that the quality of lab-grown meat, improved using GMS cell sheets, is similar in tissue characteristics to both soy-based meat and chicken breast (Park et al., 2021).

b. Polysaccharide-Based Scaffold Biomaterial (Animal or Plant-Derived Polysaccharide Biopolymer)

Certain plant polysaccharides, including alginate, pectin, konjac gum, and cellulose, possess the potential to serve as valuable biomaterials due to their physiological roles and excellent cellular adherence. In addition to plants, specific types of bacteria and algae also synthesize cellulose. Cass Materials, an Australian start-up company, is now investigating the application of fermented bacterial nanocellulose as scaffolding for CM. Initial studies have shown promising results, suggesting that muscle cells can attach to the very porous scaffolds and develop fibres (Le, 2020). Nevertheless, most of the cellulose-based scaffolds exhibited a porous structure, except for the green algae Cladophora, which was predominantly fibrous (Bar-Shai et al., 2021). Another potential scaffolding material for CM is alginate, a polymer made from brown algae. Pluripotent stem cells can be cultured in alginate-derived tubes, which are compatible with differentiation techniques and enable high cell densities and growth rates (Li et al., 2018). An edible three-dimensional scaffold (CS-SA-Col/Gel) composed of chitosan, sodium alginate, collagen, and gelatin were created by Li et al., 2022a. A robust cultured cell meat (CCM) model with strong adhesion sites was constructed utilizing a 3D 2-CS-SA-Col1-Gel scaffold that was created via freeze-drying and electrostatic interactions. This scaffold successfully promotes the growth of pig muscle cells. Not only that, but the look and texture qualities (such as chewiness and resilience) of this structured CCM model were quite like those of fresh pork.

2. Synthetic Polymers

Polyglycolic acid (PGA), polylactic acid (PLA), poly DL-lactic co-glycolic acid (PLGA), polycaprolactone (PCL), and polyethylene glycol (PEG) are synthetic polymers (Biswal, 2021). In contrast to natural polymers, synthetic polymers provide exceptional mechanical strength, which is essential for supporting tissue growth. However, these materials inhibit cell growth, generate hazardous chemicals after degradation, and have poor cell adherence (Tessmar and Gopferich, 2007). Furthermore, the hydrophobic nature and lack of cell recognition sites limit synthetic biopolymers, such as the arginyl glycyl aspartic acid (RGD) peptide motif (Tallawi et al., 2015), which is not yet approved for human consumption, limiting its use in in vitro meat production (Singh et al., 2023). Biomaterials in the food sector are limited because of their non-edible nature and susceptibility to deterioration, potentially toxic to tissue (Bomkamp et al., 2022b). The General Standard for Food Additives allows for a maximum amount of 1-70 g/kg of polyethylene glycol to be added (Codex Alimentarius Commission, 2021). Moreover, FDA-approved edible scaffolds include gelatin, chitosan, pectin, cellulose, starch, gluten, alginate, and glucomannan, all of which are natural biopolymers (Singh et al., 2023).

3. Self-Assembling Peptides (SAPs)

Self-assembling peptides (SAPs) have been investigated and utilized for tissue engineering scaffolds and 3D bioprinting materials due to their versatility and ECM-mimicking properties (Grey et al., 2022). SAPs are made up of monomers that can conform into structures according to the environmental features around them, allowing for use in a variety of functions (Lee et al., 2019). Self-assembly can be tailored for specific applications by changing the nature of peptide sequences, while more robust and complex materials with advanced design features are feasible by simple crosslinking with biological macromolecules (Hao et al., 2022). Amino acid side chains offer sites for chemical alterations, producing diverse supramolecular structures and adaptable hydrogels. These hydrogels can gain properties like shear-thinning, bioactivity, self-healing, and shape memory, expanding self-assembling peptide material applications. Supramolecular peptides can structurally assemble into nanofiber hydrogels based on distinctive building blocks. These hydrogels serve as nanomorphology-mimetic scaffolds for tissue engineering. Biochemically, peptide nanofiber hydrogels can have bioactive motifs and factors either covalently tethered or physically absorbed into them, providing various functions based on physiological and pharmacological needs (Hao et al., 2022). Self-assembling peptides known as CH-01 and CH-02 have been used to produce hydrogels that can act as scaffolds. The hydrogel was found to successfully mimic ECM and display a nanofibrous structure like that of collagen in natural meat. The hydrogels were able to support the adherence and proliferation of muscle myoblasts (Arab et al., 2018), suggesting a viable option in cultivated meat scaffolding. The utilization of SAPs in cultivated meat remains unexplored in the existing literature, despite their use in tissue engineering. This may be attributed to the high cost of conventional peptide synthesis, which could restrict the conduct of further research. Potential strategies to reduce the cost of SAP production for cultivated meat scaffolding include the optimization of current approaches by recombinant organisms. Additionally, cell-free systems (Zhao & Wang, 2022), which eliminate the necessity for microbial hosts, present another potential method for SAP production.

4.2.3. Scaffold Fabrication Methods

During the design and fabrication of the scaffolds, the practicality and requirements of the mechanical, biological, and physicochemical features are considered. Pore interconnectivity, form, pore size, porosity, strength, and degradation rate are critical factors that influence the manufacturing of scaffolds. The top-down and bottom-up approaches are the two main techniques employed in the production of scaffolds. The top-down technique involves the initial construction of scaffolds, followed by the subsequent embedding of cells within their microstructure. The top-down strategy for fabricating three-dimensional (3D) bio-scaffolds involves many processes, including electrospinning (Lannutti et al., 2007), phase separation (Liu and Ma, 2009), lyophilization (Eltomet et al., 2019), and self-assembly (Nie et al., 2017). Conversely, the bottom-up method emphasizes the creation of small-scale tissue components with precise micro-architecture, which are then combined to produce larger tissue structures (Lu et al., 2013). These building blocks can be created utilizing many methods, such as generating cell sheets (Lee et al., 2018), self-assembling cell aggregates (Napolitano et al., 2007), encapsulating cells in hydrogels (Wodarczyk-Biegun et al., 2016), and bio-printing cells (Park et al., 2017).

1. Solvent Casting

The simplest and most common method for scaffold synthesis is solvent casting, in which biopolymers are dissolved in an organic solvent, poured into a mould, and allowed to evaporate to form a thin, sheet-like scaffold (Deb et al. 2018). In the food sector, food-grade alcohol is the organic solvent of choice (Mancuso 2021).

2. Electrospinning

The electrospinning (ESP) technique creates a fibrous structure with fibre diameters ranging from 10 nm to microns, which could be exploited to build edible scaffolds for in vitro meat production (Seah et al., 2022). Nanofiber scaffolds in this approach are highly porous, have a large surface area, and mimic natural extracellular matrix properties. In addition to forming aligned fibres that may facilitate muscle fibre development, these nanofibers have the capacity to facilitate cell adhesion and oxygen and nutrient diffusion (via the spaces between fibers). Here, an electrically charged jet applies force to the tip of the needle, causing the polymeric droplet to form. The spinneret droplets burst and are stretched when they travel through the grounded collector from the spinneret tip when the charging solvent is subjected to high voltage, where an interplay between electrostatic repulsion and surface tension takes over. As the solvent began to evaporate, the jet finally hardened into nanofibers (Gañán-Calvo et al., 1997). A variety of scaffold biomaterials, such as polylactic acid (PLA), poly (lactic-co-glycolic acid) (PLGA), polycaprolactone (PCL), gelatin methacryloyl (GelMA), fibronectin, albumin, and gelatin, can be processed using spinning techniques (Bomkamp et al. 2022b). However, collagen, gelatin, whey protein, chitosan, cellulose, and starch are edible materials that can be considered in the food industry (Levi et al., 2022). Using immersion rotation jet spinning technology, MacQueen et al., 2019 cultured rabbit skeletal muscle cells and bovine aortic smooth muscle cells on a fibre scaffold composed of porcine gelatin,

TG enzyme, and the chemical crosslinking agent EDC/NHS to produce meat-like products. The mature alignment of both types of muscle cells within anisotropic 3D muscle structure was confirmed by their adhesion to the gelatin fibers. This introduces a novel concept for the large-scale production of cultured meat.

3. Three-Dimensional Bioprinting (3D Bioprinting)

3D bioprinting is distinctive because it generates intricate and customizable structures in layers using 3D digital models created with computer-aided design (CAD) software. 3D bioprinting technology could be useful for both small-scale and large-scale production of customised cultured meat (Stephens et al., 2019). Additionally, light-assisted printing, inkjet printing, and extrusion printing are prevalent bioprinting techniques (Kacarevic et al., 2018). The majority of bioprinting processes use extrusion printing, which dispenses bioink as continuous filaments (fibres) instead of droplets using pneumatic pressure or a mechanical screw plunger. Bioink, a combination of various scaffold biomaterials, plays an essential role in 3D printing by creating the scaffolding necessary for the differentiation of stem cells into meat (Sun et al., 2018). Moreover, the printed scaffold offers a micro-milieu and a habitat for the developed muscle cells, which are usually cultivated in bioreactors that enable the transportation of nutrients on a large scale (Bishop et al., 2017). Vat photopolymerization-based, extrusion-based, and jetting-based bioprinting are the primary methods in 3D bioprinting. The bioink is deposited with high precision in extrusion-based bioprinting, resulting in customized 3D structures with excellent structural integrity. This is achieved through the continuous deposition of filaments. In extrusion-based bioprinting, the bioink is deposited with high precision, obtaining customized 3D structures with good structural integrity due to the continuous deposition of filaments. The entire process of bioprinting is carried out under the control of a computer (Ozbolat and Hospodiuk, 2016). Jetting-based bioprinting can produce ink droplets with a controllable size and low volume, depositing the ink in specific locations with high precision and without contact. Employing this technique, it is possible to use a variety of biomaterials as well as the incorporation of living cells (Li et al., 2020). Finally, Vat polymerization-based bioprinting is an emerging technology in the biofabrication of scaffolds applied in tissue engineering, used for its high resolution compared to other bioprinting technologies. Ink particles with a controlled size and low volume can be generated through jetting-based bioprinting, which deposits the ink in particular locations with high precision and without contact. This method enables the incorporation of living cells and a diverse array of biomaterials (Li et al., 2020). Lastly, vat polymerization-based bioprinting is a new technology that is being used in the biofabrication of scaffolds for tissue engineering. This technology is valued for its superior resolution in comparison to other bioprinting technologies. A cereal prolamin ink for 3D printing was generated using zein, a protein derived from barley and rye, and utilized to produce a fibrous framework that enables the attachment and growth of C2C12 and pig skeletal muscle satellite cells (Su et al., 2023). Moreover, Dutta et al. (2022 developed a bioink from alginate and gelatine-based hydrogel scaffolds with plant- or insect-derived hydrolysates for bovine myosatellite cells and produced cell-based pepperoni meat prototypes $20 \times 20 \times 5$ mm in size. Also, Liu et al. (2021b) created a 15×15 mm 3D-printed structure for porcine skeletal muscle satellite cells using sodium alginate-gelatine and gelatine-methacrylate (GelMA)-silk fibroin. In a gelatine-based gel, Xu et al. (2023) produced a cell-based fish fillet measuring $20 \times 12 \times 4$ mm utilizing piscine adipocytes and satellite cells. Prior studies highlight the benefits of integrating 3D bioprinting with biomimetic scaffolds derived from authentic tissue architectures to produce cell-based meat products. Recently, companies have used 3D bioprocessing technology for the biofabrication of cultured meat. 3D Bioprinting Solutions, in collaboration with KFC (Kentucky Fried Chicken), intends to manufacture and promote lab-grown nuggets. 3D Bio-Tissues Ltd. (3DBT) has formed partnerships with CPI (Independent Centre for Technological Innovation) and the United Kingdom government's High-Value Manufacturing Catapult to enhance cell culture media for the cultured meat business. Nissin Food Holdings, a Japanese company, is collaborating with the University of Tokyo to create printed meat cubes (State of the Industry Report, 2022). 3D printing technology has been used to successfully make several food varieties, such as chocolate, cakes, and breads, in addition to cultured meat (Li et al., 2022).

4. Freeze-Drying (Lyophilization)

Freeze drying is a method of drying polymeric solutions in which a substance is frozen to an extremely low temperature and then the surrounding pressure is lowered, allowing the frozen water to sublime (Deb et al., 2018). The process is distinguished by three distinct steps: the initial step involves the preparation of the polymer solution; the second step involves the casting or molding of the polymer solution; and the final step involves the freezing and drying of the polymer solution at low pressure. Sublimation and desorption are employed to extract the ice and unfrozen water, respectively, during the third phase. Freeze drying can generate scaffolds with pore sizes ranging from 20 to 200 μm and a porosity of approximately 90%. Temperature, polymer concentration, and freeze rate regulate the size of pores (Deb et al., 2018; Seah et al., 2022). So far, the freeze-drying technique has been used for preparing the scaffold using synthetic non-edible and edible polymers, which should be further investigated and replaced with plant- and biopolymer-derived edible scaffolds (Shit et al., 2014; Bomkamp et al., 2022).

4.3. Cells Scaling Up

Massive cellular proliferation is essential to producing cultured meat. There are two types of scaling up cells: the first is scaling up based on scaffolding material and the second is a scaffold-free approach.

a. Cells Scaling Up Based on Scaffold Material

Bioreactors, a fundamental cell expansion technology, are responsible for the provision of the requisite stimuli and capacity to achieve the scale-up of cell sources for cultivated meat production. They are a controlled environment that contains a nutrient-dense medium containing essential amino acids, carbohydrates, and growth factors. It can also facilitate nutrient diffusion and cell development by stimulating or agitating the cells to promote their maturation and proliferation. In the initial phases of cultivated meat production, when cell proliferation is a top priority, a bioreactor is indispensable for the facilitation of large-scale cell culture as well as the simplification of medium recycling and replacement during the proliferation stage. Optimal culture conditions can be ensured by managing the biological conditions. To guarantee that the media is saturated with dissolved oxygen, oxygen can be introduced through sparges or upstream aeration. The pH is maintained at 7.2–7.4 by monitoring the carbon dioxide concentration as detected by sensors (Santos et al., 2023). Currently, the three primary bioreactor varieties are classified according to the method of medium introduction into the main vessel of the bioreactor: batch, fed batch, fed batch, and continuous (Spier et al., 2011). Also, the bioreactors can be classified according to how they mix their contents. Adding mixing to the bioreactor system promotes cell growth and development. Mechanical bioreactors mix with agitators or impellers. Currently, stirred tank bioreactors are the most common bioprocess scale-up bioreactors (Martin et al., 2004). Stirred tank systems are a viable bioreactor type for scaling up cultured meat production due to their proven reliability and scalability. The fundamental challenge in these societies is to ensure efficient delivery of nutrients and the elimination of waste at high cell concentrations. Bioreactor yield and operating cost are both affected by cell density, making it an important operational aim. Consequently, developing media formulations that sustain fast growth while reducing the formation of growth-inhibitory compounds such as ammonia is a significant biological problem (Kolkmann et al., 2022). Interventions to metabolically remodel cells for improved bioprocess adaptability can be better designed with an understanding of the mechanisms involved in nutrition and waste metabolite sensing (Shyh-Chang and Ng, 2017). While glucose is essential for anabolic activity, too much of it might inhibit cell growth (Furuichi et al., 2021). When glucose is present, it activates mTORC1, which in turn increases cellular anabolic responses (Leprivier and Rotblat, 2020). To keep growth rates constant, amino acid supplementation is essential, as it converges on mTOR signaling as well. Bioprocess optimization could be achieved by tinkering with the mTOR signaling network, which regulates cell size, proliferation rate, and glucose intake rate, among other cellular responses (Dreesen and ussenegger, 2011). Therapeutic options to target abnormally growing cells have emerged via inhibitor screening methods (Brüggenthies et al., 2022). Improving mTOR signalling in cultured meat through genetic or pharmacological means is an uncharted territory that could lead to fruitful discoveries in the long run, since cell proliferation is the end goal.

b. Scaffold-Free Approach (Cell Layering and Self-Assembly)

Layer-by-layer (LbL) assembly is a versatile and straightforward method for creating multilayer structures by self-assembly. It is feasible to create multilayer coatings with a precise structure and composition using a wide range of materials that are readily accessible. These coatings have various uses in the field of biomedicine (Zhang et al., 2018). This production process is rapid and easily expandable, and it has the capability to fabricate densely packed, multicellular, and textured tissues using standard culture plates without the need for a bioreactor. The three fundamental techniques for cell layering are stacking cell sheets, rolling a cohesive tissue sheet, and in situ deposition of cell-laden biomaterials. The initial method involves the utilization of a culture dish coated with a temperature-responsive polymer to create a tissue with many layers. The second approach involves enveloping an entire section of a slender tissue sheet around a cylindrical support and cultivating it until tissue fusion occurs. The third strategy involves the utilization of a handheld device to apply cell-laden biomaterials (Jo et al., 2021). Biomaterials are employed to enhance or inhibit cell adhesion, regulate cellular phenotypes, and offer three-dimensional structures for cell culture or co-culture. The biomaterials used for LbL assembly encompass a variety of substances, such as biomolecules, polyelectrolytes, particles, and colloids (Zhang et al., 2018). The successful co-culture of myoblasts and preadipocytes has already proven the possibility of using this method to construct meat-like tissues of varying dimensions and thicknesses. Scaffolds are unnecessary as the cells generate their own extracellular matrix (ECM), which remains intact and forms strong layers (Shahin-Shamsabadi and Selvaganapathy, 2022).

Figure 2. Steps of processing and technology of cultured meat

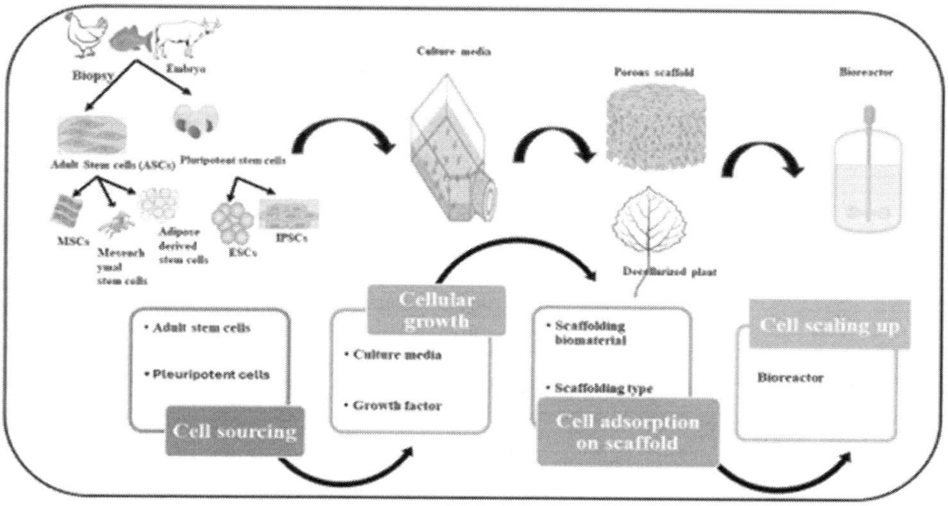

MSCs: myosatallite stem cells, ESCs: Embryonic stem cells, IPSCs: induced pluripotent stem cells.

5. ALTERNATIVES TO CULTURED MEAT

There are several alternatives to existing animal products in terms of food protein and energy sources:

5.1 Insects

In terms of converting biomass into protein or calories, edible insects can be produced more efficiently than conventional animals and potentially become a significant source of nourishment for humans (Tabassum-Abbasi et al., 2016). They include a lot of vitamins, protein, and fat (Nowak et al., 2016). Insects are more efficient at converting feed into consumable food than conventional meat, which only consumes 40% of the live animal weight. The fact that insects consume up to 100% of their feed partially explains this. As they are poikilothermic, insects consume less energy because they do not use their metabolism to heat or cool themselves. Compared with traditional cattle, they often have increased fertility, potentially yielding thousands of offspring (Premalatha et al., 2011). Rapid development rates and the fact that insects can achieve maturity in days, as opposed to months or years also contribute to efficiency (Alexander et al., 2017). A bone isotope study showed that insects have been a staple of human evolution (de Magistris et al., 2015), and a variety of species are currently consumed (> 2000 species) (Rumpold and Schlüter, 2013). However, the problem of limited consumer acceptance is widespread, especially in Western countries. The countries where a shift from eating animal products to eating insects would have the biggest effects are also the ones with high rates of animal product consumption per person. There are already indications that consumer perceptions may be beginning to shift in developed nations like the United States and the United Kingdom (Alexander et al., 2017). For instance, in the European Union, laws about novel foods and the permissible status of foods derived from insects dictate that insects cannot be processed and must instead be sold whole (de-Magistris et al., 2015). Insects are parasites or scavengers that primarily feed on grains. Their ability to scavenge indicates that they probably carry a variety of pathogens. Therefore, in addition to the standard food safety regulations for the development of insect-based foods, perceived pathogen risk overshadows the nutritional benefits of insects. The pathogens that could emerge in colonies that have been raised are not well understood. A recent study found that industrially raised mealworm and cricket samples, primarily composed of members of the Bacillus cereus group, contained a bacterial endospore fraction. Additionally, results indicated that norovirus genogroup II, hepatitis A virus, and hepatitis E virus were not found in the sample collection, suggesting a low risk of these viral pathogens affecting food safety (Vandeweyer et al., 2020). Inadequate handling, inappropriate culinary treatment, and eating insects at the wrong developmental stages are additional risks associated with eating edible insects (Kouřimská & Adámková, 2016). The EFSA states that the type of insect, the substrates utilised, the feed added to the rearing colonies, the production process, and the stage of harvesting of the insects all have a significant impact on the frequency and degree of contamination in insects and insect-based food products. Foods derived from insects and edible insects collected in the wild are also concerned about pesticide residues. Because the kinds of organic materials that wild insects eat are not regulated, they may eat pesticide-treated vegetation or crops, which could cause the pesticides to bioaccumulate in their tissues (FESA, 2015).

5.2 Microalgae

With an estimated 200,000 to 800,000 species, microalgae are a varied group that may grow quickly in a variety of settings when photoautotrophic conditions are met (Koyande et al., 2019) and can be cultivated in both warm and cold areas because of their remarkable resilience to harsh climatic conditions. Microalgae have demonstrated superior yields compared to those of conventional crops. By consuming nutrients found in wastewater, microalgae lessen their reliance on chemicals and freshwater (Brennan & Owende, 2010). Another significant feature that improves the use of microalgae as a continuous source of protein is their broad resistance to high pH and salt concentrations. Moreover, microalgae exhibit a remarkable capacity to concentrate vital nutrients and useful chemicals necessary for human well-being (Wells et al., 2017). These qualities have identified microalgae as one of the most reliable sources of protein. These were the subjects of studies conducted in the latter part of the 20th century that dealt with alternative agriculture and food. Up to 50–70% of the dry weight of microalgae is protein; this contrasts with 17.4% of beef, 19.2–20.6% of fish, 19–24% of chicken, 27% of wheat germ, 36% of soybean flour, and 47% of eggs (Koyande et al., 2019). The protein content of some high-protein microalgae (as % dry mass) has been reported as 60–71% (*Spirulina maxima*); 63% (*Synechoccus sp.*): 42–63% (*Spirulina platensis*), 53% (*Chlorella pyrenoidosa*), 57% (*Dunaliella salina*), 49% (*Dunaliella bioculata*), 52% (*Tetraselmis maculata*), 50–56% (*Scenedesmus obliquus*), 43–56% (Anabaena cylindrica), and 48% (*Chlamydomonas rheinhardii*) (Wang et al. 2021). Challenges: Lack of knowledge about the health benefits of microalgae and limited incentives for producers are major barriers to the successful use of this alternative protein source. Microalgae production has a minimal carbon footprint; if these challenges are successfully overcome, the use of microalgae in the food and nutraceutical industries could help meet the dietary protein needs of the growing world population and address climate change (Koyande et al., 2019). The unappealing taste, smell, and colour of microalgae alter the organoleptic qualities of processed foods, which presents a challenge for food scientists attempting to use them as food or as a source of ingredients (Verni et al., 2023). Another problem that needs to be solved to maximize the nutrient content of algae is the extraction of nutrients from microalgae. Bioactivity, bioavailability, and bioaccessibility are crucial factors to consider. The extracellular matrix (ECM), which is made up of pectin, cellulose, alginic, fibrillary peptidoglycan layers, and some polysaccharides (i.e., cellulose, hemicelluloses such as xyloglucan, mannans, glucuronan, beta-glucan, and lignin), is a multilayered, highly complex cell wall found in microalgae (Quesada-Salas et al., 2021). Sorting out the species that can react to cell wall disruption techniques is important because different microalgal species have varying degrees of cell wall complexity. Consequently, when selecting the microalgae species to be used as protein extraction subjects, it is essential to delve into the parameters that represent ease of cell disruption and protein recovery (Fatima et al., 2023).

5.3 Imitation Meat

Without utilizing meat products, imitation meat or meat analogues aim to replicate particular types of meat, including nutrients and aesthetic characteristics (such as texture, flavour, and appearance). The most popular fake meats, such as tempeh or tofu, are probably those made from soy (Malav et al., 2015), which has been producing and consuming tofu or soybean curd from coagulated soy milk for centuries. It can be further cooked to more closely resemble meat products in terms of flavor and texture. For example, flavoring can be added to make it taste like gammon, sausage, chicken, beef, or lamb (Malav et al., 2015). Soy and tofu contain high levels of protein while being low in fat (Sahirman & Ardiansyah, 2014). The Protein Digestibility-Corrected Amino Acid Score (PDCAAS) of soy and beef is similar, suggesting that their protein values in human nutrition are comparable (Reeds et al., 2000).

5.4 Aquaculture

Salmon and other carnivorous fish can eat up to five times as much fish as they eventually produce as feed (Ma et al., 2009), which poses a challenge to the growth of farmed carnivorous fish (Diana, 2009), thereby lowering the likelihood of significant replacement with current animal products. Because herbivorous and omnivorous species have much lower "fish-to-fish" conversion ratios—carp, for example, currently has a ratio of 0.1, and further reductions are predicted, this problem is less severe for these species (Tacon & Metian, 2008) because consuming feed derived from fish is not necessary for their nutrition—two-thirds of aquaculture's total output comes from freshwater systems, which dominate production. The primary species are either herbivorous or omnivorous, with carp producing the most, though more recently, tilapia and catfish have also become more popular (Bostock et al., 2010).

6. OPPORTUNITIES AND CHALLENGES

6.1. Opportunities for Cultured Meat

With the introduction of enriched and functional foods, consumers are more open to trying goods modified to have specific nutritional qualities (Burdock et al., 2006). There are numerous methods available for producing designer meat in vitro. It is possible to modify the culture medium's composition to affect the flavor of cultured meat and fatty acid makeup (Bhat et al., 2013), and the health benefits of meat can be increased by incorporating elements such as specific vitamins into the culture medium, which may have a positive impact on health, and co-culturing with different cell types may improve the quality of the meat. Supplementing with fats after production allows for better control over the ratio of saturated to polyunsaturated fatty acids and overall fat content (Bhat et al., 2013), and in vitro production systems have significant environmental potential. Because the circumstances in an in vitro meat production system are controlled and manipulable, it will not only significantly lessen ecological hazards but also ensure sustainable production of designer, chemical-safe, and disease-free meat (Bhat & Fayaz, 2011). In vitro, meat has the potential to significantly lessen animal suffering and eliminate the need for animal consumption (Hopkins & Dacey, 2008). Since a product from a bioreactor is not subject to the same environmental fluctuations as animal products and is not location- or soil-specific, it presents opportunities for new production locations or alternate land uses. This makes it a more dependable alternative. Moreover, animals are notoriously unreliable as a raw material to produce meat from a commercial standpoint due to illness, stress, and uneven growth (Bhat & Bhat, 2011). In the current meat production systems, it takes several weeks rather than months (for chickens) or years (for pigs and cows) before the meat can be harvested. The growth of meat in in vitro systems requires much less time than traditional meat production methods. This implies that less time will be needed to maintain the tissue, which will result in less feed and labor being needed per kilogramme of in vitro cultured meat (Bhat et al., 2013).

6.2. Challenges of Cultured Meat

While many people support in vitro meat because of its potential benefits to the environment and climate, as well as because animal activists support it, there are also concerns and criticisms surrounding it (Welin, 2013).

1. Sensorial Characteristics

The color and appearance of in vitro meat might be challenging to compete with those of conventional meat. In 2013, a sensory panel at London's Riverside Studios reported that the cultured meat produced and tasted was colorless. A small amount of red beetroot juice and saffron were added to the meat to improve its color (Bhat et al., 2015). Therefore, it is necessary to develop new meat processing technologies to improve the flavor and appearance of in vitro meat products. Initially, scaffold-derived tissue monolayers and yolk-like blobs of self-assembling muscle fibres were used to produce in vitro meat, which was subsequently used to prepare communal meat products. Nonetheless, numerous attempts have been made to employ tissue engineering techniques to create more enticing meat products by seeding scaffolds with muscle cells to produce the final product. Additionally, there have been attempts and proposals to develop scaffolds that enable 3-D tissue culture and complex meat structuring using edible and natural biomaterials like collagen (Hopkins & Dacey, 2008).

2. Alienation to Nature

The fact that the in vitro meat production system might make us more aloof from animals and the natural world and contribute to our urbanization is another issue. Cultured meat is compatible with our growing reliance on technology, which raises concerns about our growing alienation from nature (Welin, 2013). Less land will be impacted by human activity if livestock farming is abandoned, which is beneficial for the environment but may also drive humans away from it (Bhat et al., 2015).

3. Cost of Production and Economic Disturbances

The primary potential barrier is the extraordinarily high cost of cultured meat, although market penetration and large-scale production are typically linked to sharp price reductions. Industrial-scale in vitro meat production is only possible if a reasonably affordable method for producing a product that is qualitatively comparable to already available meat products is developed and given government support similar to that given to other agribusinesses (Bhat & Fayaz, 2011), and the economies of these countries that depend on the export of meat to other countries and engage in large-scale conventional meat production will undoubtedly be impacted by the production of meat in vitro. Employment in the agricultural sector will be affected by this technology in nations where cultured meat production has been widely adopted. These production centers will lessen environmental pollution because they are close to cities, which will save transportation costs, but perhaps this will not be so good for the countryside (Bhat et al., 2015).

4. Social Acceptance

One of the biggest obstacles to the public's acceptance of cultured meat is its unnaturalness (Welin, 2013). In vitro meat's unnatural nature worries some potential customers, but as Hopkins and Dacey point out, just because something is natural does not mean it is good for you (Hopkins & Dacey, 2008). Furthermore, consumers may view in vitro meat as fake meat rather than the real thing, which would make them devalue it similarly to how they would artificially flowers or synthetic diamonds (Hopkins & Dacey, 2008). Many opponents of the idea of producing meat in vitro are concerned that because this technology can culture human muscle tissue, it may lead to cannibalism with fewer victims (Hopkins & Dacey, 2008). Another argument is that original cells obtained from an animal in a morally dubious manner must be used to create in vitro meat and that doing so will morally contaminate all subsequent generations of tissue (Hopkins & Dacey, 2008).

7. LEGISLATIVE CONSIDERATION

7.1. Regulations Consideration

In numerous countries, the production and sale of cultured meat (CM) lacks clearly defined rules. So, food regulatory agencies should develop rules to promote the acceptance and commercialization of cell-based meat. In November 2018, the FDA, USDA, and Good Food Institute released a joint statement on CM rules (Bryant, 2020), indicating that a collaborative agreement might alleviate consumer trust concerns and promote future confidence. In 2019, the USDA and FDA announced a formal agreement to regulate cell-cultured meat products from livestock and poultry cell lines, adding clarity to the US regulatory approach. According to the agreement, the FDA will have regulatory authority over components of the manufacturing process that occur prior to cell harvest and product development. These stages include cell line isolation, selection, and banking, as well as cell proliferation and differentiation into specific tissue types. The USDA will oversee processing items downstream from harvest, including product testing, inspections, labelling, and safety evaluations (Fish et al., 2020). Regulatory approval has been granted to new products, including poultry breasts, one year later (Kantono et al., 2022). However, CM-producing countries have implemented thorough legislative measures to ensure consumer safety. The Singapore Food Agency (SFA) has released guidance on its safety assessment requirements for novel foods, outlining specific information submission requirements for the approval of cultured meat products. The sale of cultured chicken meat from Eat Just Inc. was authorized by the SFA on December 1, 2020, marking the first-ever approval of cultivated meat worldwide (Barbosa et al., 2023). In Europe, the EU Novel Food Regulation (EUNFR) specifically covers food products that are created using tissue culture techniques. Consequently, CM must obtain formal clearance from the European Food Safety Authority (EFSA) before it can be available for sale. Nevertheless, the precise details regarding the required tests and certifications for this product, such as food safety, disclosed components, and nutritional value, are still uncertain (Chodkowska et al., 2022). Moreover, Regulation (EC) No. 178/2002, General Food Law (GFL), outlines the process for awarding food permits in the European Union. The Novel Food Regulation (Regulation (EU) No. 2015/2283) governs the pre-marketing authorizations for foods derived from animal cells or tissue culture. Genetic engineering in the production of cultured beef may trigger the application of the Regulation on Genetically Modified Food and Feed (Regulation (EC) No. 1829/2003). Nevertheless, the United Kingdom (UK) has ceased to participate in the EU's common food authorization procedures. To sell its products in the United Kingdom, any cultured meat company must apply for authorization from the UK Food Standards Agency (FSA) starting in May 2021

(Barbosa et al., 2023). Food Standards Australia New Zealand (FSANZ), a legislative authority responsible for developing food standards in Australia and New Zealand, has recently suggested the inclusion of cell-based meat in their current Food Standards Code. Nevertheless, it is imperative to obtain specialized premarket certification (Tingwei et al., 2019). To sell their products in any country, cultured meat manufacturers must submit for inclusion in the approved new foods list, as per FSANZ regulations regarding novel foods. This necessitates an evaluation of the safety of the production process by FSANZ, which is anticipated to last for a minimum of 14 months. The purpose of the safety assessment is to verify and establish that the product does not pose a health risk (Barbosa et al., 2023). In Canada, the submission of comprehensive information in a pre-market approval application is required for cultured meats, which are classified as novel foods. This information must include molecular characterization, nutritional composition, toxicology, allergenicity, and the types and levels of chemical contaminants, all of which serve as evidence that the food is safe for consumption. However, in Japan, cultured meat has an existing regulatory structure and may not require pre-market approval, depending on the production procedure. The Japanese government is building a regulatory framework to ensure food safety and customer acceptance.

7.2. Religious Consideration

Religion may influence how people perceive CM. In Judaism, some rabbis would consider CM Kosher if the cells came from a Kosher slaughtered animal (Kenigsberg and Zivotofsky, 2020). According to Islamic views, the use of CM is permitted (halal) if the cells are obtained from an animal slaughtered in accordance with Islamic dietary laws, and the growth medium used to produce the cells is also halal (Verbeke et al., 2015). Contemporary jurists believe that cultured meat is Halal only if the stem cells are sourced from a Halal animal and no blood or serum is used (da Silva & Conte-Junior, 2024). However, Jews, Muslims, and Buddhists showed less preference for CM compared to traditional meat. Moreover, Hindus are likely to consider cultured meat as a way of avoiding harming animals, and some may decide it is permissible to consume as long as it is not beef, as cows are considered sacred animals (Kenigsberg and Zivotofsky, 2020). According to religious standards, Christian people can consume any type of meat because it is considered clean. Consequently, cultured meat may be consumed if it is available on the market (Jagadeesan & Salem, 2020).

8. CULTURED MEAT MARKET

8.1. Customer Preference

The major obstacle to the development of CM is the lack of customer preference. Furthermore, consumer demographics, including gender, age, country of origin, eating patterns, and other social parameters, also influence meat consumption (Liu et al., 2023a). Cultural attitudes towards CM will exhibit substantial variation, as indicated by certain studies. Nevertheless, CM effectively reduces greenhouse gas emissions (GHG), ensuring that consumers recognize CM as an environmentally sustainable option (Bryant et al. 2019). Moreover, as the occurrence of zoonotic infections such as salmonella and *E. coli*, which are linked to conventionally produced meat (Bryant et al., 2019), has decreased in the CM, certain customers consider the CM a safe choice. Furthermore, with a lower fat level in CM (Bouvard et al., 2015) than traditionally farmed meat, approximately 28.6% of consumers in Europe judged CM to be healthy (Kantono et al., 2022). Also, consumer attitudes towards the prospect of cell-cultured meat differ greatly among regions. In Europe, more attention is being paid to meat production's environmental, sustainability, and animal welfare issues, but it is uncertain how much of this attention will change consumer behaviours or receptivity to meat alternatives (Santeramo et al., 2018). In 2013, a questionnaire poll was conducted on the first cultured beef burger manufactured in the United Kingdom. About two-thirds of consumers expressed an interest in consuming it (Guardian, 2013). Surveys done in the United States and Italy revealed that two-thirds and 54% of respondents, respectively, were willing to try (WTT) cultured meat (Mancini and Antonioli, 2019). In a study conducted to assess consumer attitudes towards cultured beef in Germany (n = 1000) and France (n = 1000), most consumers expressed a willingness to purchase cultured meat as an alternative to traditional meat when it became available (Bryant & Barnett, 2020). A global sample of customers (n = 3,091) from numerous countries, including China, the United States, the United Kingdom, France, Spain, the Netherlands, New Zealand, Brazil, and the Dominican Republic, were surveyed to determine their tendency to try and purchase cultured meat. Food curiosity, the importance of meat in a diet, and consumers' realistic perception of cultured meat as a viable option all had a positive impact on their willingness to test, buy, and pay more for it (Rombach et al., 2022). In South Africa, Tsvakirai and Nalley (2023) examined how psychological motivators and deterrents affect customers' desire to sample cultured meat. The study found that implicit views influence neophobic and neophilic attitudes, while concerns about social, cultural, and economic disturbances may prevent adoption. Additionally, several customers studied expressed reluctance to support sustainable lifestyles due to associated costs. However, they argue that the government should coordinate these efforts. Despite this, Morais-da-Silva et al. (2022) found that 58.8% of respondents preferred plant-based protein over cultured meat. Furthermore, based on their perception of cultured meat as being unnatural compared to traditional meat, 204 customers showed limited acceptance of it (Siegrist et al., 2018). In general, the willingness to try, eat, or pay is regulated by the respondents' age, gender, degree of education, and countries of origin. However, we found several interactions between these factors. African respondents from the richest and most educated countries were more WTT cultured meat (Kombolo Ngah et al., 2023). In terms of health, taste, and naturalness, according to Francekovi'c et al. (2021), the term cultured meat elicits curiosity but can also cause emotional resistance, particularly in Croatia, Greece, and Spain, where the proportion is lower. Surveys conducted in the United States and Italy showed that 2/3rds and 54% of respondents, respectively, were willing to try cultured meat. Additionally, vegetarians believe that cultured meat is morally justifiable and could serve as a healthy substitute for meat. Vegetarians feel cultured meat is morally acceptable and a healthy meat substitute. In addition, research by Chuah et al. (2024) suggests that customers' preferences and willingness to pay (WTP) can be greatly influenced by education on the benefits of cell-based seafood. This, in turn, could increase the marketability of these products. Several factors influence customer preferences, including the following:

1. Gender

In comparison to women, men exhibit a higher level of interest, a willingness to try (WTT) (Bryant et al., 2020), and a WTP (Kantor & Kantor, 2021) for cultured meat. Nevertheless, in Germany and, particularly, in France, women exhibit a lower level of willingness to consume cultured meat than males (Bryant et al., 2020). On the other hand, gender did not influence the willingness to eat and willingness to try (WTE and WTT) of cultured meat among African respondents (Kombolo Ngah et al., 2023).

2. Country of Origin

The likelihood of purchasing cultured meat is higher among Chinese citizens (59%), followed by Indians (56%), than among Americans (30%), which implies that government incentives for investment are in place (Bryant et al., 2019). Most French consumers believed that cultured meat lacked health benefits, taste, and natural features (Hocquette et al., 2022). Cattle are considered sacred animals in India, where cultured meat is widely consumed. While countries with large Muslim populations, including Malaysia, Qatar, and Indonesia, consider cultured beef to be Halal (Hocquette et al., 2024). Moreover, approximately 54% of respondents expressed an interest in trying cultured meat in a study conducted for Italian consumers (n = 525) (Mancini and Antonioli, 2019).

3. Age

Younger respondents exhibit higher WTT, WTE, and WTP than older respondents (>31 years of age) in Brazil (Chriki et al., 2021). However, in Europe (Grasso et al., 2019) as well as France and Germany (Bryant et al., 2020), elder respondents (65 years of age or older) show a low acceptability of cultured meat.

4. Income

In Brazil, respondents with the lowest monthly income (<3000 BRL) showed higher acceptability (WTT, WTE, and WTP) of cultured meat compared to those with the highest income (>15,000 BRL) (Chriki et al., 2021). Moreover, the initial cost of "artificial meat" will be higher, making it unaffordable for many consumers, particularly those in Africa (McKinsey, 2022).

5. Education

Cultured meat, a novel technique, will necessitate more trained and competent personnel. Education will thus play a critical role in ensuring that cultured meat is produced globally. Additionally, education had an interaction effect with income on WTP and WTE. This is since education is more readily available to individuals with higher incomes, particularly in certain developing countries (Kombolo Ngah et al., 2023). Food neophobia is more prevalent among respondents with lower levels of education (van den Heuvel et al., 2019).

8.2. Market Capacity

The cultured meat market is gaining traction as a sustainable, long-term alternative to food production. The expansion of this sector is indicative of the pursuit of more ethical solutions as well as the potential economic impact on countries that integrate this technology into their production models in the short and long term (da Silva & Conte-Junior, 2024). Animal cell-based meat, poultry, and seafood products that are analogous to traditional products are ready to be introduced into the market (Dolgin, 2020). Currently, over 150 organizations globally are engaged in the development of technology, either by contributing resources or manufacturing final products. The total amount of committed capital in this endeavor is projected to reach $2.8 billion by 2022 (Good Food institute, 2022). The initial cultured chicken nugget product has obtained regulatory permission for commercialization in Singapore (Singapore Food Agency, 2020), and the development of regulatory procedures for these goods is underway in numerous other regions. The United States Food and Drug Administration (FDA) and the United States Department of Agriculture (USDA) Food Safety and Inspection Service (USDA-FSIS) have officially agreed to collaborate in the regulation of cell-cultured beef and poultry products. The FDA will have exclusive jurisdiction over the regulation of seafood products (Post et al., 2020). Sustainable alternatives to traditional beef meet the increasing demand for ethical and environmentally sustainable diets. To date, there are 32 emerging start-up companies in the cultured meat production sector. Out of these, 25% are dedicated to producing cultured cattle, 22% to cultured chicken, 19% to cultured pigs and shellfish, and 15% to cultured exotic meat. North America accounts for 40% of the companies, followed by Asia (31%), and Europe 25% (Kumar et al., 2021). Startup companies and research groups worldwide are exploring cultured meat technologies to make them more accessible to customers, with the majority of groups based in the United States and Europe (Rubio et al., 2020). Mumbai, the first city in the world, hosted the first laboratory-grown meat research Centre, the Centre for Excellence in Cellular Agriculture, which cultured animal cells extracted painlessly. Their research focuses on developing and optimizing the most important cell lines to improve this sector (Porto and Berti, 2022). Australia, on the other hand, is the most recent nation to have entered the emerging laboratory-cultured meat industry, with two producers. Vow Food, a biotechnology start-up headquartered in Sydney, is one such producer that has secured approximately USD 20 million in venture funding. Vow Food's objective is to replicate meats that are currently produced and marketed on a large scale in Australia, with an emphasis on premium quality (Young et al., 2022). Glen Neal, the general manager of risk management and intelligence at FSANZ (Food Standards Australia and New Zealand), has indicated that cultured meat may be available for purchase in 2023 (Bowling, 2022). Singapore stands out in the global cultural meat landscape due to its limited production and export of traditional meat. Moreover, in 2023, cultured meat will become an essential part of the U.S. food chain. Brazil, a major meat producer and exporter, is currently engaged in innovative research on the production of cultured poultry meat, with sensory and nutritional analyses anticipated to be fulfilled by 2024 (EMBRAPA, 2023a). UPSIDE Foods, founded in 2015, was the inaugural CM startup. Since that time, there has been a rapid and significant expansion, resulting in the establishment of numerous enterprises in over 20 countries. In 2021, at least 21 new companies debuted, representing tremendous growth, as there had previously only been 86 CM enterprises (Santos et al., 2023). In Europe, the Netherlands, the origin of CM, committed EUR 60 million in financing in April 2022 to promote the construction of a national cellular agriculture ecosystem through the National Growth Fund. Spain made a large investment of EUR 5.2 million in a CM project managed by BioTech Foods in 2021 (Vegaconomist,2022). The Good Food Institute cites an increase in investment in cultured meat enterprises, demonstrating the production sector's confidence in this emerging market. In 2022, governments around the world offered significant funding and grants to cultured meat businesses, primarily for research and development initiatives. Global investments in cultured beef totaled $896 million. Several countries spearheaded the investment push, including Australia, China, the European Union, India, Japan, New Zealand, Qatar, Singapore, South Korea, Spain, the United Kingdom, and the United States (Bomkamp et al., 2022a). In 2021, invested capital increased by approximately 336% from 2020, reaching USD 410 million (GFI, 2021). The global cultured meat industry is anticipated to grow to $0.20 and $0.39 billion by 2023 and 2027, respectively, up from $0.16 billion in 2022 (da Silva and Conte-Junior, 2024). Moreover, the United States has made substantial investments in the production of cultured meat, with 43 companies currently engaged in the research of this technology, as indicated by data from the Good Food Institute (GFI) (Bomkamp et al., 2022a). Scaling remains a prominent obstacle in the industry, as it is crucial for lowering prices in the commercialization of cultured meat. In 2022, multiple collaborations were established with the aim of enhancing cultured meat production by facilitating the exchange of technologies, infrastructure, and resources among enterprises. Industries assure that they will be able to provide a restricted level of demand soon once cultured meat is globally regulated (Bomkamp et al., 2022a). Nevertheless, with the continuous growth of the cultured meat industry, there is an anticipated rise in the global production and distribution of cultured meat alongside traditional meat. Future Meat recently inaugurated its inaugural factory, which is dedicated to the production of meat that is derived from poultry, pork, and lamb cells (Porto and Berti, 2022). This achievement is a significant milestone in the technological advancement of the cultivated meat market because it serves as a catalyst for product industrialization. "MeaTech 3D Ltd." has successfully printed a 104 g cultured steak using its proprietary 3D bioprinting technology. They derived the steak from adipose and muscle cells. It is reputed to be one of the largest cultivated steaks to have been produced in recent years (Dadhania, 2022). Since 2015, various private cultured meat enterprises have formed in many countries, including the United States (Memphis Meats, now known as Upside Foods) and the Netherlands (Mosa Meat) (Chriki et al., 2020), encouraging cultured meat production within the next five years (Zhang et al., 2021). In 2016, the Good Food Institute, a nonprofit organization, was founded to promote new meat alternatives, such as cultured meat. As of early 2022, 60 of the 112 global enterprises were working in cultured meat processing. In October 2023, Cell MEAT submitted a request for certification from the Ministry of Food and Drug Safety (MFDS) of the Republic of Korea for the provisional use of Dokdo prawns

(*Lebbeus groenlandicus*) cell culture as a food ingredient (Cell MEAT, 2023). New amendments to an application received from Vow seeking approval of cultured quail were announced by Food Standards Australia New Zealand (FSANZ) in December 2023 (FSANZ, 2023). Cell-based chicken, beef, and seafood mixed products like burgers and nuggets cost $66.4/kg to $2200.5/kg (Guan et al., 2021). These costs are significantly higher than the retail price of conventional seafood products, such as salmon, tuna, and shrimp, which are priced at approximately $10.17/lb, $10.29/lb, and $9.05/lb, respectively (USDA, 2022).

Poultry Meat Products

Memphis Meats, a food technology company, effectively manufactured and launched cultured meat products in 2016 (Newman, 2020). In 2019, a cell-cultured poultry nugget manufactured by JUST was priced at $50 USD (Van Loo et al., 2020). In late 2020, the Singapore Food Agency (SFA) approved Eat Just Inc.'s sale of cultured chicken meat, making it the first government to approve the commercialization of cultured meat. A restaurant sold the product for approximately US$23. In 2013, Mosa Meat, a Dutch startup company, produced its first beef burger, which cost roughly $330,000 (Luiz Morais-da-Silva et al., 2022). Eat Just's cultivated chicken (GOOD MeatTM) was authorised for sale by the Singapore Food Agency (SFA) in December 2020. Moreover, Memphis Meats, a California-based startup, created cultured duck meat *(Hocquette et al., 2022)*. Gourmey, a French startup company, used duck egg cells and adjusted nutrients to produce artificial foie gras (ethical foie gras) (Guan et al., 2021). In 2020, JUST used cultured duck cells to produce duck pate and chorizo (Profeta et al., 2021).

Beef Meat Products

Mosa Meat, a Dutch startup company, generated cultured beef from cow stem cells in a medium without bovine serum, resulting in cost-effective CM (Bryant and Barnett, 2020). In 2016, Memphis Meats, a California-based startup, created the first cultured meatballs with cell-cultured beef (Stephens, 2022).

Pork Meat Products

In 2018, Meatable, a Dutch startup company, utilized stem cell technology to readily extract particular cells and create cell-cultured pork meat. New Age Meats, a San Francisco startup company, has produced prototype pork sausages using muscle and fat cells from live pigs (Profeta et al., 2021).

Table 2. Start-up companies for cultured meat production (beef products, pork products, poultry products and fish and shellfish products)

Species	Name of Company	Country	Name of Product
Cattle	Mosa meat	Netherland	Beef burger
	Modern meadow	USA	Meat steak
	Memphis meat	USA	Meat ball
	Aleph farms	Israel	Cultured steaks using proprietary 3-D technology
Pork	New age meat	USA	Pork sausage
	Higher steaks	UK	Pork belly and bacon
Poultry	Memphis meat	USA	Chicken tender
	JUST	USA	Chicken nuggets
	SuperMeat	Israel	chicken
	Memphis meats	USA	Duck meat nuggets
	Peace of meat	Belgium	Chicken meat
	JUST	USA	Duck pâté & chorizo
Fish and shellfish	Finless Food	USA	Bluefin tuna
	Bluefin Foods	USA	Bluefin tuna
	Blue Nalu	USA	Tuna, mahi mahi, red snapper
	Memphis Meats	USA	Coho salmon
	Wildtype	USA	Salmon
	Cultured Decadence	USA	Lobster
	Sound Eats	USA	Whitefish, zebrafish
	Shiok Meats	Singapore	Crab, lobster, shrimp
	Umami Meats	Singapore	Japanese eel, red snapper, grouper, yellowfin tuna
	Magic Caviar	Netherland	Caviar
	Bluu Biosciences	Germany	Salmon, trout, carp
	Cell Ag Tech	Canada	Whitefish
	Another Fish	Canda	Whitefish
	Avant Meats	China	Fish maw, sea cucumber, whitefish

9. CONCLUSION

In conclusion, in vitro-cultured meat represents a promising solution to the challenges faced by traditional meat production systems. Through advancements in tissue engineering, cell culture techniques, and sustainable practices, lab-grown meat offers a sustainable, ethical, and environmentally friendly alternative to conventional meat. The promise of cultured meat lies in its ability to provide a sustainable and ethical alternative to conventional animal agriculture. It offers solutions to critical issues such as environmental degradation, animal welfare concerns, and the increasing demand for protein from a growing global population. Cultured meat production requires significantly less land, water, and energy, and it has the potential to dramatically reduce greenhouse gas emissions, making it a key player in the fight against climate change. Moreover, in vitro meat can enhance food security by reducing dependency on traditional livestock and mitigating the risks of foodborne pathogens. The controlled environment in which cultured meat is produced minimizes contamination risks, ensuring a safer and more consistent product. However, the path to widespread adoption of cultured meat is not without obstacles. Technical challenges remain, including the need to improve the efficiency and scalability of production processes and to develop cost-effective culture media. Additionally, consumer acceptance and regulatory frameworks will play crucial roles in the successful commercialization of cultured meat. Public perception and education are vital to overcoming skepticism and fostering acceptance. Efforts to communicate the benefits and address misconceptions about cultured meat will be essential. Moreover, regulatory bodies must establish clear guidelines to ensure the safety, labelling, and marketing of cultured meat products. While challenges remain, continued research, innovation, and collaboration among scientists, policymakers, and industry stakeholders will be key to realizing its full potential. As we move forward, the development and adoption of cultured meat could lead to a more sustainable, ethical, and resilient food system for future generations.

REFERENCES

Abitbol, T., Rivkin, A., Cao, Y., Nevo, Y., Abraham, E., Ben-Shalom, T., & Shoseyov, O. (2016). Nanocellulose, a tiny fiber with huge applications. *Current Opinion in Biotechnology*, 39, 76–88. 10.1016/j.copbio.2016.01.00226930621

Alexander, P., Brown, C., Arneth, A., Dias, C., Finnigan, J., Moran, D., & Rounsevell, M. D. A. (2017). Could consumption of insects, cultured meat or imitation meat reduce global agricultural land use? In *Global Food Security* (Vol. 15, pp. 22–32). Elsevier B.V. 10.1016/j.gfs.2017.04.001

Ali, A., & Ahmed, S. (2018). Recent advances in edible polymer-based hydrogels as a sustainable alternative to conventional polymers. *Journal of Agricultural and Food Chemistry*, 2018(66), 6940–6967. 10.1021/acs.jafc.8b0105229878765

Allan, S. J., Ellis, M. J., & De Bank, P. A. (2021). Decellularized grass as a sustainable scaffold for skeletal muscle tissue engineering. *Journal of Biomedical Materials Research. Part A*, 109(12), 2471–2482. 10.1002/jbm.a.3724134057281

Anomaly, J., Browning, H., Fleischman, D., & Veit, W. (2023). Flesh without blood: The public health benefits of lab-grown meat. *Journal of Bioethical Inquiry*, 21(1), 167–175. 10.1007/s11673-023-10254-737656382

Arab, W., Rauf, S., Al-Harbi, O., & Hauser, C. A. (2018). Novel ultrashort self-assembling peptide bioinks for 3D culture of muscle myoblast cells. *International Journal of Bioprinting*, 4(2). Advance online publication. 10.18063/ijb.v4i1.12933102913

Asioli, D., Bazzani, C., & Nayga, R.Jr. (2021). Are consumers willing to pay for in-vitro meat? an investigation of naming effects. *Journal of Agricultural Economics*, 73(2), 356–375. 10.1111/1477-9552.12467

Bakhsh, A., Lee, E.-Y., Ncho, C. M., Kim, C.-J., Son, Y.-M., Hwang, Y.-H., & Joo, S.-T. (2022). Quality characteristics of meat analogs through the incorporation of textured vegetable protein: A systematic review. *Foods*, 11(9), 1242. 10.3390/foods1109124235563965

Bar-Nur, O., Gerli, M. F., Di Stefano, B., Almada, A. E., Galvin, A., Coffey, A., Huebner, A. J., Feige, P., Verheul, C., Cheung, P., Payzin-Dogru, D., Paisant, S., Anselmo, A., Sadreyev, R. I., Ott, H. C., Tajbakhsh, S., Rudnicki, M. A., Wagers, A. J., & Hochedlinger, K. (2018). Direct reprogramming of mouse fibroblasts into functional skeletal muscle progenitors. *Stem Cell Reports*, 10(5), 1505–1521. 10.1016/j.stemcr.2018.04.00929742392

Bar-Shai, N., Sharabani-Yosef, O., Zollmann, M., Lesman, A., & Golberg, A. (2021). Seaweed cellulose scaffolds derived from green macroalgae for tissue engineering. *Scientific Reports*, 11(1), 11843. 10.1038/s41598-021-90903-234088909

Barbosa, W., Correia, P., Vieira, J., Leal, I., Rodrigues, L., Nery, T., Barbosa, J., & Soares, M. (2023). Trends and Technological Challenges of 3D Bioprinting in Cultured Meat: Technological Prospection. *Applied Sciences (Basel, Switzerland)*, 13(22), 12158. 10.3390/app132212158

Bartholet, J. (2011). Inside the meat lab. *Scientific American*, 304(6), 64–69. 10.1038/scientificamerican0611-6421608405

Ben-Arye, T., & Levenberg, S. (2019). Tissue engineering for clean meat production. *Frontiers in Sustainable Food Systems*, 3, 46. 10.3389/fsufs.2019.00046

Ben-Arye, T., Shandalov, Y., Ben-Shaul, S., Landau, S., Zagury, Y., Ianovici, I., Lavon, N., & Levenberg, S. (2020). Textured soy protein scaffolds enable the generation of three-dimensional bovine skeletal muscle tissue for cell-based meat. *Nature Food*, 1(4), 210–220. 10.1038/s43016-020-0046-5

Bhat, Z. F., & Bhat, H. (2011). Animal-free meat biofabrication. *American Journal of Food Technology*, 6(6), 441–459. 10.3923/ajft.2011.441.459

Bhat, Z. F., Bhat, H., & Pathak, V. (2013). Prospects for In Vitro Cultured Meat - A Future Harvest. In Principles of Tissue Engineering: Fourth Edition (pp. 1663–1683). Elsevier Inc. 10.1016/B978-0-12-398358-9.00079-3

Bhat, Z. F., & Fayaz, H. (2011). Prospectus of cultured meat - Advancing meat alternatives. In Journal of Food Science and Technology (Vol. 48, Issue 2, pp. 125–140). 10.1007/s13197-010-0198-7

Bhat, Z. F., Kumar, S., & Fayaz, H. (2015). In vitro meat production: Challenges and benefits over conventional meat production. In Journal of Integrative Agriculture (Vol. 14, Issue 2, pp. 241–248). Editorial Department of Scientia Agricultura Sinica. 10.1016/S2095-3119(14)60887-X

Bishop, Mostafa, Pakvasa, Luu, Lee, Wolf, Ameer, He, & Reid. (2017). 3-D bioprinting technologies in tissue engineering and regenerative medicine: current and future trends. Genes Dis 4:185–195. https://doi.org/.2017.10.00210.1016/j.gendis

Biswal, T. (2021). Biopolymers for tissue engineering applications: A review. *Materials Today: Proceedings*, 41, 397–402. 10.1016/j.matpr.2020.09.628

Bodiou, V., Moutsatsou, P., & Post, M. J. (2020). Microcarriers for upscaling cultured meat production. *Frontiers in Nutrition*, 7, 10. 10.3389/fnut.2020.0001032154261

Bomkamp, C., Carter, M., Cohen, M., Gertner, D., Ignaszewski, E., Murray, S., O'Donnell, M., Pierce, B., Swartz, E., & Voss, S. (2022a), Cultivated meat state of the industry report. https://gfi.org/resource/cultivated-meat-eggs-and-dairy-state-of-the-industry-report/

Bomkamp, C., Skaalure, S. C., Fernando, G. F., Ben-Arye, T., Swartz, E. W., & Specht, E. A. (2022b). Scaffolding biomaterials for 3D cultivated meat: Prospects and challenges. *Advancement of Science*, 9(3), 2102908. 10.1002/advs.20210290834786874

Bostock, J., McAndrew, B., Richards, R., Jauncey, K., Telfer, T., Lorenzen, K., Little, D., Ross, L., Handisyde, N., Gatward, I., & Corner, R. (2010). Aquaculture: Global status and trends. In Philosophical Transactions of the Royal Society B: Biological Sciences (Vol. 365, Issue 1554, pp. 2897–2912). Royal Society. 10.1098/rstb.2010.0170

Bouvard, V., Loomis, D., Guyton, K. Z., Grosse, Y., El Ghissassi, F., Benbrahim-Tallaa, L., Guha, N., Mattock, H., & Straif, K. (2015): Carcinogenicity of consumption of red and processed meat. Lancet Oncol., 16, 1599.

Brennan, L., & Owende, P. **(2010).** Biofuels from microalgae-A review of technologies for production, processing, and extractions of biofuels and co-products. In Renewable and Sustainable Energy Reviews (Vol. 14, Issue 2, pp. 557–577). https://doi.org/10.1016/j.rser.2009.10.009

Brennan, T., Katz, J., Quint, Y., & Spencer, B. (2021). Cultivated meat: Out of the lab, into the frying pan. McKinsey & Company.

Brüggenthies, J. B., Fiore, A., Russier, M., Bitsina, C., Brötzmann, J., Kordes, S., Menninger, S., Wolf, A., Conti, E., Eickhoff, J. E., & Murray, P. J. (2022). A cell-based chemical-genetic screen for amino acid stress response inhibitors reveals torins reverse stress kinase GCN2 signaling. *The Journal of Biological Chemistry*, 298(12), 102629. 10.1016/j.jbc.2022.10262936273589

Bryant, C. (2020). *Exploring the Nature of Consumer Preferences between Conventional and Cultured Meat*. University of Bath.

Bryant, C., & Barnett, J. (2020). Consumer Acceptance of Cultured Meat: An Updated Review (2018–2020). *Applied Sciences (Basel, Switzerland)*, 10(15), 5201. 10.3390/app10155201

Bryant, C., Szejda, K., Parekh, N., Desphande, V., & Tse, B. (2019). A survey of consumer perceptions of plant-based and clean meat in the USA, India, and China. *Frontiers in Sustainable Food Systems*, 3, 11. Advance online publication. 10.3389/fsufs.2019.00011

Bryant, C., van Nek, L., & Rolland, N. C. M. (2020). European Markets for Cultured Meat: A Comparison of Germany and France. *Foods*, 2020(9), 1152. 10.3390/foods909115232825592

Burdock, G. A., Carabin, I. G., & Griffiths, J. C. (2006). The importance of GRAS to the functional food and nutraceutical industries. *Toxicology*, 221(1), 17–27. 10.1016/j.tox.2006.01.01216483705

Carletti, E., Motta, A., & Migliaresi, C. (2011). *Scaffolds for tissue engineering and 3D cell culture*. Methods Mol Bio. 10.1007/978-1-60761-984-0_2

Carrington, D. (2020). No-kill, lab-grown meat to go on sale for first time. Retrieved on September 18, 2021 from The Guardian. website: https:// www.theguardian.com/environment/2020/dec/02/nokill-lab-grown-meat-to-go-on-sale-for-first-time

Catts, O., & Zurr, I. (2002). Growing semi-living sculptures: The tissue culture project. *Leonardo*, 35(4), 365–370. 10.1162/002409402760181123

CellMEAT. (2023). Cell MEAT begins approval process for cell-cultured meat from the Ministry of Food and Drug Safety of Korea. Available from: https://www.thecellmeat.com/bbs/board.php?bo_table=gallery&wr_id=44

Chang, K. H., Liao, H. T., & Chen, J. P. (2013). Preparation and characterization of gelatin/hyaluronic acid cryogels for adipose tissue engineering: In vitro and in vivo studies. *Acta Biomaterialia*, 9(11), 9012–9026. 10.1016/j.actbio.2013.06.04623851171

Chen, Y., Zhang, W., Ding, X., Ding, S., Tang, C., Zeng, X., Wang, J., & Zhou, G. (2024). Programmable scaffolds with aligned porous structures for cell cultured meat. *Food Chemistry*, 430, 137098. Advance online publication. 10.1016/j.foodchem.2023.13709837562260

Chodkowska, K. A., Wódz, K., & Wojciechowski, J. (2022). Sustainable future protein foods: The challenges and the future of cultivated meat. *Foods*, 11(24), 4008. 10.3390/foods1124400836553750

Choi, K., Yoon, J., Kim, M., Jeong, J., Ryu, M., Park, S., & Lee, C. (2020). Optimization of culture conditions for maintaining pig muscle stem cells in vitro. *Food Science of Animal Resources*, 40(4), 659–667. 10.5851/kosfa.2020.e3932734272

Chriki, S., Ellies-Oury, M.-P., Fournier, D., Liu, J., & Hocquette, J.-F. (2020). Analysis of scientific and press articles related to cultured meat for a better understanding of its perception. *Frontiers in Psychology*, 11, 1845. 10.3389/fpsyg.2020.0184532982823

Chriki, S., Payet, V., Pflanzer, S. B., Ellies-Oury, M. P., Liu, J., Hocquette, É., Rezende-de-Souza, J. H., & Hocquette, J. F. (2021). Brazilian consumers' attitudes towards so-called "cell-based meat". *Foods*, 10(11), 2588. 10.3390/foods1011258834828869

Chuah, S. X. Y., Gao, Z., Arnold, N. L., & Farzad, R. (2024). Cell-Based Seafood Marketability: What Influences United States Consumers' Preferences and Willingness-To-Pay? *Food Quality and Preference*, 113, 105064. 10.1016/j.foodqual.2023.105064

Chui, C. Y., Odeleye, A., Nguyen, L., Kasoju, N., Soliman, E., & Ye, H. (2019). Electrosprayed genipin cross-linked alginate–chitosan microcarriers for ex vivo expansion of mesenchymal stem cells. *Journal of Biomedical Materials Research. Part A*, 107(1), 122–133. 10.1002/jbm.a.3653930256517

Codex Alimentarius Commission (CAC). (2021). *General Standard for Food Additives*. CAC.

Contessi Negrini, N., Tofoletto, N., Farè, S., & Altomare, L. (2020). Plant tissues as 3D natural scaffolds for adipose, bone and tendon tissue regeneration. *Frontiers in Bioengineering and Biotechnology*, 723, 723. Advance online publication. 10.3389/fbioe.2020.0072332714912

Cuadri, A. A., Bengoechea, C., Romero, A., & Guerreroet, A. (2016). A natural based polymeric hydrogel based on functionalized soy protein. *European Polymer Journal*, 85, 164–174. 10.1016/j.eurpolymj.2016.10.026

da Silva, B. D., & Conte-Junior, C. A. (2024). Perspectives on cultured meat in countries with economies dependent on animal production: A review of potential challenges and opportunities. *Trends in Food Science & Technology*, 149, 104551. 10.1016/j.tifs.2024.104551

Dadhania, S. (2022). 3D Printing Meets Meat in the Largest Cultured Steak Ever Made. Available online: https://www.idtechex.com/en/research-article/3d-printing-meets-meat-in-the-largest-cultured-steak-ever-made/25517

de-Magistris, T., Pascucci, S., & Mitsopoulos, D. (2015). Paying to see a bug on my food: How regulations and information can hamper radical innovations in the European Union. *British Food Journal*, 117(6), 1777–1792. 10.1108/BFJ-06-2014-0222

Deb, P., Deoghare, A. B., Borah, A., Barua, E., & Lala, S. D. (2018). Scaffold development using biomaterials: A review. *Materials Today: Proceedings*, 5(5), 12909–12919. 10.1016/j.matpr.2018.02.276

Del Bakhshayesh, A. R., Mostafavi, E., Alizadeh, E., Asadi, N., Akbarzadeh, A., & Davaran, S. (2018). Fabrication of Three-Dimensional Scaffolds Based on Nano-biomimetic Collagen Hybrid Constructs for Skin Tissue Engineering. *ACS Omega*, 3(8), 8605–8611. 10.1021/acsomega.8b0121931458990

Di Micco, R., Krizhanovsky, V., Baker, D., & d'Adda di Fagagna, F. (2021). Cellular senescence in ageing: From mechanisms to therapeutic opportunities. *Nature Reviews. Molecular Cell Biology*, 22(2), 75–95. 10.1038/s41580-020-00314-w33328614

Diana, J. S. (2009). Aquaculture production and biodiversity conservation. *Bioscience*, 59(1), 27–38. 10.1525/bio.2009.59.1.7

Dolgin, E. (2020). Will cell-based meat ever be a dinner staple? Nature, 588, S64–S64.

Dong, J., Sun, Q., & Wang, J.-Y. (2004). Basic study of corn protein, zein, as a biomaterial in tissue engineering, surface morphology and biocompatibility. *Biomaterials*, 25(19), 4691–4697. 10.1016/j.biomaterials.2003.10.08415120515

Dreesen, I. A. J., & Fussenegger, M. (2011). Ectopic expression of human mTOR increases viability, robustness, cell size, proliferation, and antibody production of Chinese hamster ovary cells. *Biotechnology and Bioengineering*, 108(4), 853–866. 10.1002/bit.2299021404259

Dutta, S. D., Ganguly, K., Jeong, M. S., Patel, D. K., Patil, T. V., Cho, S. J., & Lim, K. T. (2022). Bioengineered lab-grown meat-like constructs through 3D bioprinting of antioxidative protein hydrolysates. *ACS Applied Materials & Interfaces*, 14(30), 34513–34526. 10.1021/acsami.2c1062035849726

Edelman, P., McFarland, D., Mironov, V., & Matheny, J. (2005). Commentary: In Vitro-Cultured Meat Production. *Tissue Engineering*, 11(5-6), 659–662. 10.1089/ten.2005.11.65915998207

Eltom, A., Zhong, G., & Muhammad, A. (2019). Scaffold techniques and designs in tissue engineering functions and purposes: A review. *Advances in Materials Science and Engineering*, 2019(1), 3429527. 10.1155/2019/3429527

Enrione, J., Blaker, J. J., Brown, D. I., Weinstein-Oppenheimer, C. R., Pepczynska, M., Olguin, Y., Sanchez, E., & Acevedo, C. A. (2017). Edible Scaffolds Based on Non-Mammalian Biopolymers for Myoblast Growth. *Materials (Basel)*, 10(12), 1404. 10.3390/ma1012140429292759

Farms, A. (2024). Aleph Farms granted world's first regulatory approval for cultivated beef. Available from: https://alephfarms.com/journals/aleph-farms-granted-worlds-first-regulatory-approval-for-cultivated-beef/

Farwa, M., Zulfiqar, A. R., Syeda, R. B., Muhammad, Z., Ozgun, C. O., & Ammara, R. (2022). Preparation, properties, and applications of gelatin-based hydrogels (GHs) in the environmental, technological, and biomedical sectors. *International Journal of Biological Macromolecules*, 218, 601–633. 10.1016/j.ijbiomac.2022.07.16835902015

Fatima, N., Emambux, M. N., Olaimat, A. N., Stratakos, A. C., Nawaz, A., Wahyono, A., Gul, K., Park, J., & Shahbaz, H. M. (2023). Recent advances in microalgae, insects, and cultured meat as sustainable alternative protein sources. *Food and Humanity*, 1, 731–741. 10.1016/j.foohum.2023.07.009

FESA. (2015). Risk profile related to production and consumption of insects as food and feed. *EFSA Journal*, 13(10), 4257. Advance online publication. 10.2903/j.efsa.2015.4257

Fish, K. D., Rubio, N. R., Stout, A. J., Yuen, J. S., & Kaplan, D. L. (2020). Prospects and challenges for cell-cultured fat as a novel food ingredient. *Trends in Food Science & Technology*, 98, 53–67. 10.1016/j.tifs.2020.02.00532123465

Flycatcher. (2013). Kweekvlees Cultured Meat. Netherlands. Available online: http://www.flycatcherpanel.nl/news/item/nwsA1697/media/images/Resultaten_onderzoek_kweekvlees.pdf (accessed on 8 August 2020)

Food Standards Australia New Zealand. (2023). A1269: Cultured quail as a novel food. Available from: https://www.foodstandards.gov.au/food-standards-code/applications/A1269-Cultured-Quail-as-a-Novel-Food

Freeman, F. E., & Kelly, D. J. (2017). Tuning alginate bioink stiffness and composition for controlled growth factor delivery and to spatially direct MSC fate within bio printed tissues. *Scientific Reports*, 7(1), 1–2. 10.1038/s41598-017-17286-129213126

Furuichi, Y., Kawabata, Y., Aoki, M., Mita, Y., Fujii, N. L., & Manabe, Y. (2021). Excess glucose impedes the proliferation of skeletal muscle satellite cells under adherent culture conditions. *Frontiers in Cell and Developmental Biology*, 9, 640399. 10.3389/fcell.2021.64039933732705

Gañán-Calvo, A. M., Dávila, J., & Barrero, A. (1997). Current and droplet size in the electro spraying of liquids. Scaling laws. *Journal of Aerosol Science*, 28(2), 249–275. 10.1016/S0021-8502(96)00433-8

GFI (Good Food Institute). (2021). State of the Industry Report. Cultivated Meat and Seafood. Available online: https://gfi.org/resource/cultivatedmeat-eggs-and-dairy-state-of-the-industry-report/

Ghanbari, M., Sadjadinia, A., Zahmatkesh, N., Mohandes, F., Dolatyar, B., Zeynali, B., & Salavati-Niasari, M. (2022). Synthesis and investigation of physicochemical properties of alginate dialdehyde/gelatin/ZnO nanocomposites as injectable hydrogels. *Polymer Testing*, 110, 107562. 10.1016/j.polymertesting.2022.107562

Ghanbari, M., Salavati-Niasari, M., & Mohandes, F. (2021a). Thermosensitive alginate-gelatin-nitrogen-doped carbon dots scaffolds as potential injectable hydrogels for cartilage tissue engineering applications. *RSC Advances*, 11(30), 18423–18431. 10.1039/D1RA01496J35480940

Ghanbari, M., Salavati-Niasari, M., Mohandes, F., Dolatyarb, B., & Zeynalib, B. (2021b). In vitro study of alginate–gelatin scaffolds incorporated with silica NPs as injectable, biodegradable hydrogels. *RSC Advances*, 11(27), 16688–16697. 10.1039/D1RA02744A35479165

Good Food Institute. (2022). state of the industry report: Cultivated meat and Older Consumers' Readiness to Accept Alternative, More Sustainable Protein Sources in the European Union. *Nutrients*, 11, 1904.

Gray, V. P., Amelung, C. D., Duti, I. J., Laudermilch, E. G., Letteri, R. A., & Lampe, K. J. (2022). Biomaterials via peptide assembly: Design, characterization, and application in tissue engineering. *Acta Biomaterialia*, 140, 43–75. 10.1016/j.actbio.2021.10.03034710626

Guan, X., Lei, Q., Yan, Q., Li, X., Zhou, J., Du, G., & Chen, J. (2021). Trends and Ideas in Technology, Regulation and Public Acceptance of Cultured Meat. *Future Foods: a Dedicated Journal for Sustainability in Food Science*, 3, 100032. 10.1016/j.fufo.2021.100032

Guardian. (2013). Would you eat a synthetic beefburger? https://www.theguardian.com/commentisfree/poll/2013/aug/0 5/eat-synthetic-beefburger-poll

Guardian. (2022). All sizzle, no steak: how Singapore became the center of the plant-based meat industry. https://www.theguardian.com/e nvironment/2022/nov/06/all-sizzle-no-steak-how-singapore-became-the-centre-of-the-plant-based-meat-industry

Hao, Z., Li, H., Wang, Y., Hu, Y., Chen, T., Zhang, S., Guo, X., Cai, L., & Li, J. (2022). Supramolecular peptide nanofiber hydrogels for bone tissue engineering: From multi hierarchical fabrications to comprehensive applications. *Advancement of Science*, 9(11), 2103820. 10.1002/advs.20210382035128831

Harris, A. F., Lacombe, J., & Zenhausern, F. (2021). The emerging role of decellularized plant-based scafolds as a new biomaterial. *International Journal of Molecular Sciences*, 22(22), 12347. 10.3390/ijms22221234734830229

Hocquette, É., Liu, J., Ellies-Oury, M.-P., Chriki, S., & Hocquette, J.-F. (2022). Does the Future of Meat in France Depend on Cultured Muscle Cells? Answers from Different Consumer Segments. *Meat Science*, 188, 108776. 10.1016/j.meatsci.2022.10877635245709

Hocquette, J. F., Chriki, S., Fournier, D., & Ellies-Oury, M. P. (2024). Will "cultured meat" transform our food system towards more sustainability? Animal, 101145.

Hopkins, P. D., & Dacey, A. (2008). Vegetarian meat: Could technology save animals and satisfy meat eaters? *Journal of Agricultural & Environmental Ethics*, 21(6), 579–596. 10.1007/s10806-008-9110-0

Jagadeesan, P., & Salem, S. (2020). Religious and Regulatory concerns of animal free meat and milk. Science Open *Preprints*, 10.14293/S2199-1006.1.SOR-.PP7B21S.v1

Jin, G., & Bao, X. (2024). Tailoring the taste of cultured meat. *eLife*, 13, e98918. Advance online publication. 10.7554/eLife.9891838813866

Jo, B., Nie, M., & Takeuchi, S. (2021). Manufacturing of animal products by the assembly of microfabricated tissues. *Essays in Biochemistry*, 65(3), 611–623. 10.1042/EBC2020009234156065

Jones, J. D., Rebello, A. S., & Gaudette, G. R. (2021). Decellularized spinach: An edible scaffold for laboratory-grown meat. Food Bioscience, 41,1100986 12. 10.1016/j.cofs.2015.08.002

Kačarević, Ž. P., Rider, P. M., Alkildani, S., Retnasingh, S., Smeets, R., Jung, O., Ivanišević, Z., & Barbeck, M. (2018). An introduction to 3D bioprinting: Possibilities, challenges and future aspects. *Materials (Basel)*, 11(11), 2199. 10.3390/ma1111219930404222

Kantono, K., Hamid, N., Malavalli, M. M., Liu, Y., Liu, T., & Seyfoddin, A. (2022). Consumer acceptance and production of in vitro meat: A review. *Sustainability (Basel)*, 14(9), 4910. 10.3390/su14094910

Kantor, B. N., & Kantor, J. (2021). Public attitudes and willingness to pay for cultured meat: A cross-sectional experimental study. *Frontiers in Sustainable Food Systems*, 5, 1–7. 10.3389/fsufs.2021.594650

Kenigsberg, J. A., & Zivotofsky, A. Z. (2020). A Jewish religious perspective on cellular agriculture. *Frontiers in Sustainable Food Systems*, 128, 128. 10.3389/fsufs.2019.00128

Khan, F., & Tanaka, M. (2018). Designing Smart Biomaterials for Tissue Engineering. *International Journal of Molecular Sciences*, 19(1), 17. 10.3390/ijms1901001729267207

Kolkmann, A. M., van Essen, A., Post, M. J., & Moutsatsou, P. (2022). Development of a chemically defined medium for in vitro expansion of primary bovine satellite cells. *Frontiers in Bioengineering and Biotechnology*, 10, 895289. 10.3389/fbioe.2022.89528935992337

Kolodkin-Gal, I., Dash, O., & Rak, R. (2023). Probiotic cultivated meat: Bacterial-based scaffolds and products to improve cultivated meat. *Trends in Biotechnology*.37805297

Kombolo Ngah, M., Chriki, S., Ellies-Oury, M. P., Liu, J., & Hocquette, J. F. (2023). Consumer perception of "artificial meat" in the educated young and urban population of Africa. *Frontiers in Nutrition*, 10, 1127655. 10.3389/fnut.2023.112765537125051

Kouřimská, L., & Adámková, A. (2016). Nutritional and sensory quality of edible insects. Elsevier GmbH., 10.1016/j.nfs.2016.07.001

Koyande, A. K., Chew, K. W., Rambabu, K., Tao, Y., Chu, D. T., & Show, P. L. (2019). Microalgae: A potential alternative to health supplementation for humans. In Food Science and Human Wellness (Vol. 8, Issue 1, pp. 16–24). Elsevier B.V. 10.1016/j.fshw.2019.03.001

Kumar, P., Sharma, N., Sharma, S., Mehta, N., Verma, A. K., Chemmalar, S., & Sazili, A. Q. (2021). In-vitro meat: A promising solution for sustainability of meat sector. *Journal of Animal Science and Technology*, 63(4), 693–724. 10.5187/jast.2021.e8534447949

Lannutti, J., Reneker, D., Ma, T., Tomasko, D., & Farson, D. (2007). Electrospinning for tissue engineering scaffolds. *Materials Science and Engineering C*, 27(3), 504–509. 10.1016/j.msec.2006.05.019

Le, B. (2020). Australian Startup Develops Novel Edible Scaffold to Modernize Meat Production, https://www.proteinreport.org/australian-startup-develops-novel-edible-scaffold-modernize-meat-production

Lee, J. D., Shin, D., & Roh, J.-L. (2018). Development of an in vitro cell-sheet cancer model for chemotherapeutic screening. *Theranostics*, 8(14), 3964–3973. 10.7150/thno.2643930083273

Lee, M., Park, S., Choi, B., Kim, J., Choi, W., Jeong, I., Han, D., Koh, W.-G., & Hong, J. (2022). Tailoring a gelatin/agar matrix for the synergistic effect with cells to produce high-quality cultured meat. *ACS Applied Materials & Interfaces*, 14(33), 38235–38245. 10.1021/acsami.2c1098835968689

Lee, S., Trinh, T. H., Yoo, M., Shin, J., Lee, H., Kim, J., Hwang, E., Lim, Y., & Ryou, C. (2019). Self-assembling peptides and their application in the treatment of diseases. *International Journal of Molecular Sciences*, 20(23), 5850. 10.3390/ijms2023585031766475

Lei, Z. L., Li MingSheng, L. M., Ma ZhongRen, M. Z., & Feng YuPing, F. Y. (2017). Preparation of DEAE-soybean starch microspheres for anchorage-dependent mammal cell culture.

Leprivier, G., & Rotblat, B. (2020). How does mTOR sense glucose starvation? AMPK is the usual suspect. *Cell Death Discovery*, 6(1), 27. 10.1038/s41420-020-0260-932351714

Lerman, M. J., Lembong, J., Muramoto, S., Gillen, G., & Fisher, J. P. (2018). The evolution of polystyrene as a cell culture material. *Tissue Engineering. Part B, Reviews*, 24(5), 359–372. 10.1089/ten.teb.2018.005629631491

Levi, S., Yen, F. C., Baruch, L., & Machluf, M. (2022). Scaffolding technologies for the engineering of cultured meat: Towards a safe, sustainable, and scalable production. *Trends in Food Science & Technology*, 126, 13–25. 10.1016/j.tifs.2022.05.011

Li, G., Hu, L., Liu, J., Huang, J., Yuan, C., Takaki, K., & Hu, Y. (2022a). A Review on 3D Printable Food Materials: Types and Development Trends. *International Journal of Food Science & Technology*, 2022(57), 164–172. 10.1111/ijfs.15391

Li, L., Chen, L., Chen, X., Chen, Y., Ding, S., Fan, X., Liu, Y., Xu, X., Zhou, G., Zhu, B., Ullah, N., & Feng, X. (2022Chitosan-sodium alginate-collagen/gelatin three-dimensional edible scaffolds for building a structured model for cell cultured meat. *International Journal of Biological Macromolecules*, 209, 668–679. 10.1016/j.ijbiomac.2022.04.05235413327

Li, Q., Lin, H., Du, Q., Liu, K., Wang, O., Evans, C., Christian, H., Zhang, C., & Lei, Y. (2018). Scalable and physiologically relevant microenvironments for human pluripotent stem cell expansion and differentiation. *Biofabrication*, 10(2), 025006. 10.1088/1758-5090/aaa6b529319535

Li, W., Han, Y., Yang, H., Wang, G., Lan, R., & Wang, J. Y. (2016). Preparation of microcarriers based on zein and their application in cell culture. *Materials Science and Engineering C*, 58, 863–869. 10.1016/j.msec.2015.09.04526478381

Li, X., Liu, B., Pei, B., Chen, J., Zhou, D., Peng, J., Zhang, X., Jia, W., & Xu, T. (2020). Inkjet Bioprinting of Biomaterials. *Chemical Reviews*, 120(19), 10793–10833. 10.1021/acs.chemrev.0c0000832902959

Li, Y., Liu, W., Li, S., Zhang, M., Yang, F., & Wang, S. (2021). Porcine skeletal muscle tissue fabrication for cultured meat production using three-dimensional bioprinting technology. *J Future Foods.*, 1(1), 88–97. 10.1016/j.jfutfo.2021.09.005

Liu, F., Zhang, S. Y., Chen, K. X., & Zhang, Y. (2023). Fabrication, in-vitro digestion and pH-responsive release behavior of soy protein isolate glycation conjugates-based hydrogels. *Food Research International*, 169, 112884. 10.1016/j.foodres.2023.11288437254332

Liu, J., Chriki, S., Kombolo, M., Santinello, M., Pflanzer, S. B., Hocquette, É., Ellies-Oury, M.-P., & Hocquette, J.-F. (2023). Consumer perception of the challenges facing livestock production and meat consumption. *Meat Science*, 200, 109144. 10.1016/j.meatsci.2023.10914436863253

Liu, X., & Ma, P. X. (2009). Phase separation, pore structure, and properties of nanofibrous gelatin scaffolds. *Biomaterials*, 30(25), 4094–4103. 10.1016/j.biomaterials.2009.04.02419481080

Liu, Y., Wang, R., Ding, S., Deng, L., Zhang, Y., Li, J., Shi, Z., Wu, Z., Liang, K., Yan, X., Liu, W., & Du, Y. (2022). Engineered meatballs via scalable skeletal muscle cell expansion and modular micro-tissue assembly using porous gelatin microcarriers. *Biomaterials*, 287, 121615. 10.1016/j.biomaterials.2022.12161535679644

Lu, T. Y., Li, Y., & Chen, T. (2013). Techniques for fabrication and construction of threedimensional scaffolds for tissue engineering. *International Journal of Nanomedicine*, 337, 337. Advance online publication. 10.2147/IJN.S3863523345979

Luiz Morais-da-Silva, R., Glufke Reis, G., Sanctorum, H., & Forte Maiolino Molento, C. (2022). The social impacts of a transition from conventional to cultivated and plant- **based** meats: Evidence from Brazil. *Food Policy*, 111, 102337. 10.1016/j.foodpol.2022.102337

Ma, L., Seager, M., Wittmann, M., Jacobson, M., Bickel, D., Burno, M., Jones, K., Graufelds, V. K., Xu, G., Pearson, M., McCampbell, A., Gaspar, R., Shughrue, P., Danziger, A., Regan, C., Flick, R., Pascarella, D., Garson, S., Doran, S., & Ray, W. J. (2009). Selective activation of the M1 muscarinic acetylcholine receptor achieved by allosteric potentiation. *Proceedings of the National Academy of Sciences of the United States of America*, 106(37), 15950–15955. 10.1073/pnas.090090310619717450

MacQueen, L. A., Alver, C. G., Chantre, C. O., Ahn, S., Cera, L., Gonzalez, G. M., ... Parker, K. K. (2019). Muscle tissue engineering in fibrous gelatin: implications for meat analogs. NPJ Science of Food, 3(1), 20.

Malav, O. P., Talukder, S., Gokulakrishnan, P., & Chand, S. (2015). Meat Analog: A Review. *Critical Reviews in Food Science and Nutrition*, 55(9), 1241–1245. 10.1080/10408398.2012.68938124915320

Mancini, M. C., & Antonioli, F. (2019). Exploring consumers' attitude towards cultured meat in Italy. Meat Science, 150, 101–110. https://doi.org/.Meatscience,12.01410.1016/J

Mancuso, J. (2021). What are food grade solvents. Accessed: https:// ecolink.com/info/what-are-food-grade-solvents/

Martin, I., Wendt, D., & Heberer, M. (2004). The role of bioreactors in tissue engineering. *Trends in Biotechnology*, 22(2), 80–86. 10.1016/j.tibtech.2003.12.00114757042

MBRAPA. (2023). Brasil está na vanguarda no desenvolvimento de carne cultivada - Portal Embrapa. https://www.embrapa.br/busca-de-noticias/-/noticia/77704192/brasil-esta-na-vanguarda-no-desenvolvimento-de-carne-cultivada

Melzener, L., Verzijden, K. E., Buijs, A. J., Post, M. J., & Flack, J. E. (2021). Cultured beef: From small biopsy to substantial quantity. *Journal of the Science of Food and Agriculture*, 101(1), 7–14. 10.1002/jsfa.1066332662148

Mendes, A. C., Stephansen, K., & Chronakis, I. S. (2017). Electrospinning of food proteins and polysaccharides. *Food Hydrocolloids*, 68, 53–68. 10.1016/j.foodhyd.2016.10.022

Modulevsky, D. J., Cuerrier, C. M., & Pelling, A. E. (2016). Biocompatibility of subcutaneously implanted plant-derived cellulose biomaterials. *PLoS One*, 11(6), e0157894. 10.1371/journal.pone.015789427328066

Morais, A. Í., Vieira, E. G., Afewerki, S., Sousa, R. B., Honorio, L. M., Cambrussi, A. N., Santos, J. A., Bezerra, R. D. S., Furtini, J. A. O., Silva-Filho, E. C., Webster, T. J., & Lobo, A. O. (2020). Fabrication of polymeric microparticles by electrospray: The impact of experimental parameters. *Journal of Functional Biomaterials*, 11(1), 4. 10.3390/jfb1101000431952157

Morais-da-Silva, R. L., Villar, E. G., Reis, G. G., Sanctorum, H., & Molento, C. F. M. (2022). The expected impact of cultivated and plant-based meats on jobs: The views of experts from Brazil, the United States and europe. Humanities and Social Sciences Communications, 9(1), 1–14. 10.1057/s41599-022-01316-z

Mridul, A. (2021). Cultured meat to hit UK menus by 2023, says cell-based startup Ivy Farm. The Vegan Review. https://theveganreview.com/cultured-meat-ukmenus-2023-cell-based-startup-ivy-farm/

Napolitano, A. P., Chai, P., Dean, D. M., & Morgan, J. R. (2007). Dynamics of the self-assembly of complex cellular aggregates on micromolded nonadhesive hydrogels. *Tissue Engineering*, 13(8), 2087–2094. 10.1089/ten.2006.019017518713

Newman, L. (2020). *The Promise and Peril of "Cultured Meat"; McGill-Queen's Uuniversity Press*. QU, Canada.

Nie, W., Peng, C., Zhou, X., Chen, L., Wang, W., Zhang, Y., Ma, P. X., & He, C. (2017). Three dimensional porous scaffold by self-assembly of reduced graphene oxide and nano-hydroxyapatite composites for bone tissue engineering. *Carbon*, 116, 325–337. 10.1016/j.carbon.2017.02.013

Nowak, V., Persijn, D., Rittenschober, D., & Charrondiere, U. R. (2016). Review of food composition data for edible insects. *Food Chemistry*, 193, 39–46. 10.1016/j.foodchem.2014.10.11426433285

Ogrodnik, M. (2021). Cellular aging beyond cellular senescence: Markers of senescence prior to cell cycle arrest in vitro and in vivo. *Aging Cell*, 20(4), e13338. 10.1111/acel.1333833711211

Ozbolat, I. T., & Hospodiuk, M. (2016). Current Advances and Future Perspectives in Extrusion-Based Bioprinting. *Biomaterials*, 76, 321–343. 10.1016/j.biomaterials.2015.10.07626561931

Padilha, L., Malek, L., & Umberger, W. (2021). Food choice drivers of potential lab-grown meat consumers in australia. *British Food Journal*, 123(9), 3014–3031. 10.1108/BFJ-03-2021-0214

Park, J. A., Yoon, S., Kwon, J., Kim, Y. K., Kim, W. J., Yoo, J. Y., & Jung, S. (2017). Freeform micropatterning of living cells into cell culture medium using direct inkjet printing. *Scientific Reports*, 7(1), 14610. 10.1038/s41598-017-14726-w29097768

Park, S., Jung, S., Choi, M., Lee, M., Choi, B., Koh, W. G., Lee, S., & Hong, J. (2021). Gelatin MAGIC powder as nutrient-delivering 3D spacer for growing cell sheets into cost-effective cultured meat. *Biomaterials*, 278, 121155. 10.1016/j.biomaterials.2021.12115534607049

Phelan, M. A., Kruczek, K., Wilson, J. H., Brooks, M. J., Drinnan, C. T., Regent, F., Gerstenhaber, J. A., Swaroop, A., Lelkes, P. I., & Li, T. (2020). Soy protein nanofiber scaffolds for uniform maturation of human induced pluripotent stem cell-derived retinal pigment epithelium. *Tissue Engineering. Part C, Methods*, 26(8), 433–446. 10.1089/ten.tec.2020.007232635833

Porto, L., & Berti, F. (2022). Carne Cultivada: Perspectivas e Oportunidades Para o Brasil. Available online: https://gfi.org.br/wp-content/uploads/2022/06/WP-Carne-Cultivada-no-Brasil-GFI-Brasil-05_2022_.pdf

Post, M. J. (2014). Cultured beef: Medical technology to produce food. *Journal of the Science of Food and Agriculture*, 94(6), 1039–1041. 10.1002/jsfa.647424214798

Post, M. J., Levenberg, S., Kaplan, D. L., Genovese, N., Fu, J., Bryant, C. J., Negowetti, N., Verzijden, K., & Moutsatsou, P. (2020). Scientific, sustainability and regulatory challenges of cultured meat. *Nature Food*, 1(7), 403–415. 10.1038/s43016-020-0112-z

Premalatha, M., Abbasi, T., Abbasi, T., & Abbasi, S. A. (2011). Energy-efficient food production to reduce global warming and ecodegradation: The use of edible insects. In Renewable and Sustainable Energy Reviews (Vol. 15, Issue 9, pp. 4357–4360). 10.1016/j.rser.2011.07.115

Profeta, A., Siddiqui, S. A., Smetana, S., Hossaini, S. M., Heinz, V., & Kircher, C. (2021). The Impact of Corona Pandemic on Consumer's Food Consumption. *Journal für Verbraucherschutz und Lebensmittelsicherheit*, 16(4), 305–314. 10.1007/s00003-021-01341-134421498

Quesada-Salas, M. C., Delfau-Bonnet, G., Willig, G., Préat, N., Allais, F., & Ioannou, I. (2021). Article optimization and comparison of three cell disruption processes on lipid extraction from microalgae. *Processes (Basel, Switzerland)*, 9(2), 1–20. 10.3390/pr9020369

Radaei, P., Mashayekhan, S., & Vakilian, S. (2017). Modeling and optimization of gelatin- chitosan micro-carriers preparation for soft tissue engineering: Using Response Surface Methodology. 10.1016/j.msec.2017.02.108

Reddy, M. S. B., Ponnamma, D., Choudhary, R., & Sadasivuni, K. K. (2021). A comparative review of natural and synthetic biopolymer composite scaffolds. *Polymers*, 13(7), 1105. 10.3390/polym1307110533808492

Reeds, P., Schaafsma, G., Tomé, D., & Young, V. (2000). Criteria and Significance of Dietary Protein Sources in Humans Summary of the Workshop with Recommendations 1.

Rehman, N., Edkins, V., & Ogrinc, N. (2024). Is sustainable consumption a sufficient motivator for consumers to adopt meat alternatives? a consumer perspective on plant-based, cell-culture-derived, and insect-based alternatives. *Foods*, 13(11), 1627. 10.3390/foods1311162738890856

Rojas-Tavara, A. (2023). Microalgae in lab-grown meat production. *Czech Journal of Food Sciences*, 41(6), 406–418. 10.17221/69/2023-CJFS

Rombach, M., Dean, D., Vriesekoop, F., de Koning, W., Aguiar, L. K., Anderson, M., Mongondry, P., Oppong-Gyamfi, M., Urbano, B., Gómez Luciano, C. A., Hao, W., Eastwick, E., Jiang, Z. V., & Boereboom, A. (2022). Is cultured meat a promising consumer alternative? Exploring key factors determining consumer's willingness to try, buy and pay a premium for cultured meat. *Appetite*, 179, 106307. 10.1016/j.appet.2022.10630736089124

Rubio, N. R., Xiang, N., & Kaplan, D. L. (2020). Plant-based and cell-based approaches to meat production. Nature Communications, 11(1), 1–11. 10.1038/s41467-020-20061-y

Rumpold, B. A., & Schlüter, O. K. (2013). Potential and challenges of insects as an innovative source for food and feed production. In *Innovative Food Science and Emerging Technologies* (Vol. 17, pp. 1–11). Elsevier Ltd., 10.1016/j.ifset.2012.11.005

Sahirman, S., & Ardiansyah. (2014). Assessment Of Tofu Carbon Footprint In Banyumas, Indonesia - Towards 'Greener' Tofu. Proceeding of International Conference on Research, Implementation And Education Of Mathematics And Sciences, 18–20.

Sahoo, D. R., & Biswal, T. (2021). Alginate and its application to tissue engineering. *SN Applied Sciences*, 3(1), 30–90. 10.1007/s42452-020-04096-w

Sandvig, I., Karstensen, K., Rokstad, A. M., Aachmann, F. L., Formo, K., Sandvig, A., Skjåk-Bræk, G., & Strand, B. L. (2015). RGD-peptide modified alginate by a chemoenzymatic strategy for tissue engineering applications. *Journal of Biomedical Materials Research. Part A*, 103(3), 896–906. 10.1002/jbm.a.3523024826938

Santeramo, F. G., Carlucci, D., De Devitiis, B., Seccia, A., Stasi, A., Viscecchia, R., & Nardone, G. (2018). Emerging trends in European food, diets and food industry. *Food Research International*, 104, 39–47. 10.1016/j.foodres.2017.10.03929433781

Santos, A. C. A., Camarena, D. E. M., Roncoli Reigado, G., Chambergo, F. S., Nunes, V. A., Trindade, M. A., & Stuchi Maria-Engler, S. (2023). Tissue engineering challenges for cultivated meat to meet the real demand of a global market. International Journal of Molecular Sciences, 24(7), 6033. https://gfi.org/wp-content/uploads/2023/01/2022-Cultivated-MeatState-of-the-Industry-Report.pdf(2023)

Segovia, M., Yu, N., & Loo, E. (2022). The effect of information nudges on online purchases of meat alternatives. *Applied Economic Perspectives and Policy*, 45(1), 106–127. 10.1002/aepp.13305

Shahin-Shamsabadi, A., & Selvaganapathy, P. R. (2022). Engineering Murine Adipocytes and Skeletal Muscle Cells in Meat-like Constructs Using Self-Assembled Layer-by-Layer Biofabrication: A Platform for Development of Cultivated Meat. *Cells, Tissues, Organs*, 2022(211), 304–312. 10.1159/00051176433440375

Sharma, S., Thind, S. S., & Kaur, A. (2015). In vitro meat production system: Why and how? *Journal of Food Science and Technology*, 2015(52), 7599–7607. 10.1007/s13197-015-1972-326604337

Shen, C. R., Chen, Y. U. S., Yang, C. J., Chen, J. K., & Liu, C. L. (2010). Colloid chitin azure is a dispersible, low-cost substrate for chitinase measurements in a sensitive, fast, reproducible assay. *Journal of Biomolecular Screening*, 15(2), 213–217. 10.1177/108705710935505720042532

Shi, Z., Zhang, Y., Phillips, G. O., & Yang, G. (2014). Utilization of bacterial cellulose in food. *Food Hydrocolloids*, 35, 539–545. 10.1016/j.foodhyd.2013.07.012

Shyh-Chang, N., & Ng, H. H. (2017). The metabolic programming of stem cells. *Genes & Development*, 31(4), 336–346. 10.1101/gad.293167.11628314766

Siegrist, M., Sütterlin, B., & Hartmann, C. (2018). Perceived naturalness and evoked disgust influence acceptance of cultured meat. *Meat Science*, 139, 213–219. 10.1016/j.meatsci.2018.02.00729459297

Singapore Food Agency (SFA). (2020). How are alternative proteins regulated in Singapore? https://www.sfa.gov.sg/food-information/risk-at-a-glance/safety-of-alternativeprotein(2020)

Singh, A., Kumar, V., Singh, S. K., Gupta, J., Kumar, M., Sarma, D. K., & Verma, V. (2023). Recent advances in bioengineered scaffold for in vitro meat production. *Cell and Tissue Research*, 391(2), 235–247. 10.1007/s00441-022-03718-636526810

Singh, A., Singh, S. K., Kumar, V., Gupta, J., Kumar, M., Sarma, D. K., Singh, S., Kumawat, M., & Verma, V. (2023). Derivation and characterization of novel cytocompatible decellularized tissue scaffold for myoblast growth and differentiation. *Cells*, 13(1), 41. 10.3390/cells1301004138201245

Smetana, S., Mathys, A., Knoch, A., & Heinz, V. (2015). Meat alternatives: Life cycle assessment of most known meat substitutes. *The International Journal of Life Cycle Assessment*, 20(9), 1254–1267. 10.1007/s11367-015-0931-6

Spier, M., Vandenberghe, L., Medeiros, A., & Soccol, C. (2011). Application of different types of bioreactors in bioprocesses. In *Bioreactors: Design, Properties, and Applications* (Vol. 1, pp. 53–87). Nova Science Publishers, Inc.

State of the Industry Report—Cultivated Meat. (2022). Available online: https://gfi.org/wp-content/uploads/2021/04/COR-SOTIRCultivated-Meat-2021-0429.pdf

Stephens, N. (2022). Join Our Team, Change the World: Edibility, Producibility and Food Futures in Cultured Meat Company Recruitment Videos. *Food, Culture, & Society*, 25(1), 32–48. 10.1080/15528014.2021.188478735177960

Stephens, N., Di Silvio, L., Dunsford, I., Ellis, M., Glencross, A., & Sexton, A. (2018). Bringing cultured meat to market: Technical, socio-political, and regulatory challenges in cellular agriculture. *Trends in Food Science & Technology*, 78, 155–166. 10.1016/j.tifs.2018.04.01030100674

Stephens, N., Sexton, A. E., & Driessen, C. (2019). Making sense of making meat: Key moments in the first 20 years of tissue engineering muscle to make food. *Frontiers in Sustainable Food Systems*, 45, 45. Advance online publication. 10.3389/fsufs.2019.0004534250447

Stout, A. J., Mirliani, A. B., Rittenberg, M. L., Shub, M., White, E. C., Yuen, J. S.Jr, & Kaplan, D. L. (2022). Simple and effective serum-free medium for sustained expansion of bovine satellite cells for cell cultured meat. *Communications Biology*, 5(1), 466. 10.1038/s42003-022-03423-835654948

Su, L., Jing, L., Zeng, X., Chen, T., Liu, H., Kong, Y., Wang, X., Yang, X., Fu, C., Sun, J., & Huang, D. (2023). 3D-Printed prolamin scaffolds for cell-based meat culture. *Advanced Materials*, 35(2), 2207397. 10.1002/adma.20220739736271729

Su, X., Xian, C., Gao, M., Liu, G., & Wu, J. (2021). Edible materials in tissue regeneration. *Macromolecular Bioscience*, 21(8), 2100114. 10.1002/mabi.20210011434117831

Sun, J., Zhou, W., Yan, L., Huang, D., & Lin, L. Y. (2018). Extrusion-based food printing for digitalized food design and nutrition control. *Journal of Food Engineering*, 220, 1–1. 10.1016/j.jfoodeng.2017.02.028

Surya, S., Smarak, B., & Bhat, R. (2023). Sustainable polysaccharide and protein hydrogel-based packaging materials for food products: A review. *International Journal of Biological Macromolecules*, 248, 125845. 10.1016/j.ijbiomac.2023.12584537473880

Tabassum-Abbasi & Abbasi, S. A. (2016). Reducing the global environmental impact of livestock production: The minilivestock option. In *Journal of Cleaner Production* (Vol. 112, pp. 1754–1766). Elsevier Ltd. 10.1016/j.jclepro.2015.02.094

Tacon, A. G. J., & Metian, M. (2008). Global overview on the use of fish meal and fish oil in industrially compounded aquafeeds: Trends and future prospects. *Aquaculture (Amsterdam, Netherlands)*, 285(1–4), 146–158. 10.1016/j.aquaculture.2008.08.015

Tan, J., & Joyner, H. S. (2020). Characterizing wear behaviors of edible hydrogels by kernel-based statistical modeling. J. Food Eng. 275, 109850. doi:.jfoodeng.2019.10985010.1016/j

Tessmar, J. K. A. M., & Göpferich, A. M. (2007). Customized PEG-derived copolymers for tissue- € engineering applications. *Macromolecular Bioscience*, 7(1), 23–39. 10.1002/mabi.20060009617195277

Thyden, R., Perreault, L. R., Jones, J. D., Notman, H., Varieur, B. M., Patmanidis, A. A., Dominko, T., & Gaudette, G. R. (2022). An edible, decellularized plant derived cell carrier for lab grown meat. *Applied Sciences (Basel, Switzerland)*, 12(10), 5155. 10.3390/app12105155

TingWei, JingWen, XinRui, GuoQiang, XueLiang, GuoCheng, Jian, & XiuLan. (2019). Research progress on lab-grown meat risk prevention and safety management norms. *Shipin Yu Fajiao Gongye*, 45, 254–258.

Tsvakirai, C. Z., & Nalley, L. L. (2023). The coexistence of psychological drivers and deterrents of consumers' willingness to try cultured meat hamburger patties: Evidence from South Africa. *Agricultural and Food Economics*, 11(1), 1–17. 10.1186/s40100-023-00293-4

USDA. (2022). ERS - Meat Price Spreads.

van den Heuvel, E., Newbury, A., & Appleton, K. M. (2019). The psychology of nutrition with advancing age: Focus on food Neophobia. *Nutrients*, 11(1), 151. 10.3390/nu1101015130642027

Van Loo, E. J., Caputo, V., & Lusk, J. L. (2020). Consumer Preferences for Farm-Raised Meat, Lab-Grown Meat, and Plant-Based Meat alternatives: Does Information or Brand Matter? *Food Policy*, 95, 101931. 10.1016/j.foodpol.2020.101931

Vandeweyer, D., Lievens, B., & Van Campenhout, L. (2020). Identification of bacterial endospores and targeted detection of foodborne viruses in industrially reared insects for food. *Nature Food*, 1(8), 511–516. 10.1038/s43016-020-0120-z37128070

Vegaconomist. (2022). Cultured Meat in Europe: Which Country Is Leading the Race? 2022. Available online: https://vegconomist. com/cultivated-cell-cultured-biotechnology/cultured-meat-in-europe-which-country-is-leading-the-race/

Verbeke, W., Sans, P., & Van Loo, E. J. (2015). Challenges and prospects for consumer acceptance of cultured meat. *Journal of Integrative Agriculture*, 14(2), 285–294. 10.1016/S2095-3119(14)60884-4

Verni, M., Demarinis, C., Rizzello, C. G., & Pontonio, E. (2023). Bioprocessing to Preserve and Improve Microalgae Nutritional and Functional Potential: Novel Insight and Perspectives. In Foods (Vol. 12, Issue 5). MDPI. 10.3390/foods12050983

Waltz, E. (2021). Club-goers take first bites of lab-made chicken. *Nature Biotechnology*, 39(3), 257–258. 10.1038/s41587-021-00855-133692516

Wang, Y., Tibbetts, S. M., & McGinn, P. J. (2021). Microalgae as sources of high-quality protein for human food and protein supplements. In Foods (Vol. 10, Issue 12). MDPI. 10.3390/foods10123002

Welin, S. (2013). Introducing the new meat. Problems and prospects. *Etikk i Praksis*, 7(1), 24–37. 10.5324/eip.v7i1.1788

Wells, M. L., Potin, P., Craigie, J. S., Raven, J. A., Merchant, S. S., Helliwell, K. E., Smith, A. G., Camire, M. E., & Brawley, S. H. (2017). Algae as nutritional and functional food sources: revisiting our understanding. In Journal of Applied Phycology (Vol. 29, Issue 2, pp. 949–982). Springer Netherlands. 10.1007/s10811-016-0974-5

Wilks, M., & Phillips, C. J. C. (2017). Attitudes to in Vitro Meat: A Survey of Potential Consumers in the United States. *PLoS One*, 2017(12), e0171904. 10.1371/journal.pone.017190428207878

Williams, L. A., Davis-Dusenbery, B. N., & Eggan, K. C. (2012). Snapshot: Directed differentiation of pluripotent stem cells. *Cell*, 149(5), 1174–1174. 10.1016/j.cell.2012.05.01522632979

Włodarczyk-Biegun, M. K., Farbod, K., Werten, M. W., Slingerland, C. J., De Wolf, F. A., Van Den Beucken, J. J., Leeuwenburgh, S. C. G., Cohen Stuart, M. A., & Kamperman, M. (2016). Fibrous hydrogels for cell encapsulation: A modular and supramolecular approach. *PLoS One*, 11(5), e0155625. 10.1371/journal.pone.015562527223105

Xiang, N., Yuen, J. S.Jr, Stout, A. J., Rubio, N. R., Chen, Y., & Kaplan, D. L. (2022). 3D porous scaffolds from wheat glutenin for cultured meat applications. *Biomaterials*, 285, 121543. Advance online publication. 10.1016/j.biomaterials.2022.12154335533444

Xu, E., Niu, R., Lao, J., Zhang, S., Li, J., Zhu, Y., ... Liu, D. (2023). Tissue-like cultured fish fillets through a synthetic food pipeline. NPJ Science of Food, 7(1), 17.

Yang, Z., Yuan, S., Liang, B., Liu, Y., Choong, C., & Pehkonen, S. O. (2014). Chitosan Microsphere Scaffold Tethered with RGD-Conjugated Poly (methacrylic acid) Brushes as Effective Carriers for the Endothelial Cells. *Macromolecular Bioscience*, 14(9), 1299–1311. 10.1002/mabi.20140013624895289

Ye, L., Yao, F., & Li, J. (2023). Chapter 6 - peptide and protein-based hydrogels. In Sustainable hydrogels. Elsevier.

Yegappan, R., Selvaprithiviraj, V., Amirthalingam, S., & Jayakumar, R. (2018). Carrageenan based hydrogels for drug delivery, tissue engineering and wound healing. *Carbohydrate Polymers*, 198, 385–400. 10.1016/j.carbpol.2018.06.08630093014

Young, S. (2022). Celling Meat—Is Cultivated Meat Really Here to Stay? Available online: https://thefarmermagazine.com.au/cellingmeat-is-cultivated-meat-here-to-stay/

Zeltinger, J., Sherwood, J. K., Graham, D. A., Müeller, R., & Griffith, L. G. (2001). Effect of pore size and void fraction on cellular adhesion, proliferation, and matrix deposition. *Tissue Engineering*, 7(5), 557–572. 10.1089/10763270175321318311694190

Zhang, J., Liu, L., Liu, H., Shi, A., Hu, H., & Wang, Q. (2017). Research advances on food extrusion equipment, technology and its mechanism. *Nongye Gongcheng Xuebao (Beijing)*, 2017(33), 275–283.

Zhang, L., Hu, Y., Badar, I. H., Xia, X., Kong, B., & Chen, Q. (2021). Prospects of artificial meat: Opportunities and challenges around consumer acceptance. *Trends in Food Science & Technology*, 116, 434–444. 10.1016/j.tifs.2021.07.010

Zhang, S., Xing, M., & Li, B. (2018). Biomimetic Layer-by-Layer Self-Assembly of Nanofilms, Nanocoatings, and 3D Scaffolds for Tissue Engineering. *International Journal of Molecular Sciences*, 2018(19), 1641. 10.3390/ijms1906164129865178

Zhao, Y., & Wang, S. (2022). Experimental and biophysical modeling of transcription and translation dynamics in bacterial-and mammalian-based cell-free expression systems. *SLAS Technology*.35231628

Zheng, Y.-Y., Zhu, H.-Z., Wu, Z.-Y., Song, W.-J., Tang, C.-B., Li, C.-B., Ding, S.-J., & Zhou, G.-H. (2021). Evaluation of the effect of smooth muscle cells on the quality of cultured meat in a model for cultured meat. *Food Research International*, 150, 110786. 10.1016/j.foodres.2021.11078634865801

Zhou, X. H., Yin, L., Yang, B. S., Chen, C. Y., Chen, W. H., Xie, Y., Yang, X., Pham, J. T., Liu, S., & Xue, L. J. (2021). Programmable local orientation of micropores by mold-assisted ice templating. *Small Methods*, 5(2), 2000963. Advance online publication. 10.1002/smtd.20200096334927890

Chapter 9
Future of Cellular Agriculture

Idris Adewale Ahmed
Lincoln University College, Malaysia

Ibrahim Bello
North Dakota State University, USA

Abdullateef Akintunde Raji
Federal University of Kashere, Nigeria

Maryam Abimbola Mikail
Mimia Sdn Bhd, Malaysia

ABSTRACT

Cellular agriculture, a transformative field at the intersection of biotechnology and food production, is poised to revolutionize the global food system. This review delves into the multifaceted landscape of cellular agriculture, examining key aspects such as technological advancements driving innovation, emerging market trends and lucrative opportunities, regulatory frameworks shaping industry development, and the environmental impact of transitioning to cellular agriculture. Moreover, it explores consumer acceptance and perception, crucial for mainstream adoption, alongside economic viability considerations and the evolving supply chain and infrastructure requirements. Ethical considerations, particularly concerning animal welfare, are scrutinized, highlighting the importance of addressing these concerns for industry sustainability. Furthermore, the review also evaluates the roles of international collaboration and partnerships in fostering growth and overcoming challenges.

DOI: 10.4018/979-8-3693-4115-5.ch009

Copyright © 2024, IGI Global. Copying or distributing in print or electronic forms without written permission of IGI Global is prohibited.

INTRODUCTION

With the global population being predicted to grow to about 9-11 billion people by 2050, the potential challenges that would confront the current food production systems amidst the threat of climate change and limited arable land are unimaginable (Eibl et al., 2021; Moritz et al., 2022). Such challenges include but are not limited to malnutrition, food insecurity, insufficient food availability for the growing population, and the pressure to reduce the environmental footprint of conventional agriculture (Nyika et al., 2021). Furthermore, the increased urbanization and economic development in low- and middle-income countries are continuously expected to bring certain changes in consumption patterns and lifestyles such as increased consumption of protein-dense animal-sourced foods. Thus, there will be an expected rise in concern over the negative environmental impact of meat production as its demand rises (Talwar et al., 2024).

The field of agriculture has thus become very challenging, especially in the era of increasing advocacy for sustainability. The adverse megatrends in agriculture such as increasing deforestation, land usage, climate change, human health issues, pollution of water bodies, and the ethical aspect of rearing and eating animals, sustainability challenges, and global population growth have also become a global concern for both crop farming and livestock production (Moritz et al., 2022; Räty et al., 2023). Current agricultural policies in most industrialized and Western countries favor specialized large-scale farms which are associated with negative impacts on food safety, animal welfare and the environment. In other words, conventional large-scale farming is one of the major contributors to climate change, water pollution as well as loss of biodiversity and soil degradation. Furthermore, the increased competition in livestock farming has greatly led to an unfavorable situation for rural development. In a transition towards sustainable food systems, it is imperative to ensure healthy and sufficient as well as sustainable nutrition for the growing global population despite the limited natural resources, thereby necessitating several changes in agriculture and food production (Räty et al., 2023).

A novel food production method that is based on in vitro cell cultivation techniques, otherwise known as "cellular agriculture," is being proposed as a solution for organizing future sustainable food systems and the post-farmed-animal bioeconomy (Moritz et al., 2022; Räty et al., 2023).

Cellular agriculture is an emerging technology that involves the production of cell-based plant and animal products with little to no plant or animal involvement to meet the increasing demand for food and nutrition and reduce environmental burdens (Nyika et al., 2021).

Cellular agriculture also involves the use of stem cell biology, tissue engineering, and in some cases, genetic engineering and synthetic biology, to produce animal products without using living animals (Helliwell & Burton, 2021; Stephens et al., 2018). It has thus been advocated as the only means of addressing the sustainability and ethical dilemmas of animal agriculture while meeting and maintaining new demand for animal proteins projected to rise by 70%–60% (Helliwell & Burton, 2021). Cultured meat (also known as in vitro meat or clean meat), produced by cultivating animal cells in a bioreactor in a nutrition medium, is an example of tissue-based cellular agriculture, yet, fermentation-based, where products are fermented by using algae, and bacteria, or yeast rather than animal cells, is another form of cellular agriculture production (Moritz et al., 2022; Stephens et al., 2018). While tissue engineering-based cellular agriculture involves the use of cells or cell lines taken from living animals or a genetically modified cell line for the production of useable tissue with minimal quantities of animal tissue input the cells, fermentation-based uses genetically modified microorganisms through recombinant DNA rather than tissue from a living animal (Stephens et al., 2018).

Depending on the type of products, cellular agriculture is usually categorized into broad divisions, namely, cellular and acellular products (Figure 1). Cellular products are the cultivated cells themselves while acellular products are only the substances produced by the cultivated cells. Cellular products include cultivated plant cells that are being used in cosmetics, cultured coffee, and pharmaceuticals or cultivated animal cells that are being used for the production of cultured meat. On the other hand, acellular products include materials such as egg white protein, milk proteins, and fatty acids that are synthesized by microbes, algae, or yeast (Räty et al., 2023).

Figure 1. Categories of Cellular Agriculture; Cellular and Acellular Products

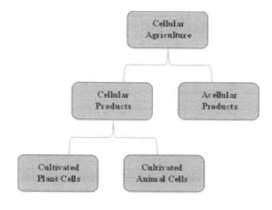

Cultured animal products promise an efficient, safe, humane system of animal protein production, and more environmentally sustainable as against conventional animal agriculture's perpetuation of animal suffering and slaughter, the potential for prolific antibiotic use and zoonotic disease, as well as broad environmental impacts such as greenhouse gas emissions, land and water pollution, and energy use (Helliwell & Burton, 2021; Nyika et al., 2021).

The world's first laboratory-grown burger was unveiled in London at a press conference on the 5th of August, 2013 by Mark Post (a scientist from a research group at Maastricht University). Though the first food product from synthesized animal proteins (an ice cream) was made available by Perfect Day in the US as a limited-run product less than 6 years later, Eat Just in Singapore was the first restaurant to receive regulatory approval from the Singapore Food Agency to sell its cell-cultured chicken to the public in November/December 2020 (Helliwell & Burton, 2021).

The U.S. Food and Drug Administration, in June 2023, also approved the sale of cultured meat in the USA. Though the complex technologies involved in the production of cultured meat products and the associated challenges, such as the scaling of production, cost-effectiveness, and animal-free growth media continue to hinder the marketability of cultured meat products, simpler acellular products as well as plant-based cellular products have emerged in the cosmetics and food markets. Nevertheless, cellular agriculture technologies hold great potential and may not only be successful but also more easily accepted if they can improve animal welfare, provide higher profitability, and more importantly reduce the environmental footprint of food systems (Räty et al., 2023).

On the other hand, potential threats such as the increased consumer alienation from food production systems and the monopolization of cellular agriculture technologies by large multinational companies are some of the growing concerns about cellular agriculture (Räty et al., 2023). Therefore, in this review study, we aimed to present the current status of the multifaceted landscape of cellular agriculture by examining key aspects such as technological advancements, emerging market trends and lucrative opportunities, regulatory frameworks shaping the industry development, the environmental impact of transitioning to cellular agriculture, consumer acceptance and perception, ethical considerations, particularly concerning animal welfare, the roles of international collaboration and partnerships in fostering growth and overcoming challenges, and cellular agriculture's prospects and challenges.

MARKET OPPORTUNITIES

The global market opportunities for cellular agriculture and alternative proteins vary according to cultural, economic, environmental, and social factors in both high-income countries and low- and medium-income countries.

Though the market is growing globally across regions, the Asia-Pacific region accounts for the fastest-growing alternative proteins market with an expected compound annual growth rate of 18.5% from 2021 to 2027 as compared to >17.5% in the United States even though the US market exceeded 50 billion USD in 2020 (Talwar et al., 2024).

Although still modest in comparison to traditional agribusiness and the well-established plant-based food sector, cellular agriculture has undoubtedly drawn significant investment from Asia, Europe, and the Americas in recent years (Ye et al., 2022). Investors are recognizing the potential of cellular agriculture to revolutionize the food industry by providing sustainable and ethical alternatives to conventional meat production. Certain businesses with long-term strategic goals have gradually shifted their focus from improving traditional meat production techniques to investigating meat replacements and building healthy meat manufacturing and supply chains. This shift is underscored by the projected growth of the global market for meat replacements, from $1.9 billion in 2021 to $4 billion in 2027 (Ye et al., 2022). This substantial market potential is attracting a diverse array of stakeholders, including new startups, established food firms, SMEs, large corporations from other industries, financial institutions, and even governments.

ANIMAL WELFARE CONSIDERATION

The global demand for animal products continues to rise, placing immense pressure on traditional agricultural systems (Godfray and Garnett, 2014). In response to this challenge, cellular agriculture has emerged as a promising alternative approach to producing animal-derived foods (Bapat et al., 2022; Rischer et al., 2020). Through cellular agriculture, the need for raising and slaughtering animals is bypassed. While this technology offers potential benefits in terms of sustainability, environmental impact, and human health, it also raises ethical questions and welfare concerns that necessitate careful examination. Proponents of cellular agriculture advocate for its widespread adoption by highlighting its utilitarian benefits (Dutkiewicz and Abrell, 2021). They argue that embracing this technology could mitigate the negative impacts of animal agriculture, such as reducing the number of animals slaughtered for food, while also enhancing animal welfare. By cultivating animal tissues in controlled lab environments, cellular agriculture aims to minimize or even eradicate the suffering endured by animals in conventional farming setups. This ethical stance prioritizes the reduction of harm and suffering inflicted upon sentient beings, aligning with principles of compassion and welfare.

However, cellular agriculture's ethical framework is shaped by a comprehensive examination of society's widespread acceptance of animal agriculture, which inflicts significant harm on animal populations each year (Pond et al., 2011). In the United States alone, nearly ten billion chickens are slaughtered annually (Welty, 2007), with the global figure surpassing 70 billion (Ilea, 2009), not to mention the vast number of marine creatures killed to meet human consumption demands (Dutkiewicz and Abrell, 2021). Critics of animal agriculture argue that this acceptance of mass violence is largely due to consumers' lack of understanding regarding the methods used in meat production (Rowe, 2011).

Beneath the surface of consumerism, the lives and deaths of animals occur unnoticed by the majority of individuals. The transformation of human diets into a meat-centric culture, driven by political and economic factors as well as sophisticated marketing tactics, has largely taken place out of public view, particularly in developed regions like the west (Chiles and Fitzgerald, 2018). This has led to a culture of ignorance regarding the true impact of meat production.

However, as awareness grows regarding the numerous harms associated with animal agriculture—such as animal suffering, environmental degradation, and labor exploitation—some consumers are beginning to shift away from traditional meat consumption (Dutkiewicz and Abrell, 2021; Graça et al., 2019). The emerging plant-based meat alternative industry, valued at $1 billion in retail sales, is attracting interest from both omnivores and vegetarians (Safdar et al., 2022). Despite their appeal, plant-based options are limited by their inability to fully replicate the taste and texture of animal-derived meat (Arora et al., 2023).

In contrast, cellular agriculture offers a promising solution that transcends these limitations. Through the cultivation of animal fat and muscle from stem cells, cellular agriculture offers a method to produce animal meat without resorting to traditional slaughter practices This transformative and promising technology can alleviate ethical concerns related to animal welfare, presenting consumers with a sustainable and humane alternative to conventional meat production techniques.

ENVIRONMENTAL ETHICS

Furthermore, the use of antibiotics in traditional animal agriculture promotes the spread of antibiotic-resistant microorganisms, creating serious public health concerns (Manyi-Loh et al., 2018). Cellular agriculture, with its regulated and sterile laboratory conditions, offers the ability to reduce the need for antibiotics in food production, thereby mitigating the threats of antibiotic resistance (Hanley et al., 2024). This innovative approach also offers opportunities to mitigate other environmental impacts, such as curbing deforestation and habitat destruction, due to its requirement of significantly less land than traditional livestock farming. According to a life-cycle assessment by Tuomisto and Teixeira De Mattos (2011), producing a thousand kilograms of cultured meat requires approximately 99% less land and 82%–96% less water compared to conventionally produced European livestock. By utilizing smaller land footprints, cellular agriculture can help preserve vital ecosystems and biodiversity. Additionally, the design of cultured meat inherently reduces food waste, as only the necessary prime cuts are produced rather than the entire carcass (Stephens et al., 2018). Thus, proponents argue that embracing cellular agriculture not only aligns with ethical imperatives to protect the environment but also conserves natural resources for future generations.

MORAL ETHICS

The debate over the moral status of cultivated cells in cellular agriculture raises fundamental problems concerning the ethical treatment of living organisms (Stephens et al., 2018). Some claim that grown cells have no inherent moral value and hence deserve less ethical consideration than conscious animals (Franklin, 2004). Others, however, express concern about the possible manipulation and commercialization of living organisms, such as cells cultivated for food production (Anthis et al., 2023; Schaefer and Savulescu, 2014). These concerns underline the importance of carefully considering the ethical implications of using any form of life solely as a means to an end.

CULTURAL PERSPECTIVE

Certain civilizations have enduring customs and rituals concerning the raising and consumption of animals (Franklin, 1999), and transitioning to lab-grown meat may result in the loss of cultural history and identity (Volden and Wethal, 2021). The Maasai Community in East Africa (Kenya and Tanzania) is well-known for its cattle-rearing practices, which involve livestock not only providing food but also playing an important role in social and cultural rites (King et al., 1984; Quinlan et al., 2016). Cattle are seen as a symbol of prosperity and class, and many ceremonies, such as weddings or adulthood rituals, involve the exchange or sacrifice of cattle. The transition to lab-grown meat has the potential to undermine deeply rooted cultural traditions and destroy the social fabric of Maasai society. Similarly, in many regions of rural India, small-scale farming and livestock raising, such as cows, goats, and poultry, are tightly linked to religious and cultural traditions (Batra, 1981; Jha, 2002; Valpey, 2020). For example, cows are valued in Hinduism, and their care and usage in agriculture and dairy production are regarded as holy acts (Jha, 2002). Despite its potential benefits, lab-grown meat may be perceived as culturally insensitive or even offensive.

CULINARY TRADITIONS

Food is a fundamental component of cultural identity, and conventional meals often incorporate specialized ways of animal rearing, slaughtering, and processing (Anderson, 2014; Montanari, 2006). In Japan, for example, Wagyu beef, which is known for its quality and taste, is the result of meticulous breeding and specialized feeding practices that have been refined over centuries (Matsuishi et al., 2001; Motoyama et al., 2016). The historical significance of Wagyu beef extends beyond its consumption, demonstrating a commitment to craftsmanship and tradition. While lab-grown Wagyu can replicate the flavor and texture of conventionally produced beef, it may lack cultural and culinary value. Similarly, in Spain, the manufacture of Jamón Ibérico, a famous cured ham, includes feeding Iberian pigs' acorns and allowing them to wander freely. This traditional process offers distinct tastes, flavors, and characteristics to the meat, which are highly valued in Spanish culture (Kosmatko and Węglarz, 2018; Szyndler-Nedza et al., 2019). Lab-grown imitations of such products may be seen to lack originality and fail to reflect the region's cultural heritage.

RELIGIOUS CONSIDERATIONS

Religious dietary regulations, as well as the spiritual significance of animal rearing and consumption, play a crucial role in shaping cultural attitudes toward food (Alonso, 2015). Halal dietary standards in Islam require specific methods of slaughtering animals to authenticate the meat as permissible for consumption (Farouk et al., 2016; Nakyinsige et al., 2012). The ethical and spiritual dimensions of Halal slaughter are deeply embedded in Islamic culture and Islamic clerics have continued to debate the acceptance and alignment of lab-grown meat with religious principles (Kashim et al., 2023). Similarly, the Kosher food regulations in Judaism also demand precise slaughtering techniques to ensure the meat is appropriate for consumption. The Rabbinical authorities are actively debating the status of lab-grown meat in light of these regulations and the results of these discussions will have a profound impact on Jewish communities' acceptance of lab-grown meat.

DRIVERS OF GROWTH, DEMAND AND CONSUMPTION

Some of the factors driving the growth, demand, and consumption of cellular agriculture include religious and cultural considerations, the increasing consumption of healthy food ingredients, the growing health consciousness as well as the availability of low-wage labor and raw materials in countries such as China and India. Despite the projected growth of the cellular agriculture market globally the main obstacles to its entry and expansion in low- and medium-income countries include factors such as affordability, cultural barriers, education, regulations, and supply chain scalability (Talwar et al., 2024).

For instance, in high-income countries, though there is a growing awareness regarding the downsides of high animal protein consumption, the adoption of a more plant-based diet is being hindered not only by the limited availability of plant-based products but also majorly by meat consumption habits. Furthermore, the integration of cellular agriculture into existing culinary and traditions requires the adaptation of the products to local flavors and textures (Stephens et al., 2018). Other motivators for the acceptance of cultured meat in Germany include the assurance that cultured meat is free from antibiotics, then the importance of food safety, animal welfare, and environmental benefits (Moritz et al., 2022). In other words, environmental and sustainability factors are the major drivers of transition in high-income countries while food preferences, nutritional value, and sociocultural factors such as habits, market food prices, and traditions are the major drivers of transition in low-income countries

CONSUMER ACCEPTANCE

Despite the potential of cellular agriculture in meeting and coping with some of the challenges associated with conventional agriculture, it is also very important to address the question of its acceptability by consumers and the general public. Cellular agriculture's viability depends on consumer acceptance of meat produced in laboratories (Stephens et al., 2018). The market's acceptance of lab-grown meat can be strongly influenced by public perception and cultural attitudes. According to the literature, there are several criteria determining consumer acceptance or rejection of any novel agro-food technologies as well as their resulting end-products. The first set of determinants includes the objective and subjective perceived personal and societal benefits, the risks of the technology, and the perceived differences in who essentially benefits and who bears the risks that are associated with the technology and its end products (Table 1). The second set of determinants is related to the technology itself. Such technology-related perceptions are mostly related to the objective and subjective perceived scientific knowledge or uncertainty, perceived controllability of the technological processes (such as safety monitoring and quality control), and perceived naturalness of the technology and its end products. Other determinants include but are not limited to the role, content, and quantity of media coverage, public awareness and familiarity with the technology, general trust in science, the perceived efficacy of the regulatory framework and regulation in the food domain, and the level of consumer or public involvement in the technology development process (Verbeke et al., 2015).

Table 1. Determinants of Consumer Acceptance of Cellular Agriculture

Categories of Determinants	
First Category	• Objective and subjective perceived personal and societal benefits. • The risks of the technology. • Perceived differences in who essentially benefits and who bears the risks that are associated with the technology and its end products.
Second Category is about technology itself.	• Objective and subjective perceived scientific knowledge or uncertainty. • Perceived controllability of the technological processes. • Perceived naturalness of the technology and its end products
Other determinants	• Role, content, and quantity of media coverage • Public awareness and familiarity with the technology. • General trust in science. • Perceived efficacy of the regulatory framework and regulation in the food domain. • Level of consumer or public involvement in the technology development process.

For instance, the perceived unnaturalness of cellular agriculture could lead to strong reticence among consumers and the general public, considering that the technology represents yet another manipulation of nature to the advantage of man. Though the perceived (un)naturalness is expected to be one of the most challenging issues for cellular agriculture, it has also been argued that the unnatural or natural status of a process or its end product does not necessarily translate to the goodness or otherwise of the product. Besides, a product may be natural even if produced in an unnatural way (Kadim et al., 2015; Verbeke et al., 2015).

Public-private collaborations can be extremely helpful in addressing consumer concerns regarding safety, ethics, and taste, as well as in promoting the cultural integration of these novel goods and teaching customers about the advantages of cellular agriculture. Overcoming skepticism and fostering customer acceptability can be accomplished with the use of persuasive communication techniques and clear facts regarding the benefits to health, the environmental impact, and the production method of lab-grown beef (Cecchinato, 2021).

ECONOMIC VIABILITY

Cellular agriculture is a new trend emanating from the field of Biotechnology. It involves the development of animal tissues or organs from animal cells, using tissue culture and fermentation techniques, for the manufacturing of protein. The protein sources are then made available for commercialization. All over the world, the need for animal protein sources of food cannot be overemphasized. Therefore, with the emergence of cellular agriculture, it is highly hoped that consumer patronage, due to the inevitable demand factor, will be overwhelming. Given the above, the venture promises to be lucrative and highly worthwhile. The business can go a long way to impact national GDP, especially in developing nations. This is because the cost of the available animal source of protein is too high for an average citizen to bear, hence, the new cellular agriculture foods landing at the market at cheaper and affordable cost shall be an inevitable alternative for the citizenry. Several researchers have confirmed the potential of cellular agriculture to deliver protein, and other food products such as fats, flavors, etc., for human consumption at reasonable and bearable prices. The technique has shown to be environmentally friendly and sustainable (Smith et al., 2022).

There is no doubt that the new trend in animal sources of protein production has erupted arguments, especially on ethical considerations, the scientific procedures only needed to be clarified and simplified for adherents of faith to comprehend, once things are clear, there is likely to be no hindrance to the large-scale market targets. Furthermore, the emergence of cellular agriculture could create jobs for the teaming national youths who are jobless. It possesses the potential to increase food production, and new business opportunities (Newman et al., 2021).

Over time, other protein products from animals, such as dairy, could be subjected to production for human consumption. There is no doubt that across the world, the shortage of milk and other dairy products is reaching an alarming stage, especially in developing nations. The need for government to engage in proper funding from conception to commercialization, and proper orientation to carry the citizens along is highly inevitable. Some of the major problems encountered by conventional cattle herders, poultry farmers, and other animal farmers, such as animal feeds, diseases, drought, and challenging environmental hazards shall be overcome, thus, ultimately leading to reduced cost of production and improved risk control.

SUPPLY CHAIN AND INFRASTRUCTURE

The emergence of cellular agriculture for the production of animal sources of proteins for human consumption presents potential new business opportunities for people. The procedures entail the utilization of animal cells for tissue culture and fermentation and bioengineering techniques, for manufacturing of the proteins. The fact that the products of the project are meant to be supplied to the final consumers at large commercial scales warrants a definite supply chain for easy accessibility (Manuel Jesús & Garzón-Moreno, 2022).

Given the above, since the laboratories constitute the primary infrastructure for production, it becomes highly necessary for industries to join and collaborate as a large-scale deposit center for high-level refrigeration and preservation to prevent spoilage. The need to engage the state-of-the-art packaging system, allowing for long-distance shipping and transportation also becomes essential for a business meant to cover a large area of distribution; therefore, necessitating the involvement of some notable industries in line with the concept of `from laboratories to the market. The next infrastructure along the chain of supply after the industries are the warehouses/store rooms where wholesale marketers shall deposit the purchased products for industries/companies. The warehouses/store rooms must be well designed to serve the purpose of proper storage, refrigeration, and preservation for retailers to be able to access the protein products while retaining their qualities. The retailers shall also build suitable mini warehouses/store rooms for depositing the purchased products pending when the hawkers or consumers will come to buy (Chandramohan et al., 2023). In summary, therefore, the chain of supply will follow the trend illustrated in Figure 2.

Figure 2. The chain of supply from laboratory to consumers of cellular agricultural products

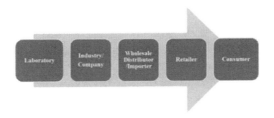

INTERNATIONAL COLLABORATION AND PARTNERSHIPS

According to Stephens and Ellis (2020), modern cellular agriculture has evolved from an academic focus to a catalyst for change that is drawing interest and cooperation from all across the world. Producing agricultural products from cell cultures as opposed to whole plants or animals is a new topic that has the potential to address important challenges including food security, animal welfare, and sustainability. Nevertheless, there are many obstacles in the way of achieving this potential, and international cooperation and strategic alliances will be crucial to the development of cellular agriculture in the future.

International collaboration and public-private partnerships are increasingly proving to be the driving forces behind the emerging cellular agriculture industry (Yuen, 2017). These collaborations have enabled the sharing and exchange of knowledge, resources, and technologies across borders, facilitating advancements that might be unattainable by individual entities working in isolation. For example, research institutions in different countries can pool their expertise to tackle scientific hurdles such as optimizing cell growth mediums and scaling up production processes. One notable instance is the cooperation between the European Union and various Asian countries in funding and researching cellular agriculture (Boden et al., 2010; Post et al., 2020). Such partnerships not only accelerate technological innovation but also help harmonize regulatory standards, ensuring that products can be safely and efficiently brought to market worldwide.

Nevertheless, the nascent field of cellular agriculture encounters numerous noteworthy obstacles, such as those related to technology, regulations, and market adoption (Nyika et al., 2021). It will take strategic alliances and concerted efforts to overcome these obstacles and other technical challenges, utilizing the diverse stakeholder groups' capabilities to promote innovation and guarantee the system's successful integration.

TECHNICAL CHALLENGES

The tissue engineering being largely explored in cellular agriculture has mostly been focused on medical applications such as regenerative medicine as well as drug discovery and toxicology from in-vitro models. While the technical principles are the same for food production such as cultured meat, the food scale is not only much larger but also the product must be affordable as a commodity. Furthermore, raw materials' grade (purity) should not be as high and the regulatory requirements should not be so stringent as biomedical applications (Stephens et al., 2018). Furthermore, cellular agriculture epitomizes a shift in the chemical systems, energy, and land that support food production. Though the technology presents opportunities for environmental and health improvements, achieving such desirable outcomes, however, will require personal (such as maintaining a healthy lifestyle) and corporate commitments (such as following and maintaining established cleanliness protocols) as well as a realistic understanding of the technology involved in guiding its development (Mattick, 2018). On the other hand, it is a difficult and expensive process to scale up manufacturing from the lab to commercial levels (Yuen, 2017; Chan et al., 2024). There are several technological challenges in moving from small-scale lab research to large-scale manufacturing, including cost-cutting measures, consistent product quality, and cell growth medium optimization. To create scalable and affordable production techniques, academic institutions, biotech firms, and food producers must work together. These collaborations can quicken the pace of technological development and move lab-grown meat closer to the general public by combining resources and knowledge.

ENVIRONMENTAL IMPACT

Cellular agriculture is an important revolutionary technology that would not only allow more food production on less land than ever before and mitigate eutrophication in some of the world's oceans but also offer other benefits such as involving animal welfare, human health, and environmental sustainability. However, none of these benefits is absolutely guaranteed without wise decision-making that is based on a realistic understanding of the possibilities and challenges of cellular agriculture presents. Additionally, the absence of livestock and the use of less land would not necessarily translate into efficient or sustainable processes, especially in terms of energy requirements. Food producers must continuously engage in realistic environmental assessment and ongoing innovation toward energy efficiency (Mattick, 2018). The industrial nature of cellular agriculture means that there is still a high uncertainty in the efficiency of the technology in terms of energy requirements. For instance, cultured meat is associated with high energy consumption thus, may lead to higher greenhouse gas emissions than conventional agriculture (Mattick et al., 2015).

REGULATORY CHALLENGES

The regulatory environment around cellular agriculture is still changing, with different nations implementing different regulatory frameworks. This lack of consistency presents a big obstacle for businesses trying to expand into new areas.

International cooperation can aid in the development of uniform laws that guarantee product safety and ease the entry of products into various markets.

For cell-based food to be distributed globally, these laws must be harmonized to enable goods to be advertised and sold with uniform safety and quality criteria everywhere (Chan et al., 2024). Building consumer confidence in lab-grown meat products will also be aided by the establishment of clear and uniform regulatory frameworks. To create uniform regulations for cellular agriculture, regulatory agencies from several nations can engage in talks facilitated by international organizations and trade associations (Fraser et al., 2024). Companies will be able to operate more effectively internationally, and the approval process will be streamlined by the creation of a single regulatory framework.

CONCLUSION

Ensuring adequate, healthy, nutritious, environment-friendly and sustainably sourced food for all and sundry remains a priority in the midst of increasingly growing world population. The cellular agriculture is expected to play a crucial role to play in meeting these growing demands. Cellular agriculture can also help to shape a better, sustainable and more hopeful future for all with continued research and innovation as well as investment. The evaluation of the market opportunities in each region is vital. The role of cultural, economic, environmental, and social factors is also very important. The energy requirement of the cellular agriculture also calls for innovation towards energy efficiency owing to the fact the technology is industrially-drives and any use of alternative fuels would have different impacts on diverse stakeholders. Therefore, addressing the challenges facing cellular agriculture will require a multi-faceted approach, involving collaboration among various stakeholders, including governments, research institutions, industry players, and consumer advocacy groups. Furthermore, to find creative answers to technical problems, governments and businesses might collaborate to support research projects. Food producers can offer useful insights into market demands and production scalability, while academic institutions can collaborate with biotech companies to conduct innovative research. In addition, educational campaigns regarding the advantages of lab-grown meat can be created by public-private partnerships. These campaigns can address common misconceptions and highlight the benefits of cellular agriculture on health, animal welfare, and the environment through product demonstrations, taste tests, and information sessions. To guarantee uniformity in quality and safety, industry standards for cell-based meat manufacturing should be established. Establishing a strong and reliable cellular agricultural sector will be facilitated by cooperative efforts to specify best practices and standards.

Thus, the future of cellular agriculture rests on its ability to successfully navigate obstacles related to technology, regulations, and market acceptance. Stakeholders can develop creative solutions, establish favorable regulatory settings, and promote consumer acceptance through strategic collaboration and partnerships. Through collaboration, the international community can fully realize the potential of cellular agriculture, laying the groundwork for an ethical and sustainable food system that can satisfy the needs of an expanding population while reducing its negative effects on the environment.

REFERENCES

Alonso, E. B. (2015). The impact of culture, religion and traditional knowledge on food and nutrition security in developing countries. Retrieved from https://ageconsearch.umn.edu/record/285169/files/30_briones.pdf

Anderson, E. N. (2014). Everyone eats: Understanding food and culture. NYU Press. Retrieved from https://books.google.com/books?hl=en&lr=&id=OKEUCgAAQBAJ&oi=fnd&pg=PP9&dq=Food+is+a+fundamental+component+of+cultural+identity,+and+conventional+meals+often+incorporate+specialized+ways+of+animal+rearing,+slaughtering,+and+processing.+&ots=bJwX_OqW0X&sig=BaqSQGblYkeIAMTXHx25cJ-p48E

Anthis, J. R., Berenstein, P., Beurrier, C., Biehl, C., Birdie, P., Burley, J., . . . d'Origny, G. (2023). Modern meat: the next generation of meat from cells. Retrieved from https://ora.ox.ac.uk/objects/uuid:6dd61d4b-781a-4ccb-bcd6-9029728348fe/files/sfx719n99b

Arora, S., Kataria, P., Nautiyal, M., Tuteja, I., Sharma, V., Ahmad, F., Haque, S., Shahwan, M., Capanoglu, E., Vashishth, R., & Gupta, A. K. (2023). Comprehensive Review on the Role of Plant Protein As a Possible Meat Analogue: Framing the Future of Meat. *ACS Omega*, 8(26), 23305–23319. 10.1021/acsomega.3c0137337426217

Bapat, S., Koranne, V., Sealy, M. P., Shakelly, N., Huang, A., Sutherland, J. W., & Malshe, A. P. (2022). Cellular agriculture: An outlook on smart and resilient food agriculture manufacturing. *Smart and Sustainable Manufacturing Systems*, 6(1), 1–11. 10.1520/SSMS20210020

Batra, S. (1981). Cows and cow-slaughter in India: Religious, political and social aspects. Retrieved from https://repub.eur.nl/pub/34927/

Bierbaum, R., Leonard, S. A., Rejeski, D., Whaley, C., Barra, R. O., & Libre, C. (2020). Novel entities and technologies: Environmental benefits and risks. *Environmental Science & Policy*, 105, 134–143. 10.1016/j.envsci.2019.11.002

Boden, M., Cagnin, C., Carabias-Hütter, V., Haegemann, K., & Könnölä, T. (2010). Facing the future: time for the EU to meet global challenges. Retrieved from https://digitalcollection.zhaw.ch/handle/11475/6040

Cecchinato, A. (2021). The Plant-Based Meat: a study of packaging to communicate a sustainable food product's innovation.

Chan, D. L.-K., Lim, P.-Y., Sanny, A., Georgiadou, D., Lee, A. P., & Tan, A. H.-M. (2024). Technical, commercial, and regulatory challenges of cellular agriculture for seafood production. *Trends in Food Science & Technology*, 144, 104341. 10.1016/j.tifs.2024.104341

Chandramohan, J., Asoka Chakravarthi, R. P., & Ramasamy, U. (2023). A comprehensive inventory management system for non-instantaneous deteriorating items in supplier- retailer-customer supply chains. Supply Chain Analytics, 3, 100015. https://doi.org/https://doi.org/10.1016/j.sca.2023.100015

Chiles, R. M., & Fitzgerald, A. J. (2018). Why is meat so important in Western history and culture? A genealogical critique of biophysical and political-economic explanations. *Agriculture and Human Values*, 35(1), 1–17. 10.1007/s10460-017-9787-7

Dutkiewicz, J., & Abrell, E. (2021). Sanctuary to table dining: Cellular agriculture and the ethics of cell donor animals. Politics and Animals, 7, 1–15. Retrieved from https://journals.lub.lu.se/pa/article/view/22252

Eibl, R., Senn, Y., Gubser, G., Jossen, V., van den Bos, C., & Eibl, D. (2021). Cellular Agriculture: Opportunities and Challenges. *Annual Review of Food Science and Technology*, 12(1), 51–73. 10.1146/annurev-food-063020-12394033770467

Farouk, M. M., Pufpaff, K. M., & Amir, M. (2016). Industrial halal meat production and animal welfare: A review. *Meat Science*, 120, 60–70. 10.1016/j.meatsci.2016.04.02327130540

Franklin, A. (1999). Animals and modern cultures: A sociology of human-animal relations in modernity. Animals and Modern Cultures, 1–224. Retrieved from https://www.torrossa.com/gs/resourceProxy?an=5019341&publisher=FZ7200

Franklin, J. H. (2004). *Animal Rights and Moral Philosophy*. Columbia University Press. 10.7312/fran13422

Fraser, E. D. G., Newman, L., Yada, R. Y., & Kaplan, D. L. (2024). The foundations of cellular agriculture: science, technology, sustainability and society. In Fraser, E. D. G., Kaplan, D. L., Newman, L., & Yada, R. Y. (Eds.), *Cellular Agriculture* (pp. 3–10). Academic Press. 10.1016/B978-0-443-18767-4.00010-X

Godfray, H. C. J., & Garnett, T. (2014). Food security and sustainable intensification. *Philosophical Transactions of the Royal Society of London. Series B, Biological Sciences*, 369(1639), 20120273. 10.1098/rstb.2012.027324535385

Graça, J., Godinho, C. A., & Truninger, M. (2019). Reducing meat consumption and following plant-based diets: Current evidence and future directions to inform integrated transitions. *Trends in Food Science & Technology*, 91, 380–390. 10.1016/j.tifs.2019.07.046

Hanley, L., Zai, B., Reisiger, C., & Glaros, A. (2024). Cellular agriculture and public health, nutrition, and food security. In *Cellular Agriculture* (pp. 407–422). Elsevier. 10.1016/B978-0-443-18767-4.00007-X

Helliwell, R., & Burton, R. J. F. (2021). The promised land? Exploring the future visions and narrative silences of cellular agriculture in news and industry media. *Journal of Rural Studies*, 84, 180–191. 10.1016/j.jrurstud.2021.04.002

Ilea, R. C. (2009). Intensive livestock farming: Global trends, increased environmental concerns, and ethical solutions. *Journal of Agricultural & Environmental Ethics*, 22(2), 153–167. 10.1007/s10806-008-9136-3

Jha, D. N. (2002). The myth of the holy cow. Verso. Retrieved from https://books.google.com/books?hl=en&lr=&id=vTrKWRUOpbcC&oi=fnd&pg=PR9&dq=livestock+like+cows+tied+to+religious+practices+in+india&ots=z-i8Hl4qMe&sig=XTVOzpiXqcq0DsYGuJEQeAzmK3I

Kadim, I. T., Mahgoub, O., Baqir, S., Faye, B., & Purchas, R. (2015). Cultured meat from muscle stem cells: A review of challenges and prospects. *Journal of Integrative Agriculture*, 14(2), 222–233. 10.1016/S2095-3119(14)60881-9

King, J. M., Sayers, A. R., Peacock, C. P., & Kontrohr, E. (1984). Maasai herd and flock structures in relation to livestock wealth, climate and development. *Agricultural Systems*, 13(1), 21–56. 10.1016/0308-521X(84)90054-4

Kosmatko, J., & Węglarz, A. (2018). Production Technology and Nutritive Value of Iberian Ham (jamón ibérico). Retrieved from https://wz.izoo.krakow.pl/files/WZ_2018_4_art16_en.pdf

Manuel Jesús, H.-O., & Garzón-Moreno, J. (2022). Risk management methodology in the supply chain: A case study applied. *Annals of Operations Research*, 313(2), 1051–1075. Advance online publication. 10.1007/s10479-021-04220-y

Manyi-Loh, C., Mamphweli, S., Meyer, E., & Okoh, A. (2018). Antibiotic use in agriculture and its consequential resistance in environmental sources: Potential public health implications. *Molecules (Basel, Switzerland)*, 23(4), 795. 10.3390/molecules2304079529601469

Matsuishi, M., Fujimori, M., & Okitani, A. (2001). Wagyu beef aroma in Wagyu (Japanese Black cattle) beef preferred by the Japanese over imported beef. Nihon Chikusan Gakkaiho, 72(6), 498–504. Retrieved from https://www.jstage.jst.go.jp/article/chikusan1924/72/6/72_6_498/_article/-char/ja/

Mattick, C. S. (2018). Cellular agriculture: The coming revolution in food production. *Bulletin of the Atomic Scientists*, 74(1), 32–35. 10.1080/00963402.2017.1413059

Mattick, C. S., Landis, A. E., Allenby, B. R., & Genovese, N. J. (2015). Anticipatory Life Cycle Analysis of In Vitro Biomass Cultivation for Cultured Meat Production in the United States. *Environmental Science & Technology*, 49(19), 11941–11949. 10.1021/acs.est.5b01614 26383898

Montanari, M. (2006). *Food is culture*. Columbia University Press. Retrieved from https://books.google.com/books?hl=en&lr=&id=SRloQL52eysC&oi=fnd&pg=PR5&dq=Food+is+a+fundamental+component+of+cultural+identity,+and+conventional+meals+often+incorporate+specialized+ways+of+animal+rearing,+slaughtering,+and+processing.+&ots=8qPu4AEpw7&sig=zOjfzqFsFX9r2LmMLmNCfw1PzK0

Moritz, J., Tuomisto, H. L., & Ryynänen, T. (2022). The transformative innovation potential of cellular agriculture: Political and policy stakeholders' perceptions of cultured meat in Germany. *Journal of Rural Studies*, 89, 54–65. 10.1016/j.jrurstud.2021.11.018

Motoyama, M., Sasaki, K., & Watanabe, A. (2016). Wagyu and the factors contributing to its beef quality: A Japanese industry overview. *Meat Science*, 120, 10–18. 10.1016/j.meatsci.2016.04.026 27298198

Nakyinsige, K., Man, Y. B. C., & Sazili, A. Q. (2012). Halal authenticity issues in meat and meat products. *Meat Science*, 91(3), 207–214. 10.1016/j.meatsci.2012.02.015 22405913

Newell, R., Newman, L., & Mendly-Zambo, Z. (2021). The role of incubators and accelearators in the fourth agricultural revolution: A case study of Canada. *Agriculture*, 11(11), 1066. 10.3390/agriculture11111066

Newman, L., Newell, R., Mendly-Zambo, Z., & Powell, L. (2021). Bioengineering, telecoupling, and alternative dairy: Agricultural land use futures in the Anthropocene. *The Geographical Journal*, 188(3), 342–357. 10.1111/geoj.12392

Nyika, J., Mackolil, J., Workie, E., Adhav, C., & Ramadas, S. (2021). Cellular agriculture research progress and prospects: Insights from bibliometric analysis. *Current Research in Biotechnology*, 3, 215–224. 10.1016/j.crbiot.2021.07.001

Pond, W. G., Bazer, F. W., & Rollin, B. E. (2011). Animal welfare in animal agriculture. CRC Press. Retrieved from https://api.taylorfrancis.com/content/books/mono/download?identifierName=doi&identifierValue=10.1201/b11679&type=googlepdf

Post, M. J., Levenberg, S., Kaplan, D. L., Genovese, N., Fu, J., Bryant, C. J., Negowetti, N., Verzijden, K., & Moutsatsou, P. (2020). Scientific, sustainability and regulatory challenges of cultured meat. *Nature Food*, 1(7), 403–415. 10.1038/s43016-020-0112-z

Quinlan, R. J., Rumas, I., Naisikye, G., Quinlan, M. B., & Yoder, J. (2016). Searching for symbolic value of cattle: Tropical livestock units, market price, and cultural value of Maasai livestock. *Ethnobiology Letters*, 7(1), 76–86. https://www.jstor.org/stable/26423652. 10.14237/ebl.7.1.2016.621

Räty, N., Tuomisto, H. L., & Ryynänen, T. (2023). On what basis is it agriculture?: A qualitative study of farmers' perceptions of cellular agriculture. *Technological Forecasting and Social Change*, 196, 122797. 10.1016/j.techfore.2023.122797

Rischer, H., Szilvay, G. R., & Oksman-Caldentey, K.-M. (2020). Cellular agriculture—Industrial biotechnology for food and materials. *Current Opinion in Biotechnology*, 61, 128–134. 10.1016/j.copbio.2019.12.00331926477

Rowe, B. D. (2011). Understanding animals-becoming-meat: Embracing a disturbing education. Critical Education, 2(7). Retrieved from https://ices.library.ubc.ca/index.php/criticaled/article/view/182311

Safdar, B., Zhou, H., Li, H., Cao, J., Zhang, T., Ying, Z., & Liu, X. (2022). Prospects for plant-based meat: Current standing, consumer perceptions, and shifting trends. *Foods*, 11(23), 3770. 10.3390/foods1123377036496577

Schaefer, G. O., & Savulescu, J. (2014). The Ethics of Producing In Vitro Meat. *Journal of Applied Philosophy*, 31(2), 188–202. 10.1111/japp.1205625954058

Smith, D. J., Helmy, M., Lindley, N. D., & Selvarajoo, K. (2022). The transformation of our food system using cellular agriculture: What lies ahead and who will lead it? *Trends in Food Science & Technology*, 127, 368–376. 10.1016/j.tifs.2022.04.015

Stephens, N., Di Silvio, L., Dunsford, I., Ellis, M., Glencross, A., & Sexton, A. (2018). Bringing cultured meat to market: Technical, socio-political, and regulatory challenges in cellular agriculture. *Trends in Food Science & Technology*, 78, 155–166. 10.1016/j.tifs.2018.04.01030100674

Stephens, N., & Ellis, M. (2020). Cellular agriculture in the UK: A review. *Wellcome Open Research*, 5, 12. 10.12688/wellcomeopenres.15685.132090174

Szyndler-Nedza, M., Nowicki, J., & Malopolska, M. (2019). The production system of high-quality pork products–an example. Annals of Warsaw University of Life Sciences-SGGW. *Animal Science (Penicuik, Scotland)*, 58. https://agro.icm.edu.pl/agro/element/bwmeta1.element.agro-813b019a-1c97-4a01-b46a-1be10f806300

Talwar, R., Freymond, M., Beesabathuni, K., & Lingala, S. (2024). Current and Future Market Opportunities for Alternative Proteins in Low- and Middle-Income Countries. *Current Developments in Nutrition*, 8, 102035. 10.1016/j.cdnut.2023.10203538476721

Thompson, P. B., & Nardone, A. (1999). Sustainable livestock production: Methodological and ethical challenges. *Livestock Production Science*, 61(2–3), 111–119. 10.1016/S0301-6226(99)00061-5

Tuomisto, H. L., & Teixeira De Mattos, M. J. (2011). Environmental Impacts of Cultured Meat Production. *Environmental Science & Technology*, 45(14), 6117–6123. 10.1021/es200130u21682287

Valpey, K. R. (2020). Cow care in Hindu animal ethics. Springer Nature. Retrieved from https://library.oapen.org/handle/20.500.12657/22832

Verbeke, W., Sans, P., & Van Loo, E. J. (2015). Challenges and prospects for consumer acceptance of cultured meat. *Journal of Integrative Agriculture*, 14(2), 285–294. 10.1016/S2095-3119(14)60884-4

Volden, J., & Wethal, U. (2021). What happens when cultured meat meets meat culture. Changing Meat Cultures: Food Practices, Global Capitalism, and the Consumption of Animals, 185–206. Retrieved from https://books.google.com/books?hl=en&lr=&id=M3ZPEAAAQBAJ&oi=fnd&pg=PA185&dq=Transitioning+to+lab-grown+meat+may+result+into+loss+of+cultural+history+and+identity.+&ots=lN0v0cRUm2&sig=at7ihJOgFAMCmaCgeS_5qG8ACvk

Welty, J. (2007). Humane slaughter laws. *Law and Contemporary Problems*, 70(1), 175–206. https://www.jstor.org/stable/27592169?casa_token=EpZ-IAHFWDMAAAAA:QcnjGrlYJuGE_2lA2CpkeCoU7WSCHFltShG4CdkwhNNod4Xck5p4yxtZoRCbnR0egXNOEfccOKyGpv3ZLRCE8wOvf11y32CR9BHJpO95AzvmyP0h-mE

Ye, Y., Zhou, J., Guan, X., & Sun, X. (2022). Commercialization of cultured meat products: Current status, challenges, and strategic prospects. *Future Foods : a Dedicated Journal for Sustainability in Food Science*, 6, 100177. 10.1016/j.fufo.2022.100177

Yuen, K. K.-C. (2017). New sustainable models of open innovation to accelerate technology development in cellular agriculture (PhD Thesis). Massachusetts Institute of Technology. Retrieved from https://dspace.mit.edu/handle/1721.1/113537

Chapter 10
Addressing Global Food Security Through Cellular Agriculture

Ahmed M. Alhussaini Hamad
https://orcid.org/0000-0001-5037-9379
Benha University, Egypt

ABSTRACT

In the face of an ever-growing global population and the escalating impacts of climate change, ensuring food security has become a critical challenge of our time. This chapter explores the transformative potential of cellular agriculture as a viable solution to address global food security. Cellular agriculture, which involves producing animal products from cell cultures rather than traditional livestock farming, offers a sustainable and efficient alternative to conventional agriculture. The chapter also examines the socio-economic implications of cellular agriculture, such as its potential to create new job opportunities, support rural economies, and contribute to food equity by making nutritious food accessible to diverse populations. Through case studies and expert insights, this chapter provides a comprehensive overview of how cellular agriculture can play a pivotal role in building a resilient, sustainable, and secure global food system for the future.

1. INTRODUCTION

The organization of the food system, with its largely unsustainable and supply-driven features, is not fit for the 21st century (Pereira et al.2020; Tutundjian et al.2020). The number of people on Earth, with diverse dietary needs and preferences, is accelerating, outpacing the capability of ecosystems and farmers to increase food

DOI: 10.4018/979-8-3693-4115-5.ch010

production. To feed the world's population of 8.5 billion people by 2030 requires a substantial shift in the way we think about and organize the food system. There is also an increasing realization that the food system plays a central role, not only in the food security and health of people, but in the environmental health of the planet too (Bahar et al.2020; Bin Rahman & Zhang, 2023; Sadigov, 2022).

The food system stands at the interface of four desperately urgent challenges: climate change, which it intensifies; land degradation and desertification, as the growing demand for food fuels deforestation and over-exploitation of land; aquatic systems, with food production having a negative impact on the world's creeks and lakes, as well as the seas; and a global pandemic of obesity and non-communicable diseases (Wang & Azam, 2024; Chirwa & Adeyemi, 2020; Fasona et al.2022).

This chapter explores the use of scientific advances at the cellular level to re-imagine food production practices and the food continuum, as a means of addressing these challenges, improving welfare, and feeding a future world where food will be the centerpiece of societal, planetary, and economic issues.

2. BACKGROUND AND SIGNIFICANCE

Global food and hunger issues are complex. Agriculture and the supply of food are impacted by several challenges, including growing population rates, urbanization, shifting diets towards livestock products, climate change, deforestation, and reliance on water. (Mora et al.2020; Vermeulen et al.2020; Sekaran et al.2021). The Food and Agriculture Organization (FAO) of the United Nations projects that the global population will reach approximately 9.7 billion in 2050, with a significant portion of it living in urban areas. As population and urbanization increases, diets continue shifting towards more meat, dairy, and milk-based protein products (Bahar et al.2020; Falcon et al.2022; Gu et al., 2021).

This shift causes strain on our Earth systems such as trusted agricultural soils and new ecological systems with all of the reliance on food production, and the significant carbon and methane footprints that come from animal livestock activity. The dietary World Bank data, an international database, finds that on average, in developing countries, 95% of girls and 89% of boys over 5 years of age are consuming less than the World Health Organization minimum dietary intake recommendations (Kupka et al., 2020; De et al.2020; Bai et al., 2022; Li et al.2020).

In a fast-growing global market, it has taken considerable time for the cellular agriculture field to gain traction as a commercial platform. However, the technology has been receiving increased attention in the past few years. Although cellular agriculture, the process of taking animal cells from stem cell samples and feeding these cells in vitro to generate pet food, seafood, and more animal-free products, has not seen a product to market or a vast amount of capital investment, the technology is still in its infancy (Thyden et al.2022; Bryant & Barnett, 2020; Helliwell & Burton, 2021; Eibl et al.2021).

It is important to emphasize that while it is crucial to advocate for the potential dietary impacts of cellular agriculture before it reaches the masses, the development of these technologies requires time, capital, and research to be realized. However, these products and technologies could ultimately have an important impact in a movement towards greater global food security. With technical development, these products could have a broad societal impact (Moritz et al., 2022; Chiles et al.2021; Mendly-Zambo et al. 2021).

3. GLOBAL FOOD SECURITY CHALLENGES

Future researchers will need to investigate the cutting-edge ethical and legal areas impacting the global production and use of cellular agriculture products. However, researchers cannot assume that the global south will adopt cell-based meat or seafood technologies in the same way as higher-income countries. Platforms for public engagement with cell agriculture technologies should include concerns such as the potential benefits or risks for low- or middle-income countries, which could include both nutrition and food security considerations as well as legal, ethical, or religious dimensions (Reis et al., 2020; Hadi & Brightwell, 2021; Chandimali et al.2024).

To date, cellular agriculture policy discussions have centered on high-income countries, including Israel, the United States, and the Netherlands. Some scholars suggest that the global south could benefit from advancing the cell-based meat industry and from niche markets for cell-based dishes. As with renewable energy and rare earth mining, high-income countries could outsource many of the animal agriculture industry's externalities to low- and middle-income countries (Newman et al.2023; Soice & Johnston, 2021)(Mancini & Antonioli, 2022; Talwar et al.2024).

The cellular agriculture field also faces the challenge of thinking globally when considering cell sources and production solutions. Recent reports call for research, theory, and practices grounded in social justice; as wealthy countries advance cell-based meat and seafood industries, researchers must consider potential benefits and risks at the local level and across the broader food system. Strategic partnerships between producers, researchers, global health and development agencies, as well as governments, will ensure that research addresses global nutrition, food access, and other local food security priorities. The existing, regionalized, and place-based critical food studies literature may expand to include comparative case studies, early-stage analysis of cellular agriculture ethics and policy, and field-building associated with producing and selling cell-based meat and seafood in different countries (Newton & Blaustein-Rejto, 2021; Chiles et al.2021; Eibl et al.2021).

3.1. Population Growth and Urbanization

In cellular agriculture, eliminating the requirement for a farm or environment can facilitate the production of food. The development of cellular agriculture is also capable of alleviating population density and global climate change issues that may arise from traditional farming practices (Rubio et al., 2020; Helliwell & Burton, 2021; Eibl et al.2021).

In addition to its commercial application, cellular agriculture has the added potential to bring about changes in food and the dietary behavior of the next generation. Cellular agriculture can also enrich food supply, offer better approaches to adjust food supply, and create new alternatives to promote food security. There are basic characteristics of cellular agriculture that are in accord with the demands of population growth and urbanization (Nyyssölä et al.2022; Wikandari et al.2021; Mrabet, 2023).

Building facilities in available places within the city, the ability to consume food and address food supply at the place of production, and the shared technique applied to different areas and occasions are all elements that cellular agriculture shares with the new urbanism and big data techniques utilized in urban studies. Additionally, cellular agriculture can deliver multifaceted data through the production and consumption processes, both of which directly reflect urban behavior. Therefore, cellular agriculture is a potential way to increase social welfare in the context of population growth and urban issues (Ghandar et al.2021; Ebenso et al.2022; Andronie et al. 2021; Mohamed, 2023; Galanakis, 2024; Bhuyan et al., 2022).

Global population is projected to grow from about 7.6 billion currently to 8.6 billion in 2030 and 9.8 billion in 2050. Most of this growth is taking place in urban areas, resulting in urbanization and the continual growth and expansion of megacities. Increasing amounts of farmland are used for urbanization and industrialization. Just enough arable land will be left for food production and only limited farmland areas will be suitable for the cultivation of crops, limiting the acreage that can be used for agricultural production. Consequently, improving agricultural productivity is more significant than increasing the acreage to address this global food security concern. The increasing demand for agricultural products, together with the challenges like farmland reduction and climate change, have also drawn our attention to the need to reform and reconstruct the old agricultural system (Falcon et al.2022; Samal et al.2022; Olowe, 2021; Errera et al., 2023).

4. TRADITIONAL AGRICULTURE VS. CELLULAR AGRICULTURE

The main concept of cellular agriculture is producing meat, egg, and milk directly from the cells of an animal without raising livestock. Cultured cellular meat is a rapid emergence technology with momentum in beef, poultry, and fish produced through tissue engineering (Chen et al.2022; Hong et al.2021; Ramani et al.2021).

Such cultured meat can play a role in providing a more sustainable solution in the face of growing world population, global demand for protein, food security concerns, and food safety. To better understand this field of research, we first provide a systematic review through analysis of 413 documents collected from Web of Science, exploring the present state of the cultured meat literature. We also summarize the current modeling approaches in cultured meat related studies. Finally, we identify limitations, discuss opportunities and provide direction for future research in light of the present state of developments (Bryant & Barnett, 2020; Jairath et al.2021; Hadi & Brightwell, 2021; Hubalek et al., 2022; Treich, 2021).

Traditionally, animal meat and by-products have been acquired through the rearing and slaughter of various animals including cows, pigs, chickens, and fish for thousands of years. However, traditional agriculture is neither efficient nor productive (Resare et al.2022; Van Eenennaam & Werth, 2021; Birhanu et al.2021).

For every unit of food produced, multiple units of water, space, and energy are required. This has led to various environmental concerns. Between 80-90% of arable land is dedicated to animal agriculture, amounting to over 1 billion hectares. In comparison, just 30 million hectares are used for fruits, vegetables, and nuts. Demand for animal products is only projected to increase (Van Eenennaam & Werth, 2021; Zhichkin et al.2021; Piwowar, 2020; Jebari et al.2024).

By 2050, the world will be looking at a 69% increase in calorie and an 88% increase in protein consumption. The traditional agriculture status quo may not be up to meet these demands. Cellular agriculture presents a potential alternative (Van Dijk et al., 2021; Nadathur et al.2024; Glaros et al.2022).

4.1. Definition and Principles

Because of these damages, the global population is scheduled to reach nine billion by 2050 and due to food shortages, the world will have to double food production. At the same time, with global warming and the desire for several reforms, agriculture is the production system that emits the most greenhouse gases. Not to mention the effects of the use of tropical forests and wilderness in this activity. To meet the needs of future people, several sectors are working to redefine food production and the cellular agriculture sector is already a solution. Some large companies are investing in this area, João de Roma, to launch the first individual preparation. In addition, researchers are already developing work with other products, e.g., egg whites and fish. (Gu et al., 2021; Bahar et al.2020; Falcon et al.2022; Aryal et al., 2022; Sadigov, 2022; Vollset et al.2020; Tian et al.2021; Van Dijk et al., 2021 ; Ezeh et al., 2020).

The cellular agriculture model improves food production by using a tip of the biotechnology tool to grow animal products such as meat, eggs, and dairy without the need to raise or slaughter animals. The process consists of some steps, is the development of a cell line from an animal source, and, from that, developing a method to grow and differentiate these cells into a final product. In this model, there is no usage of animal parts for nutrient extraction and the condition of the developed products with slaughter can be found in other food production methods. The products on the market are mainly related to meat as there are more researches in this area, but there is also work with milk for baby formula and medicines. However, despite the promising future of such products, there is still industrial manufacture and popular acceptance in this area (Nyyssölä et al.2022; Helliwell & Burton, 2021; Rischer et al.2020; Eibl et al.2021; Mendly-Zambo et al.2021).

5. BIOTECHNOLOGICAL ADVANCES IN CELLULAR AGRICULTURE

Given the stated nonrenewability of these finite agricultural resources, cellular agriculture, once developed, provides an important long-term technology for disassociating the production of meat and meat-like suites from the production of the living minds and bodies of animals (Jairath et al.2021; Evans & Johnson, 2021).

The purpose of this is to motivate the research that advances the technology of cultured meat with the scale and efficiency of current meat production, by focusing proper scientific and technical efforts on improving large-scale, mammalian cell cultivation. As a discipline, these discussions belong to the broader moral question of what are the limits of in vitro organization and of well-organized metabolic activity vis-à-vis future production technologies and the current natural metaphysic. Informed, society-determined judgment is required to distinguish productive biotechnology constructs from nonproductive or contrary-to-good biotechnology constructs. (Hong et al.2021; Chen et al.2022; Hubalek et al., 2022; Ramani et al.2021).

From a strictly cellular standpoint, several engineering challenges must be tackled in order to transform cellular agriculture into a viable industry. These quantitative challenges give rise to the bulk of the research energies in this emerging field. A major quantitative obstacle is one of biological scaling. The scaling relationships between cell size, growth rate, nutrient usage, and resource availability in a bioreactor are all key determinants of the efficiency of any cultured meat production process (Chiles et al.2021; Moritz et al., 2022).

A second key research challenge is one of biological process control, and stems from the fact that all meat cells are derived from the stem cells attached to the basal lamina of skeletal muscle of origin. In cultured meat, the challenge of robust feedback and feedforward control becomes much more acute, as the nutrient and growth factor requirements of these pluripotent myogenic stem cells in culture are not well characterized. A third general challenge is the scalability of meat production. Economies of scale, commensurate with those in the global feedlot infrastructure, will be necessary for cultured meat to be able to displace substantial quantities of feedlot meat from the marketplace. The key opportunity for biological process engineering, as distinct from biological science, is to use the commodity resources as raw material for biological mass production to grow safe, clean, tasty, and healthy meat in a process that uses fewer resources and creates less waste than our current, inefficient growing process (Chiles et al.2021; Moritz et al., 2022; Post et al.2020; Djisalov et al.2021; Glaros et al.2022).

5.1. Cellular Agriculture Techniques

Public relations considerations aside, the term cellular agriculture describes food and agricultural technologies that utilize cells to make consumable products without utilizing the rest of the given species. This "cells to products," cellular-only model is an important distinction from the many other historical uses of cell lines in the agricultural and biotech industries. Likewise, it is the unique feature of using cells to make non-cell based products that enable cellular agriculture to be argued as an alternative and potential replacement for conventional agriculture, both presently and in the future. Given the unique products and processes inherent to cellular agriculture, it is only logical that a plan addressing global food security through cellular agriculture would concentrate on these unique aspects. It is the unique cellular aspects of cellular agriculture that can open up new solutions to completely bypass many limitations found in traditional meat production, making cellular agriculture a possible fix to our failure at providing food security both now and in the future (Ercili-Cura & Barth, 2021; Helliwell & Burton, 2021; Jönsson, 2020; Stephens & Ellis, 2020).

Meat, dairy, and egg products are all composed of specific types of cells. These unpleasantly-textured, unappetizing, and not particularly useful lumps of cells also happen to be the products required for traditional meat production. However, cellular agriculture bypasses the need to turn animal cells into animal products by instead turning animal cells into (ostensibly) animal products directly. It is this previously-omitted, technology-specific detail of cellular agriculture which allows cellular agriculture to potentially offer a solution to addressing the challenges associated with food security. We cannot have a serious discussion about the ability to achieve global food security with cellular agriculture if with the intention to avoid controversial language, we avoid the term, or do not understand the underlying technology. It is the product and processes unique to cellular agriculture, not simply biotechnology in general, that is relevant to its potential to address food security (Knežić et al.2022; Dupuis et al.2023; Jönsson, 2020; Ge et al.2023; Soice & Johnston, 2021; Lee et al.2023; Reis et al., 2020).

6. ENVIRONMENTAL AND ECONOMIC IMPACTS

6.1 Environmental Impacts

One of the social implications we have not accounted for is the general effect that food and ingredient related industries have on the environment. The environmental impacts of cellular agriculture are likely to be positive relative to traditional agriculture. Livestock farming has significant environmental impacts, accounting for around 18-20% of global greenhouse gas emissions. Ruminants like cattle also have high water and feed requirements. The World Wildlife Fund and the Zoological Society of London have estimated that livestock will consume 50% of the world's maize, barley, and triticale by 2050, needing an additional 60 million tons of protein feed—all while using precious resources to meet an increasing demand. Importantly, unlike traditional farming, cellular agriculture could reduce the amount of food waste both at the farm and consumer level. Over 33% of food globally is wasted, a significant contributor to GHG emissions. We have the opportunity to revolutionize the way we sustainably produce meat and meat products. (Zhang et al.2022; Eisen & Brown, 2022; MacLeod et al.2020; Panchasara et al., 2021; Twine, 2021).

6.2. Economic Impacts

Our analysis of the economic implications involves multiple elements: the implications cellular agriculture has on global food security, the implications that cellular agriculture has on upstream and downstream actors in the food supply chain (for example, chemical and equipment vendors; ingredient blenders; companies engaged in purchase and sales, storage and preserve efforts, transportation and delivery, and on-and off-line services related to food cultivation and consumption), the urbanization effect, and the external infrastructural investment. (Chiles et al.2021; Rischer et al.2020; Evans & Lawson, 2020; Hadi & Brightwell, 2021; Klerkx & Rose, 2020; Mora et al.2020).

6.3. Economic Viability

Currently, at the outset, scaling effects are against cellular agriculture by at least three orders of magnitude that need to be overcome, both in terms of engineering the bioreactors and also the biology to overcome problems such as mass transfer that do not exist for traditional livestock or plant agriculture. Currently, to give an indication of the current costs of the processes, in 2016, U.S. startup Memphis Meats funded ex-Google co-founder Sergey Brin for chicken meat at US$9,000 per pound albeit before considering scale production, before Delicious Foods of San Francisco made 4 ounces of the world's most expensive meatball, at £200,000 per pound (Poirier, 2022; Humbird, 2021; Rischer et al.2020; Chen et al.2022; Hadi & Brightwell, 2021; Bodiou et al., 2020).

One of the major challenges for cellular agriculture is to be economically viable. The starting point for this is using small molecules to manipulate cells into growing and differentiating in the right way to produce the intended final product as cheaply as possible. This challenge has been described as less adventurous but arguably more difficult than the current biological challenges requiring even greater engineering and automation precision. A major advantage of cellular agriculture as compared to traditional livestock agriculture, and perhaps even to plant agriculture, is that microbes are particularly good at this. As with existing biotechnology, metabolic engineering is key to optimizing synthetic biology and synthetic chemistry is also key (Chung et al.2022; Rischer et al.2020; Reiss et al.2021; Voigt, 2020)

7. REGULATORY AND ETHICAL CONSIDERATIONS

Questions regarding pricing and access should be addressed in the early stages of the industry. Regulatory bodies, including the U.S. Department of Labor and U.S. Department of Agriculture in the U.S., as well as industry bodies, can discuss potential labeling, definitions for advertising, and standards of identity for cellular agriculture products. Meanings and values could also be discussed in order to establish familiar expectations for consumers and manufacturers. Fortunately, it is possible to review and address the interpretation of established food security regulations before cellular agriculture products are commercially available. Planning for various cellular agriculture products to be affordable and widespread increases the likelihood these products can truly and inclusively contribute to the progress of food security and dietary diversity around the world (White & Barquera, 2020; Díaz et al., 2020; Bailey, 2020; Möhring et al.2020).

Another important concern to be addressed is who will have access to the cellular agriculture products. Recent data suggest that, in wealthy countries, approximately one-third of consumers are willing to accept cellular agriculture produced meat or seafood. However, it is uncertain who will have access to these products across the globe or if the shift from animal agriculture to cellular agriculture will leave people behind. The high price of first-to-market cellular agriculture products and regulations that do not support equal distribution of these products in a diverse food system may dramatically exacerbate existing disparities in dietary quality and countries' abilities to support their own populations. Therefore, it is critical to find mechanisms to minimize the gap between those who cannot afford cellular agriculture products and those who can (Chuah et al., 2024; Kantono et al.2022; Hubbard, 2022; Tomiyama et al.2020; Ye et al., 2022; Fraser et al., 2024).

7.1. Regulatory Frameworks

The cellular agriculture sector has already signaled its willingness to engage in regulatory conversations. Dr. Mark Post, the developer of the first cultured hamburger, is quoted as saying that "The product was entirely funded by a donor. Because it was expensive, but now, as we see the interest from the people, there's a necessity for investment and the necessity for potential regulatory guidance." Dr. Uma Valeti, of Mosa Meats, who are also working with cellular agriculture, expressed similar regulatory sentiments in an interview with formerly TechCrunch. "We're not going to serve our product until there's a regulatory pathway to get it just right," he explains. "We're working with regulators both in the United States and the European Union to come up with the best pathway, and that also establishes safety for our food." To this end, EdelmanLEVY and the Good Food Institute are already offering regulatory support to the growing cellular agriculture community. The support comes in the form of helping entrepreneurs to understand the regulatory landscape and to advocate effectively (Guan et al., 2021; Ong et al.2021; Adams et al.2023; Chodkowska et al., 2022; Kumar et al.2021).

The regulation of cellular agriculture is likely to be a highly complex and politically charged process as it involves oversight that crosses multiple existing regulatory frameworks. Cellular agriculture can already draw on regulatory experience from the biomedical industry and for products such as rennet made using GM yeasts. However, cellular agriculture will need to develop regulatory frameworks for a wide range of products including cultured meat, fish, dairy and egg replacements, leather, silk, collagen, and a wide variety of other products. There are also social, ethical, economic, and geopolitical concerns that need to be addressed as cellular agriculture matures in parallel with more established food production approaches. Such concerns are not unique to cellular agriculture but need to be addressed in this context too. (Fish et al.2020; Post et al.2020; Rischer et al.2020; Rubio et al., 2020; Reiss et al.2021; Bomkamp et al.2022; Bryant & Barnett, 2020; Chen et al.2022).

8. FUTURE PROSPECTS AND RESEARCH DIRECTIONS

There is also a paucity of studies that primarily focus on applications whose intent is to address global food security. We focus primarily on this near-term application of cellular agriculture as the central theme. It is out of the scope of this paper to undertake a detailed review of the literature and research findings, yet we aim to compile a widely accessible overview of the state of the research, applications from different funders towards food security, literature, and industry trends relevant to cellular agriculture, and propose some key research challenges that are crucial to address the gap between the envisioned potential and technology readiness level of this technology to address the spacefaring missions towards self-sufficiency (Viana et al.2022; Mbow et al.2020; Vonthron et al., 2020; Karthikeyan et al., 2020; Clapp et al., 2022; Pawlak & Kołodziejczak, 2020; Barrett, 2021).

Cellular agriculture is expected to contribute to future food security in a variety of contexts. There are already numerous studies and incubating companies working on several cellular agriculture applications such as cell-based meat and fish, cell-based milk, and cell-based egg whites, while a few exploratory flavoring and fermentation-based products are reaching the markets. Although many of these ventures are supported by substantial amounts of financial investment and are involved in an iterative process of research, development, and animal-growth simulation, most need further research and development on the technological and economic feasibility, scalability, and replication process, as well as operational and management models, and the broader implications of the applications from a macroeconomic, sociocultural, and environmental perspective (Chiles et al.2021; Nyyssölä et al.2022; Rischer et al.2020; Mora et al.2020).

8.1. Emerging Technologies in Cellular Agriculture

Cellular agriculture is an emerging, interdisciplinary field that uses bioengineering tools for the production of agriculturally derived products in cell cultures. Cellular agriculture encompasses the use of animal cells to generate products such as milk, eggs, and meat in a sustainable and scalable way. Simultaneously, it uses techniques for modulating plants and animal physiology that were developed through the evolution of other bioengineering disciplines such as dependability. Indeed, the last few years have seen important advances in genetic modification, synthetic and functional genomics, genome editing, and systems biology, and these have led to new, exciting opportunities in cellular agriculture. However, unless the logistics and ethics of food production are well managed, it is likely that our food systems will not be able to deliver the future agri-food nexus security challenges, providing abundant, nutritious, and increasingly sustainable food for a rapidly growing population. (Mendly-Zambo et al.2021; Eibl et al.2021; Dupuis et al.2023; Soice & Johnston, 2021; Fytsilis et al.2024; Helliwell & Burton, 2021; Behm et al.2022).

Cellular agriculture is an emerging, interdisciplinary field that encompasses the use of bioengineering tools for the production of agriculturally derived products from cell cultures. It offers potential food security, sustainability, and food safety benefits above and beyond genetic modification. Many of the newer bioengineering tools needed to produce products are not from primary food industries but are emerging as enabling technologies in a variety of fields. These include immunological controls or 'immune switches' which allow the removal of layer products and provide resistance to carriers of pathogens such as salmonella, the modulation of metabolic pathways in plants, and the modification of cell senescence in animals and plants. This review outlines the key opportunities and challenges and presents case studies in which cellular agriculture may offer a step change in improving economic, social, and environmental food security outcomes. (Robertson et al., 2024; Pajčin et al.2022; Knežić et al.2022; Martins et al.2024; Bomkamp et al.2022; Newman et al.2023; Azhar et al.2023).

REFERENCES

Adams, C. J., Crary, A., & Gruen, L. (2023). The empty promises of cultured meat. The good it promises, the harm it does: Critical essays on effective altruism, 149.

Andronie, M., Lăzăroiu, G., Iatagan, M., Hurloiu, I., & Dijmărescu, I. (2021). Sustainable cyber-physical production systems in big data-driven smart urban economy: A systematic literature review. *Sustainability (Basel)*, 13(2), 751. 10.3390/su13020751

Aryal, J. P., Manchanda, N., & Sonobe, T. (2022). Expectations for household food security in the coming decades: A global scenario. In *Future foods* (pp. 107–131). Academic Press.

Azhar, A., Zeyaullah, M., Bhunia, S., Kacham, S., Patil, G., Muzammil, K., Khan, M. S., & Sharma, S. (2023). Cell-based meat: The molecular aspect. *Frontiers in Food Science and Technology*, 3, 1126455. 10.3389/frfst.2023.1126455

Bahar, N. H., Lo, M., Sanjaya, M., Van Vianen, J., Alexander, P., Ickowitz, A., & Sunderland, T. (2020). Meeting the food security challenge for nine billion people in 2050: What impact on forests? *Global Environmental Change*, 62, 102056. 10.1016/j.gloenvcha.2020.102056

Bai, Y., Herforth, A., & Masters, W. A. (2022). Global variation in the cost of a nutrient-adequate diet by population group: An observational study. *The Lancet. Planetary Health*, 6(1), e19–e28. 10.1016/S2542-5196(21)00285-034998455

Bailey, R. L. (2020). Current regulatory guidelines and resources to support research of dietary supplements in the United States. *Critical Reviews in Food Science and Nutrition*, 60(2), 298–309. 10.1080/10408398.2018.152436430421981

Barrett, C. B. (2021). Overcoming global food security challenges through science and solidarity. *American Journal of Agricultural Economics*, 103(2), 422–447. 10.1111/ajae.12160

Behm, K., Nappa, M., Aro, N., Welman, A., Ledgard, S., Suomalainen, M., & Hill, J. (2022). Comparison of carbon footprint and water scarcity footprint of milk protein produced by cellular agriculture and the dairy industry. *The International Journal of Life Cycle Assessment*, 27(8), 1017–1034. 10.1007/s11367-022-02087-0

Bhuyan, B. P., Tomar, R., & Cherif, A. R. (2022). A systematic review of knowledge representation techniques in smart agriculture (Urban). *Sustainability (Basel)*, 14(22), 15249. 10.3390/su142215249

Bin Rahman, A. R., & Zhang, J. (2023). Trends in rice research: 2030 and beyond. *Food and Energy Security*, 12(2), e390. 10.1002/fes3.390

Birhanu, M. Y., Alemayehu, T., Bruno, J. E., Kebede, F. G., Sonaiya, E. B., Goromela, E. H., & Dessie, T. (2021). Technical efficiency of traditional village chicken production in Africa: Entry points for sustainable transformation and improved livelihood. *Sustainability (Basel)*, 13(15), 8539. 10.3390/su13158539

Bodiou, V., Moutsatsou, P., & Post, M. J. (2020). Microcarriers for upscaling cultured meat production. *Frontiers in Nutrition*, 7, 10. 10.3389/fnut.2020.0001032154261

Bomkamp, C., Skaalure, S. C., Fernando, G. F., Ben-Arye, T., Swartz, E. W., & Specht, E. A. (2022). Scaffolding biomaterials for 3D cultivated meat: Prospects and challenges. *Advancement of Science*, 9(3), 2102908. 10.1002/advs.20210290834786874

Bryant, C., & Barnett, J. (2020). Consumer acceptance of cultured meat: An updated review (2018–2020). *Applied Sciences (Basel, Switzerland)*, 10(15), 5201. 10.3390/app10155201

Chandimali, N., Park, E. H., Bak, S. G., Won, Y. S., Lim, H. J., & Lee, S. J. (2024). Not seafood but seafood: A review on cell-based cultured seafood in lieu of conventional seafood. *Food Control*, 162, 110472. 10.1016/j.foodcont.2024.110472

Chen, L., Guttieres, D., Koenigsberg, A., Barone, P. W., Sinskey, A. J., & Springs, S. L. (2022). Large-scale cultured meat production: Trends, challenges and promising biomanufacturing technologies. *Biomaterials*, 280, 121274. 10.1016/j.biomaterials.2021.12127434871881

Chiles, R. M., Broad, G., Gagnon, M., Negowetti, N., Glenna, L., Griffin, M. A., Tami-Barrera, L., Baker, S., & Beck, K. (2021). Democratizing ownership and participation in the 4th Industrial Revolution: Challenges and opportunities in cellular agriculture. *Agriculture and Human Values*, 38(4), 943–961. 10.1007/s10460-021-10237-734456466

Chirwa, P. W., & Adeyemi, O. (2020). Deforestation in Africa: implications on food and nutritional security. Zero hunger, 197-211.

Chodkowska, K. A., Wódz, K., & Wojciechowski, J. (2022). Sustainable future protein foods: The challenges and the future of cultivated meat. *Foods*, 11(24), 4008. 10.3390/foods1124400836553750

Chuah, S. X. Y., Gao, Z., Arnold, N. L., & Farzad, R. (2024). Cell-Based Seafood Marketability: What Influences United States Consumers' Preferences and Willingness-To-Pay? *Food Quality and Preference*, 113, 105064. 10.1016/j.foodqual.2023.105064

Chung, Y. H., Church, D., Koellhoffer, E. C., Osota, E., Shukla, S., Rybicki, E. P., Pokorski, J. K., & Steinmetz, N. F. (2022). Integrating plant molecular farming and materials research for next-generation vaccines. *Nature Reviews. Materials*, 7(5), 372–388. 10.1038/s41578-021-00399-534900343

Clapp, J., Moseley, W. G., Burlingame, B., & Termine, P. (2022). The case for a six-dimensional food security framework. *Food Policy*, 106, 102164. 10.1016/j.foodpol.2021.102164

De Vries-Ten Have, J., Owolabi, A., Steijns, J., Kudla, U., & Melse-Boonstra, A. (2020). Protein intake adequacy among Nigerian infants, children, adolescents and women and protein quality of commonly consumed foods. *Nutrition Research Reviews*, 33(1), 102–120. 10.1017/S09544224190002231997732

Díaz, L. D., Fernández-Ruiz, V., & Cámara, M. (2020). An international regulatory review of food health-related claims in functional food products labeling. *Journal of Functional Foods*, 68, 103896. 10.1016/j.jff.2020.103896

Djisalov, M., Knežić, T., Podunavac, I., Živojević, K., Radonic, V., Knežević, N. Ž., & Gadjanski, I. (2021). Cultivating multidisciplinarity: Manufacturing and sensing challenges in cultured meat production. *Biology (Basel)*, 10(3), 204. 10.3390/biology10030204 33803111

Dupuis, J. H., Cheung, L. K., Newman, L., Dee, D. R., & Yada, R. Y. (2023). Precision cellular agriculture: The future role of recombinantly expressed protein as food. *Comprehensive Reviews in Food Science and Food Safety*, 22(2), 882–912. 10.1111/1541-4337.1309436546356

Ebenso, B., Otu, A., Giusti, A., Cousin, P., Adetimirin, V., Razafindralambo, H., & Mounir, M. (2022). Nature-based one health approaches to urban agriculture can deliver food and nutrition security. *Frontiers in Nutrition*, 9, 773746. 10.3389/fnut.2022.77374635360699

Eibl, R., Senn, Y., Gubser, G., Jossen, V., Van Den Bos, C., & Eibl, D. (2021). Cellular agriculture: Opportunities and challenges. *Annual Review of Food Science and Technology*, 12(1), 51–73. 10.1146/annurev-food-063020-12394033770467

Eisen, M. B., & Brown, P. O. (2022). Rapid global phaseout of animal agriculture has the potential to stabilize greenhouse gas levels for 30 years and offset 68 percent of CO2 emissions this century. *PLOS Climate*, 1(2), e0000010. 10.1371/journal.pclm.0000010

Ercili-Cura, D., & Barth, D. (2021). *Cellular Agriculture: Lab Grown Foods* (Vol. 8). American Chemical Society.

Errera, M. R., Dias, T. D. C., Maya, D. M. Y., & Lora, E. E. S. (2023). Global bioenergy potentials projections for 2050. *Biomass and Bioenergy*, 170, 106721. 10.1016/j.biombioe.2023.106721

Evans, B., & Johnson, H. (2021). Contesting and reinforcing the future of 'meat' through problematization: Analyzing the discourses in regulatory debates around animal cell-cultured meat. *Geoforum*, 127, 81–91. 10.1016/j.geoforum.2021.10.001

Evans, J. R., & Lawson, T. (2020). From green to gold: Agricultural revolution for food security. *Journal of Experimental Botany*, 71(7), 2211–2215. 10.1093/jxb/eraa11032251509

Ezeh, A., Kissling, F., & Singer, P. (2020). Why sub-Saharan Africa might exceed its projected population size by 2100. *Lancet*, 396(10258), 1131–1133. 10.1016/S0140-6736(20)31522-132679113

Falcon, W. P., Naylor, R. L., & Shankar, N. D. (2022). Rethinking global food demand for 2050. *Population and Development Review*, 48(4), 921–957. 10.1111/padr.12508

Fasona, M. J., Akintuyi, A. O., Adeonipekun, P. A., Akoso, T. M., Udofia, S. K., Agboola, O. O., Ogunsanwo, G. E., Ariori, A. N., Omojola, A. S., Soneye, A. S., & Ogundipe, O. T. (2022). Recent trends in land-use and cover change and deforestation in south–west Nigeria. *GeoJournal*, 87(3), 1411–1437. 10.1007/s10708-020-10318-w

Fish, K. D., Rubio, N. R., Stout, A. J., Yuen, J. S., & Kaplan, D. L. (2020). Prospects and challenges for cell-cultured fat as a novel food ingredient. *Trends in Food Science & Technology*, 98, 53–67. 10.1016/j.tifs.2020.02.00532123465

Fraser, E. D., Newman, L., Yada, R. Y., & Kaplan, D. L. (2024). The foundations of cellular agriculture: science, technology, sustainability and society. In *Cellular Agriculture* (pp. 3–10). Academic Press. 10.1016/B978-0-443-18767-4.00010-X

Fytsilis, V. D., Urlings, M. J., van Schooten, F. J., de Boer, A., & Vrolijk, M. F. (2024). Toxicological risks of dairy proteins produced through cellular agriculture: Current state of knowledge, challenges and future perspectives. *Future Foods : a Dedicated Journal for Sustainability in Food Science*, 10, 100412. 10.1016/j.fufo.2024.100412

Galanakis, C. M. (2024). The future of food. *Foods*, 13(4), 506. 10.3390/foods1304050638397483

Ge, C., Selvaganapathy, P. R., & Geng, F. (2023). Advancing our understanding of bioreactors for industrial-sized cell culture: Health care and cellular agriculture implications. *American Journal of Physiology. Cell Physiology*, 325(3), C580–C591. 10.1152/ajpcell.00408.202237486066

Ghandar, A., Ahmed, A., Zulfiqar, S., Hua, Z., Hanai, M., & Theodoropoulos, G. (2021). A decision support system for urban agriculture using digital twin: A case study with aquaponics. *IEEE Access : Practical Innovations, Open Solutions*, 9, 35691–35708. 10.1109/ACCESS.2021.3061722

Glaros, A., Marquis, S., Major, C., Quarshie, P., Ashton, L., Green, A. G., Kc, K. B., Newman, L., Newell, R., Yada, R. Y., & Fraser, E. D. (2022). Horizon scanning and review of the impact of five food and food production models for the global food system in 2050. *Trends in Food Science & Technology*, 119, 550–564. 10.1016/j.tifs.2021.11.013

Gu, D., Andreev, K., & Dupre, M. E. (2021). Major trends in population growth around the world. *China CDC Weekly*, 3(28), 604. 10.46234/ccdcw2021.16034594946

Guan, X., Lei, Q., Yan, Q., Li, X., Zhou, J., Du, G., & Chen, J. (2021). Trends and ideas in technology, regulation and public acceptance of cultured meat. *Future Foods : a Dedicated Journal for Sustainability in Food Science*, 3, 100032. 10.1016/j.fufo.2021.100032

Hadi, J., & Brightwell, G. (2021). Safety of alternative proteins: Technological, environmental and regulatory aspects of cultured meat, plant-based meat, insect protein and single-cell protein. *Foods*, 10(6), 1226. 10.3390/foods1006122634071292

Helliwell, R., & Burton, R. J. (2021). The promised land? Exploring the future visions and narrative silences of cellular agriculture in news and industry media. *Journal of Rural Studies*, 84, 180–191. 10.1016/j.jrurstud.2021.04.002

Hong, T. K., Shin, D. M., Choi, J., Do, J. T., & Han, S. G. (2021). Current issues and technical advances in cultured meat production: A review. *Food Science of Animal Resources*, 41(3), 355–372. 10.5851/kosfa.2021.e1434017947

Hubalek, S., Post, M. J., & Moutsatsou, P. (2022). Towards resource-efficient and cost-efficient cultured meat. *Current Opinion in Food Science*, 47, 100885. 10.1016/j.cofs.2022.100885

Hubbard, Y. E. (2022). Addressing Public Perceptions About Cell-Based Meat and Cellular Agriculture Through Metaphors (Master's thesis, Old Dominion University).

Humbird, D. (2021). Scale-up economics for cultured meat. *Biotechnology and Bioengineering*, 118(8), 3239–3250. 10.1002/bit.2784834101164

Jairath, G., Mal, G., Gopinath, D., & Singh, B. (2021). A holistic approach to access the viability of cultured meat: A review. *Trends in Food Science & Technology*, 110, 700–710. 10.1016/j.tifs.2021.02.024

Jebari, A., Oyetunde-Usman, Z., McAuliffe, G. A., Chivers, C. A., & Collins, A. L. (2024). Willingness to adopt green house gas mitigation measures: Agricultural land managers in the United Kingdom. *PLoS One*, 19(7), e0306443. 10.1371/journal.pone.030644338976702

Jönsson, E. (2020). On breweries and bioreactors: Probing the "present futures" of cellular agriculture. *Transactions of the Institute of British Geographers*, 45(4), 921–936. 10.1111/tran.12392

Kantono, K., Hamid, N., Malavalli, M. M., Liu, Y., Liu, T., & Seyfoddin, A. (2022). Consumer acceptance and production of in vitro meat: A review. *Sustainability (Basel)*, 14(9), 4910. 10.3390/su14094910

Karthikeyan, L., Chawla, I., & Mishra, A. K. (2020). A review of remote sensing applications in agriculture for food security: Crop growth and yield, irrigation, and crop losses. *Journal of Hydrology (Amsterdam)*, 586, 124905. 10.1016/j.jhydrol.2020.124905

Klerkx, L., & Rose, D. (2020). Dealing with the game-changing technologies of Agriculture 4.0: How do we manage diversity and responsibility in food system transition pathways? *Global Food Security*, 24, 100347. 10.1016/j.gfs.2019.100347

Knežić, T., Janjušević, L., Djisalov, M., Yodmuang, S., & Gadjanski, I. (2022). Using vertebrate stem and progenitor cells for cellular agriculture, state-of-the-art, challenges, and future perspectives. *Biomolecules*, 12(5), 699. 10.3390/biom1205069935625626

Kumar, P., Sharma, N., Sharma, S., Mehta, N., Verma, A. K., Chemmalar, S., & Sazili, A. Q. (2021). In-vitro meat: A promising solution for sustainability of meat sector. *Journal of Animal Science and Technology*, 63(4), 693–724. 10.5187/jast.2021.e8534447949

Kupka, R., Siekmans, K., & Beal, T. (2020). The diets of children: Overview of available data for children and adolescents. *Global Food Security*, 27, 100442. 10.1016/j.gfs.2020.100442

Lee, D. K., Kim, M., Jeong, J., Lee, Y. S., Yoon, J. W., An, M. J., & Lee, C. K. (2023). Unlocking the potential of stem cells: Their crucial role in the production of cultivated meat. *Current Research in Food Science*, 7, 100551. 10.1016/j.crfs.2023.10055137575132

Li, L., Sun, N., Zhang, L., Xu, G., Liu, J., Hu, J., & Han, L. (2020). Fast food consumption among young adolescents aged 12–15 years in 54 low-and middle-income countries. *Global Health Action*, 13(1), 1795438. 10.1080/16549716.2020.17954 3832762333

MacLeod, M. J., Hasan, M. R., Robb, D. H., & Mamun-Ur-Rashid, M. (2020). Quantifying greenhouse gas emissions from global aquaculture. *Scientific Reports*, 10(1), 11679. 10.1038/s41598-020-68231-832669630

Mancini, M. C., & Antonioli, F. (2022). The future of cultured meat between sustainability expectations and socio-economic challenges. In *Future foods* (pp. 331–350). Academic Press. 10.1016/B978-0-323-91001-9.00024-4

Martins, B., Bister, A., Dohmen, R. G., Gouveia, M. A., Hueber, R., Melzener, L., Messmer, T., Papadopoulos, J., Pimenta, J., Raina, D., Schaeken, L., Shirley, S., Bouchet, B. P., & Flack, J. E. (2024). Advances and challenges in cell biology for cultured meat. *Annual Review of Animal Biosciences*, 12(1), 345–368. 10.1146/annurev-animal-021022-05513237963400

Mbow, C., Rosenzweig, C. E., Barioni, L. G., Benton, T. G., Herrero, M., Krishnapillai, M., & Diouf, A. A. (2020). *Food security (No. GSFC-E-DAA-TN78913)*. IPCC.

Mendly-Zambo, Z., Powell, L. J., & Newman, L. L. (2021). Dairy 3.0: Cellular agriculture and the future of milk. *Food, Culture, & Society*, 24(5), 675–693. 10.1080/15528014.2021.1888411

Mohamed, M. (2023). Agricultural Sustainability in the Age of Deep Learning: Current Trends, Challenges, and Future Trajectories. *Sustainable Machine Intelligence Journal*, 4, 2–1. 10.61185/SMIJ.2023.44102

Möhring, N., Ingold, K., Kudsk, P., Martin-Laurent, F., Niggli, U., Siegrist, M., & Finger, R. (2020). Pathways for advancing pesticide policies. *Nature Food*, 1(9), 535–540. 10.1038/s43016-020-00141-437128006

Mora, O., Le Mouël, C., de Lattre-Gasquet, M., Donnars, C., Dumas, P., Réchauchère, O., & Marty, P. (2020). Exploring the future of land use and food security: A new set of global scenarios. *PLoS One*, 15(7), e0235597. 10.1371/journal.pone.023559732639991

Moritz, J., Tuomisto, H. L., & Ryynänen, T. (2022). The transformative innovation potential of cellular agriculture: Political and policy stakeholders' perceptions of cultured meat in Germany. *Journal of Rural Studies*, 89, 54–65. 10.1016/j.jrurstud.2021.11.018

Mrabet, R. (2023). Sustainable agriculture for food and nutritional security. In *Sustainable agriculture and the environment* (pp. 25–90). Academic Press. 10.1016/B978-0-323-90500-8.00013-0

Nadathur, S., Wanasundara, J. P., Marinangeli, C. P. F., & Scanlin, L. (2024). Proteins in Our Diet: Challenges in Feeding the Global Population. In Sustainable Protein Sources (pp. 1-29). Academic Press.

Newman, L., Newell, R., Dring, C., Glaros, A., Fraser, E., Mendly-Zambo, Z., & Kc, K. B. (2023). Agriculture for the Anthropocene: Novel applications of technology and the future of food. *Food Security*, 15(3), 613–627. 10.1007/s12571-023-01356-6

Newton, P., & Blaustein-Rejto, D. (2021). Social and economic opportunities and challenges of plant-based and cultured meat for rural producers in the US. *Frontiers in Sustainable Food Systems*, 5, 624270. 10.3389/fsufs.2021.624270

Nyyssölä, A., Suhonen, A., Ritala, A., & Oksman-Caldentey, K. M. (2022). The role of single cell protein in cellular agriculture. *Current Opinion in Biotechnology*, 75, 102686. 10.1016/j.copbio.2022.10268635093677

Olowe, V. (2021). Africa 2100: How to feed Nigeria in 2100 with 800 million inhabitants. *Organic Agriculture*, 11(2), 199–208. 10.1007/s13165-020-00307-1

Ong, K. J., Johnston, J., Datar, I., Sewalt, V., Holmes, D., & Shatkin, J. A. (2021). Food safety considerations and research priorities for the cultured meat and seafood industry. *Comprehensive Reviews in Food Science and Food Safety*, 20(6), 5421–5448. 10.1111/1541-4337.1285334633147

Pajčin, I., Knežić, T., Savic Azoulay, I., Vlajkov, V., Djisalov, M., Janjušević, L., Grahovac, J., & Gadjanski, I. (2022). Bioengineering outlook on cultivated meat production. *Micromachines*, 13(3), 402. 10.3390/mi1303040235334693

Panchasara, H., Samrat, N. H., & Islam, N. (2021). Greenhouse gas emissions trends and mitigation measures in Australian agriculture sector—. *Revista de Agricultura (Piracicaba)*, 11(2), 85.

Pawlak, K., & Kołodziejczak, M. (2020). The role of agriculture in ensuring food security in developing countries: Considerations in the context of the problem of sustainable food production. *Sustainability (Basel)*, 12(13), 5488. 10.3390/su12135488

Pereira, L. M., Drimie, S., Maciejewski, K., Tonissen, P. B., & Biggs, R. (2020). Food system transformation: Integrating a political–economy and social–ecological approach to regime shifts. *International Journal of Environmental Research and Public Health*, 17(4), 1313. 10.3390/ijerph1704131332085576

Piwowar, A. (2020). Farming practices for reducing ammonia emissions in Polish agriculture. *Atmosphere (Basel)*, 11(12), 1353. 10.3390/atmos11121353

Poirier, N. (2022). On the Intertwining of Cellular Agriculture and Animal Agriculture: History, Materiality, Ideology, and Collaboration. *Frontiers in Sustainable Food Systems*, 6, 907621. 10.3389/fsufs.2022.907621

Post, M. J., Levenberg, S., Kaplan, D. L., Genovese, N., Fu, J., Bryant, C. J., & Moutsatsou, P. (2020). Scientific, sustainability and regulatory challenges of cultured meat. *Nature Food*, 1(7), 403–415. 10.1038/s43016-020-0112-z

Ramani, S., Ko, D., Kim, B., Cho, C., Kim, W., Jo, C., & Park, S. (2021). Technical requirements for cultured meat production: A review. *Journal of Animal Science and Technology*, 63(4), 681–692. 10.5187/jast.2021.e4534447948

Reis, G. G., Heidemann, M. S., Borini, F. M., & Molento, C. F. M. (2020). Livestock value chain in transition: Cultivated (cell-based) meat and the need for breakthrough capabilities. *Technology in Society*, 62, 101286. 10.1016/j.techsoc.2020.101286

Reiss, J., Robertson, S., & Suzuki, M. (2021). Cell sources for cultivated meat: Applications and considerations throughout the production workflow. *International Journal of Molecular Sciences*, 22(14), 7513. 10.3390/ijms2214751334299132

Resare Sahlin, K., Carolus, J., von Greyerz, K., Ekqvist, I., & Röös, E. (2022). Delivering "less but better" meat in practice—A case study of a farm in agroecological transition. *Agronomy for Sustainable Development*, 42(2), 24. 10.1007/s13593-021-00737-5

Rischer, H., Szilvay, G. R., & Oksman-Caldentey, K. M. (2020). Cellular agriculture—Industrial biotechnology for food and materials. *Current Opinion in Biotechnology*, 61, 128–134. 10.1016/j.copbio.2019.12.00331926477

Robertson, S., Nyman, H., & Suzuki, M. (2024). Cell source and Types for cultivated meat production. In *Cellular Agriculture* (pp. 111–123). Academic Press. 10.1016/B978-0-443-18767-4.00026-3

Rubio, N. R., Xiang, N., & Kaplan, D. L. (2020). Plant-based and cell-based approaches to meat production. *Nature Communications*, 11(1), 1–11. 10.1038/s41467-020-20061-y33293564

Sadigov, R. (2022). Rapid growth of the world population and its socioeconomic results. *TheScientificWorldJournal*, 2022(1), 8110229. 10.1155/2022/811022935370481

Samal, P., Babu, S. C., Mondal, B., & Mishra, S. N. (2022). The global rice agriculture towards 2050: An inter-continental perspective. *Outlook on Agriculture*, 51(2), 164–172. 10.1177/00307270221088338

Sekaran, U., Lai, L., Ussiri, D. A., Kumar, S., & Clay, S. (2021). Role of integrated crop-livestock systems in improving agriculture production and addressing food security–A review. *Journal of Agriculture and Food Research*, 5, 100190. 10.1016/j.jafr.2021.100190

Soice, E., & Johnston, J. (2021). How cellular agriculture systems can promote food security. *Frontiers in Sustainable Food Systems*, 5, 753996. 10.3389/fsufs.2021.753996

Stephens, N., & Ellis, M. (2020). Cellular agriculture in the UK: a review. *Wellcome Open Research*, 5. 10.12688/wellcomeopenres.15685.1

Talwar, R., Freymond, M., Beesabathuni, K., & Lingala, S. (2024). Current and Future Market Opportunities for Alternative Proteins in Low-and Middle-Income Countries. *Current Developments in Nutrition*, 8, 102035. 10.1016/j.cdnut.2023.10203538476721

Thyden, R., Perreault, L. R., Jones, J. D., Notman, H., Varieur, B. M., Patmanidis, A. A., & Gaudette, G. R. (2022). An edible, decellularized plant derived cell carrier for lab grown meat. *Applied Sciences (Basel, Switzerland)*, 12(10), 5155. 10.3390/app12105155

Tian, X., Engel, B. A., Qian, H., Hua, E., Sun, S., & Wang, Y. (2021). Will reaching the maximum achievable yield potential meet future global food demand? *Journal of Cleaner Production*, 294, 126285. 10.1016/j.jclepro.2021.126285

Tomiyama, A. J., Kawecki, N. S., Rosenfeld, D. L., Jay, J. A., Rajagopal, D., & Rowat, A. C. (2020). Bridging the gap between the science of cultured meat and public perceptions. *Trends in Food Science & Technology*, 104, 144–152. 10.1016/j.tifs.2020.07.019

Treich, N. (2021). Cultured meat: Promises and challenges. *Environmental and Resource Economics*, 79(1), 33–61. 10.1007/s10640-021-00551-333758465

Tutundjian, S., Clarke, M., Egal, F., Dixson-Decleve, S., Candotti, S. W., Schmitter, P., & Lovins, L. H. (2020). Future food systems: challenges and consequences of the current food system. The Palgrave Handbook of Climate Resilient Societies, 1-29.

Twine, R. (2021). Emissions from animal agriculture—16.5% is the new minimum figure. *Sustainability (Basel)*, 13(11), 6276. 10.3390/su13116276

Van Dijk, M., Morley, T., Rau, M. L., & Saghai, Y. (2021). A meta-analysis of projected global food demand and population at risk of hunger for the period 2010–2050. *Nature Food*, 2(7), 494–501. 10.1038/s43016-021-00322-937117684

Van Eenennaam, A. L., & Werth, S. J. (2021). Animal board invited review: Animal agriculture and alternative meats–learning from past science communication failures. *Animal*, 15(10), 100360. 10.1016/j.animal.2021.10036034563799

Vermeulen, S. J., Park, T., Khoury, C. K., & Béné, C. (2020). Changing diets and the transformation of the global food system. *Annals of the New York Academy of Sciences*, 1478(1), 3–17. 10.1111/nyas.1444632713024

Viana, C. M., Freire, D., Abrantes, P., Rocha, J., & Pereira, P. (2022). Agricultural land systems importance for supporting food security and sustainable development goals: A systematic review. *The Science of the Total Environment*, 806, 150718. 10.1016/j.scitotenv.2021.15071834606855

Voigt, C. A. (2020). Synthetic biology 2020–2030: Six commercially-available products that are changing our world. *Nature Communications*, 11(1), 1–6. 10.1038/s41467-020-20122-233311504

Vollset, S. E., Goren, E., Yuan, C. W., Cao, J., Smith, A. E., Hsiao, T., & Murray, C. J. (2020). Fertility, mortality, migration, and population scenarios for 195 countries and territories from 2017 to 2100: A forecasting analysis for the Global Burden of Disease Study. *Lancet*, 396(10258), 1285–1306. 10.1016/S0140-6736(20)30677-232679112

Vonthron, S., Perrin, C., & Soulard, C. T. (2020). Foodscape: A scoping review and a research agenda for food security-related studies. *PLoS One*, 15(5), e0233218. 10.1371/journal.pone.023321832433690

Wang, J., & Azam, W. (2024). Natural resource scarcity, fossil fuel energy consumption, and total greenhouse gas emissions in top emitting countries. *Geoscience Frontiers*, 15(2), 101757. 10.1016/j.gsf.2023.101757

White, M., & Barquera, S. (2020). Mexico adopts food warning labels, why now? *Health Systems and Reform*, 6(1), e1752063. 10.1080/23288604.2020.175206332486930

Wikandari, R., Manikharda, , Baldermann, S., Ningrum, A., & Taherzadeh, M. J. (2021). Application of cell culture technology and genetic engineering for production of future foods and crop improvement to strengthen food security. *Bioengineered*, 12(2), 11305–11330. 10.1080/21655979.2021.200366534779353

Ye, Y., Zhou, J., Guan, X., & Sun, X. (2022). Commercialization of cultured meat products: Current status, challenges, and strategic prospects. *Future Foods : a Dedicated Journal for Sustainability in Food Science*, 6, 100177. 10.1016/j.fufo.2022.100177

Zhang, L., Tian, H., Shi, H., Pan, S., Chang, J., Dangal, S. R., & Jackson, R. B. (2022). A 130-year global inventory of methane emissions from livestock: Trends, patterns, and drivers. *Global Change Biology*, 28(17), 5142–5158. 10.1111/gcb.1628035642457

Zhichkin, K., Nosov, V., Zhichkina, L., Allen, G., Kotar, O., & Fasenko, T. (2021, September). Efficiency of regional agriculture state support. *IOP Conference Series. Earth and Environmental Science*, 839(2), 022042. 10.1088/1755-1315/839/2/022042

Chapter 11
The Role of Cellular Agriculture in Mitigating Climate Change

Ahmed Hamad
https://orcid.org/0000-0001-5037-9379
Benha University, Egypt

Dina A. B. Awad
Benha University, Egypt

ABSTRACT

Cellular agriculture, a revolutionary approach to food production, holds significant potential for mitigating climate change. Cellular agriculture addresses the environmental burdens associated with conventional animal agriculture by cultivating animal products such as meat, dairy, and eggs from cell cultures rather than traditional livestock farming. The chapter explores cellular agriculture's technological advancements, economic implications, and regulatory challenges, emphasizing its role in decreasing deforestation, methane emissions, and water usage. Furthermore, it highlights the potential for cellular agriculture to reduce the reliance on antibiotics, thereby contributing to public health. Through a comprehensive analysis of current research, case studies, and future projections, this chapter underscores the critical importance of cellular agriculture in the global effort to combat climate change and foster a sustainable food system.

DOI: 10.4018/979-8-3693-4115-5.ch011

1. INTRODUCTION TO CELLULAR AGRICULTURE

Cellular agriculture is an evolving field within the life sciences where researchers and entrepreneurs work to develop products derived from cell cultures that have conventionally come from the farming of plants and animals. The first wave of such development saw the introduction of synthetic insulin for the treatment of diabetes, a pharmaceutical product that was commercialized in the late 1980s. Though much has changed and advanced, the basic principle remains the same: manufacturing products through a patented fermentation process, just as beer and cheese manufacturers do. The industry is now heading into a further phase of commercialization, which bridges both conventional animal-derived products and new product categories based on novel cell lines to bring about the next revolution in the food sector. This time, the ambition is to make products such as meat and fish products without the need for whole animals to be bred, raised, and slaughtered, harnessing some of the very cells inside these animals at the very beginning of their life (Mathieu et al., 2021; Home & Mehta, 2021; Lee & Yoon, 2021; Home & Mehta, 2021).

1.1. Definition and Principles

Regarding this perspective, cellAg has the potential to revolutionize the food system and has a plethora of environmental and social benefits, such as mitigating greenhouse emissions, conserving land and water, and removing the presence of pathogens or antibiotic resistance genes. However, the development of cell culture media for cellAg, as a major limiting component, has raised concerns among the broader society (Alsarhan et al., 2021). These concerns, emanating from reduced immunoglobulins in serum and the presence of fetal bovine serum, have stirred controversy within the academic community, ranging from the need for society to compromise and adopt these technologies for the potential benefits they cause, to the alternative of promoting balanced diets or a vegetarian lifestyle. Settling these arguments and concerns, several interviewees from policy and industry sectors noted important considerations in the development of alternative proteins; instead of addressing strict consumer choice and forcing them to change, regulatory and policy bodies should address potential environmental, social, and economic impacts of these novel advances in food science (Roe et al., 2021; Cronin et al., 2020).

The increasing demand for meat and dairy products, as well as concerns about the environmental and ethical implications of conventional livestock production, have driven the development of alternative protein sources. Cellular agriculture (cellAg) harnesses mammalian biology to enable the production of animal proteins using cell culture techniques, without the need to raise live animals (Newman et al., 2023). Research in this area is focused on the development, optimization, and scale-up of cell culture platforms that can produce high-quality cellAg products across diverse physicochemical properties. A major concern exclusive to cellAg, however, is the media with which these cells will be cultured. Current research is focused on supplementing serum—commonly used in traditional cell culture media for in vitro cultivation—in a cost-effective, environmentally friendly, and sustainable manner to ensure the long-term viability of scalable cellAg without compromising any nutritional value or taste of traditional animal meat (Ercili-Cura & Barth, 2021l; Giglio et al., 2024).

1.2. Historical Background

Cellular agriculture has been discussed by several other researchers besides those of Maastricht University and related colleagues. Research has been published or presented at Harvard, the MIT Media Lab, and other US universities. Researchers in Canada, Spain, the UK, and other countries have addressed aspects of cellular agriculture in the scientific literature or popular press. Statements about cellular agriculture have been made by public interest groups such as New Harvest, Humane Society of the United States, Good Food Institute, and others. Biologically, cellular agriculture is a subset of bioprocessing sciences and industries. Its basic competencies are highly related to those found in agriculture and fermentation as well. Its applications are numerous across food, beverage, medicine, and other chemical industries, and a review of several of these applications is provided in Artificial Meat Handbook (Ercili-Cura & Barth, 2021; Chiles et al., 2021; Soice & Johnston, 2021).

The first cellular agriculture scientific conference took place in 2016 at Maastricht University. At this conference, an academic consensus formally established the scientific field of cellular agriculture. As of April 2018, there are at least 18 companies conducting cellular agriculture research. Many are located in the United States, particularly in the San Francisco Bay Area. Since its formal founding in 2016, cellular agriculture topics have been addressed at major conferences such as New Harvest, in scientific literature, and by various members of universities and research groups. As with other ag-tech, there has been investor interest, with several cellular agriculture companies obtaining angel, venture capital, and other forms of funding through 2017 (Reiss et al., 2021; Post et al., 2020; Klerkx & Rose, 2020; Ong et al., 2020).

2. CLIMATE CHANGE AND ITS IMPACTS

By seeking political consensus for the actions needed to combat climate change (either scheduled or not) and by addressing the environmental risks associated with global production processes along with the inherent legal, social, and policy challenges that are characteristic of carbon-intensive sectors of the economy. Promoting green industries such as circular economy, bio-products, and climate services as well as biofuels and biochemicals is equally important (Rischer et al., 2020; Engler & Krarti, 2021; Helliwell & Burton, 2021; Del Buono, 2021).

Governance responses to combating climate change and promoting a transition to a low-carbon economy are gathering pace at the international level. They include institutional and macroeconomic reforms. These and other similar policy actions are meant to allow governments, companies, and communities to build resilience, to benefit from decarbonizing the economy, and to avoid lock-in effects in high-emission pathways. Research and innovation are also important, given the many knowledge gaps and the need for smart processes, innovative products and services, and breakthrough technologies. Similarly, global standards and appropriate financial and risk management tools and instruments can help save time, money, and lives (Cosens et al., 2021; Dale et al., 2019; Afrifa et al., 2020).

Climate change is fundamentally one of the most complex and urgent challenges of our time and poses a significant threat to global societal welfare, stability, and prosperity (Awad et al., 2024). Our current lifestyle and economic system are unsustainable, and we need to speed up how we reshape them not only to grow fairly and sustainably but to safeguard future generations. The rise of average global temperatures and adverse weather events such as typhoons/hurricanes, forest fires, and droughts have increased in frequency and often intensity in various parts of the world in recent years, demonstrating the accelerating impacts of climate change as well as the significant level of economic damage to society caused by such catastrophic events. Over the coming years, many scientists are expecting further changes, such as longer-lasting heatwaves, increasingly heavy rainfall, and greater flood risks as well as permafrost melting, higher Arctic temperatures, and more intense hurricanes, are predicted (Walsh et al., 2020; Clarke et al., 2022; Beillouin et al., 2020).

2.1. Mitigation of Climate Change

There are a number of initiatives that have been put forth as climate change mitigation measures in the food and agricultural system, ranging from technological to behavioral solutions. Cellular agriculture (CA) is a technological solution that has gained popularity in the past few years, but its potential for mitigating climate change has been explored only in a few studies. In this paper, we review the literature documenting the role of CA as a climate change mitigation measure and draw on the practical experience of the authors to discuss the potential role CA can play in the future regulation of food and agriculture (Chiles et al., 2021; Rischer et al., 2020; Del Buono, 2021).

Climate change is among the most pressing environmental challenges facing the planet. Evidence compiled by different groups, including the United Nations' Intergovernmental Panel on Climate Change (IPCC), has shown that the climate is changing, as global average surface temperatures have increased by about 1°C since the end of the 19th century (Morice et al., 2021). Climate change poses risks to many aspects of human life, and impacts felt due to rising global temperatures are projected to increase. To limit these impacts, urgent, large-scale measures to reduce greenhouse gas (GHG) emissions are needed. Agricultural emissions contribute up to a third of all GHG emissions. Moreover, agriculture is also exposed to numerous climate change impacts that will have a bearing on local food production (Kaufman et al., 2020).

2.2. Impacts of Cellular Agriculture on the Environment and Society

As a result, cellular agriculture can become a significant contributing solution in a world urgently seeking ways to limit climate change and air pollution. Conversely, its wide application could reduce the negative impacts society inflicts on the environment by addressing harmful land use, water pollution and scarcity, and loss of bioenergy, among other problems. The idea of meat without animals also presents an exciting bodily and ethically resonant challenge to how societies perceive non-human animals (Chiles et al., 2021; Rischer et al., 2020). If done right, it holds out the potential for large-scale societal reforms aimed at respecting and defending the dignity of animals. In short, cellular agriculture could pave the way not merely to an incremental improvement of the existing industrial system of animal production, but to a socially innovative, environmentally sustainable, and ethically supported set of practices (Agrimonti et al., 2020).

The technology of cellular agriculture has the potential to make important contributions to environmental, social, economic, and ethical issues surrounding contemporary animal agriculture. Emissions of greenhouse gases could be substantially reduced if meat from cell cultures was produced in a bioreactor instead of from livestock. Environmentally, this presents the appealing prospect of a substantial global reduction in the contribution from agriculture to anthropogenic climate change. Production of clean meat and other cellular products can be established in more controlled environments and require far less use of water and land, and much lower emissions of harmful gases and other pollutants (Tuomisto et al., 2022; Munteanu et al., 2021; Ren et al., 2022; Zhang et al., 2020).

3. TRADITIONAL AGRICULTURE AND ITS ENVIRONMENTAL FOOTPRINT

It has been estimated that if beef consumption trends continue, global temperature targets will be impossible to achieve, as agricultural meat production is expected to increase by 120-180% in 2050 (as compared with the year 2005) in order to feed the global population. In reaction, a global effort to increase the efficiency of traditional agriculture along with an increasing number of alternative protein products will be necessary. Cellular agriculture is developing as an alternative to traditional animal agriculture, and it is heralded as capable of delivering animals' many significant benefits. Cellular agriculture involves the cultivation of animal-derived food products (such as meat, milk, and eggs) from their respective cells, bypassing the raising of whole animals (Meissner et al., 2023; Arndt et al., 2022). Um and Song estimate meat products resulting from cellular agriculture would require 82-96% less water and 43-96% less land than conventional meat products. As well, the environmental impacts of climate change could be reduced by more than 50% if the growth by cultivation of animal cells and the associated cooling process were powered by renewable energy sources. Remarkably, since our demand is largely responsible for the increase in cows and hence methane pollution, cellular agriculture products are cruelty-free and produce significantly lower greenhouse gas emissions (Potopová et al., 2023).

As the world population approaches approximately 10 billion people by 2050, the global demand for animal-derived food products is rapidly increasing. The ongoing industrialization of traditional animal farming systems and unprecedented increase in human mass have led to a multitude of problems including cruel animal treatment, sustainable natural resource depletion, environmental pollution, human health issues, worker safety issues, and social problems. Importantly, intensive livestock and poultry production, which is essential to meet the global demand of animal-derived food products, is a major contributor to climate change. In general, beef and lamb tend to have the largest environmental footprint due to their higher feed conversion efficiency and metabolism. In the landscape, traditional farms use large areas to grow crops and feed livestock animals, which contributes to deforestation (Liu et al., 2021; Zhao et al., 2022; Lee et al., 2023).

3.1. Resource Intensive Practices

Crops themselves require considerable amounts of water and fertilizer. In turn, this lowers the energy and space efficiency of traditional livestock agriculture as these products are being transferred and utilized at a one-directional flow from plants to animals. A large portion of the water used in animal agriculture is also used to clean the facility and the animals themselves. The densities at which the animals are held further increase the rate of water evaporation, so that the water utilized in these systems does not compare to the symbolically accepted norm or burden the presence of animals would naturally require. In terms of fertilizer, these concentrated accumulations of animal waste are a significant source of pollution from run-offs, while the manure itself requires transportation and return from distant sites to commercial crop farms to maintain a nutrient-rich soil in which to grow new crops. Ultimately, the production of crop inputs for livestock agriculture, along with the excessive water waste and pollution from animal agriculture, reduces the sustainability and environmental integrity of operations through energy and material waste (Logan, 2020; Tyagi et al., 2022; Penuelas et al., 2023).

Within traditional livestock agriculture, many of the greenhouse gas emissions stem from the animal's digestive process. However, a significant portion of emissions also come from the clearing of natural landscapes and the burning of existing biomass to cultivate crops for animal consumption. These land use changes result in a sizable decrease in sequestration of CO_2, methane, and nitrous oxide, which all contribute to the increase in atmospheric greenhouse gases and climb in average global temperature each year (Escribano et al., 2020; Chiriacò & Valentini, 2021; Bhattacharyya et al., 2021).

3.2. Greenhouse Gas Emissions

Research has found multiple times that CBIs have the potential to produce animal products substantially more efficiently than their conventional counterparts, with fewer GHG emissions, in addition to reducing water use, land use, and dietary energy consumption for protein. The LCA conducted by CE Delft determined that the reduction ranges from 78% to 96% for CO_2, 64% to 96% for methane, and from 67% to 99% for nitrous oxide. These substantial savings are expected to remain in place throughout the ensuing commercial phase of production (Stephens & Ellis, 2020; Mendly-Zambo et al., 2021; Godfray et al., 2024).

The process of producing conventional meat, dairy, and eggs creates significant greenhouse gas (GHG) emissions, ranging from carbon dioxide emissions produced by the transport and use of fertilizer and pesticides, and up to 37% of global methane from enteric fermentation of ruminant animals. In 2017, eating animal materials contributed to a quarter of global GHG emissions. Where emissions are today is clearer than where they could be if demand for conventional animal-based products persisted and even increased along a projection for a global population of 2050. This scenario would make it challenging to keep global temperature change below the internationally agreed-upon limit of 1.5°C relative to the preindustrial era. According to the Food and Agriculture Organization of the United Nations, agricultural GHGs would reach 50% of the global total by 2050 if our demand for animal agriculture were to grow in a business-as-usual manner (Crippa et al., 2021;Twine, 2021; Xu et al., 2021; Barthelmie, 2022).

4. CELLULAR AGRICULTURE: A SUSTAINABLE ALTERNATIVE

Carbon pricing can aid in managing and directing economic reform and re-allocating national resources. Through the reduction of GHG emissions and the stabilization of climate change, the economic, social, and environmental losses throughout national borders will be reduced. The United Nations Framework Convention on Climate Change (UNFCCC) has adopted a market mechanism that allows economic agents with limits on their GHG emissions to purchase and use emission reduction credits called Certified Emission Reduction (CER) credits. The UNFCCC facilitates the development of international standards through ongoing dialogue, financial support, and institutional structure. The Clean Development Mechanism (CDM) and Joint Implementation (JI) contribute to sustainable development. An innovative and transformative oxygen transport alternative is cellular agriculture, which has the real potential to benefit the climate, environment, economic, and health dimensions. With new knowledge and transformative technology, this area has the real potential to flourish as a sustainable alternative. Given utopic unlocal and long-term sustainability, viewing oxygen as a good provides a jumpstart that encourages stakeholders, particularly capital holders, to develop this futuristic, but urgently needed, new oxygen supply chain (Hickmann et al., 2021; Naser & Pearce, 2022; Espelage et al., 2022).

The United Nations (UN) describes two main strategies for climate change: mitigation, which is reducing or preventing the emission of GHGs into the atmosphere, and adaptation, which is adjusting to the new ecosystem and challenges modulated by the effects of climate change. The adoption of either strategy or whatever mixture from them should not negate the possibility of the coexistence of societies. The World Bank Group, in its report titled "Making the Most of Markets," has argued that a combination of carbon prices (carbon taxes or cap-and-trade) with abatement processes can provide necessary incentives for all sectors of the economy. The report suggests that carbon pricing accelerates emission reduction and may provide important information related to an appropriate mitigation policy. If implemented, it is expected that resources will be allocated efficiently, including technological advancement that could support emission reduction in certain sectors (Thisted & Thisted, 2019; Venmans et al., 2020; Green, 2021).

4.1. Definition and Scope

Today, animal-based food accounts for as much as 70% of the total agricultural land worldwide, and the presence of farm animals significantly contributes to deforestation and biodiversity loss. According to the Global Footprint Network, cattle ranching represents 66% of the deforestation in the Amazon forests. Moreover, three million tons of animal bones produced by animal husbandry each year are generally discarded or utilized for animal food production, even though bones are sources of an extensive array of proteins, lipids, peptides, etc., amplifying the environmental impact of animal production through the pollution of soil and water (Xu et al., 2021; Errickson et al., 2021; Chen et al., 2021).

Disease and deteriorating animal welfare. Additional problems associated with traditional meat production include the contribution of GHG like CO_2, CH_4, and N_2O, as well as waste and nutrients to the environment. Traditional meat production is not just associated with direct GHG emissions and land, water and other natural resources utilization. Meat production also significantly contributes to deforestation, desertification, biodiversity loss, and eutrophication, acidification, and greenhouse gases from more indirect activities and processes (Stavi et al., 2021; Zerbe, 2022).

4.2. Technological Innovations

Technological innovations are indeed critical to the development of cultivated products, with this conversation being presented through two dominant perspectives. On the one hand, one approach is rooted in optimizing traditional, costly, and time-consuming mammalian cell culture techniques to the point where exponential economic decline is a matter of "when" rather than "if", with others proposing "new approaches for reducing costs, such as off-lifting, or increasing the value creation, such as incorporating functionality similar to or better than the traditional animal product (Eibl et al., 2021; (Rischer et al., 2020; Nyyssölä et al., 2022; O'Neill et al., 2020).

4.3. Technological Innovations

Technical innovations in the field of cellular agriculture have the potential to significantly reduce the environmental impact of food production new methods of cultivating animal cells in bioreactors show promise in reducing greenhouse gas emissions and resource usage. play a pivotal role in shaping the progress of our modern society. These advancements, driven by human creativity and ingenuity, have revolutionized our way of life. From the invention of the wheel to the development of the internet, each technological innovation brings forth new possibilities and opportunities. As we continue to embrace the ever-evolving digital landscape, the potential for future breakthroughs seems limitless. Emerging fields such as artificial intelligence, quantum computing, and genetic engineering hold the promise of reshaping industries, improving healthcare, and addressing the world's most pressing challenges. By pushing the boundaries of what is possible, technological innovations propel us towards a more advanced and interconnected world, fostering global collaboration and driving forward human progress. Indeed, the impact of these innovations is profound and enduring, fundamentally transforming the way we live, work, and interact with the world around us (Chen et al., 2022; Ge et al., 2023).

5. ENVIRONMENTAL BENEFITS OF CELLULAR AGRICULTURE

In terms of climate change, if scaled for a complete replacement of conventional livestock, cellular agriculture can potentially avoid the release of 4 to 8 gigatons of carbon dioxide equivalents that livestock production causes annually. Results from a recent life cycle analysis show that cellular agriculture may produce less greenhouse gas emissions compared to equal levels of livestock production. Alternatively, it can be used to begin the process of introducing or strengthening carbon taxes or emission trading schemes, as an impetus for the transition to a low-impact society and economy. In comparison to cultured meat, conventional livestock requires more water, a commodity which is scarce in several parts of the world. Equally, the fact that cellular agriculture does not use any antibiotics reduces the disposal of residues, avoiding the generation of highly polluting waste and minimizes collateral impacts, i.e. salinization of soils, air, etc. (Glaros et al., 2022)

The research on the environmental impacts of cellular agriculture is scarce. But with the existing data, cellular agriculture is likely to present several environmentally friendly solutions compared to conventional livestock. Firstly, cellular agriculture can reduce the land requirements for meat production. Currently, over 40% of the usable land area worldwide is used for agriculture, with the livestock sector covering one-third of the total cultivable surface. Studies show that cultured meat production can decrease land use by 99%, thereby reducing the pressure on natural land areas and forests. In addition, the pressure on wildlife habitats, which face the brunt of the demand for land for livestock, will be reduced (Treich, 2021; Rodríguez Escobar et al., 2021; Munteanu et al., 2021; Chen et al., 2022).

5.1. Reduced Land Use

Shifting to a totally agricultural land area, we can see how cellular agriculture does reduce our land use and the potential mitigation of climate change and provide other benefits. Large-scale releases of domesticated animals into the wild should they escape en masse from manufacturing bioreactors are not an issue to health and food security where they harm ecosystems, but because the vast majority of cultivated animals suffer injuries, pain, and distress through their farming processes (Helliwell & Burton, 2021; Newman et al., 2021). Nor are cellular material spills the potential source of catastrophic damage to ecosystems through the release of pathogens and foreign DNA. However, we also need to consider that the energy needs for cellular agriculture need to be met from low carbon options in addition to providing food (Chiles et al., 2021; Sinke et al., 2023). New strategies were produced for a closed-loop food-energy-water nexus for plants, resulting in accelerated crop production while treating waste, creating useful resources, and producing clean, sustainable energy. Their circular strategy virtually eliminates the solid waste stream, while purifying the process water and nutrients secured for clean energy. Indeed, circularity is one potential solution that could be pursued. However, attention also needs to be given to local impacts on communities with existing links to agriculture, as it is these members of society who are going to be most impacted by a shift in the food industry's structure (Zhang et al., 2021; Fouladi & Al-Ansari, 2021; Caputo et al., 202; Ren et al., 2022;).

Current intensive agricultural models require vast amounts of space. This has resulted in serious deforestation in Latin America, the conversion of the natural grasslands of North and South America, Africa, and Asia, and the consequent loss of biodiversity and increased sedimentation of rivers, coastal areas, and coral reefs. Non-intensive agriculture has displaced indigenous people into marginal lands, and the subsequent overuse of these lands has seen the degradation and loss of them. In addition to meeting our food needs, farmers have historically been seen as the "stewards of the land," but contemporary agriculture places the need to produce food above all else, a system where environmental impact is marginalized, the complete opposite of what we require if future generations are to have a planet that still provides ecosystem services (Zeraatpisheh et al., 2020; Guerrero-Pineda et al., 2022; Goulart et al., 2023).

5.2. Lower Greenhouse Gas Emissions

For example, to only incentivize and increase the number of richer food biodiversity environments (rural spaces and urban spot markets) can enhance the carceral environment of the urban fragile, otherwise stopping them from living a life of dignity and diversity within the urban environment (Helliwell & Burton, 2021; Godfray et al., 2024).

There are food security consequences from attempting to instigate massive changes to current livestock-intensive systems. In particular, ensuring food availability and affordability for the urban fragile, i.e., those who are already malnourished, food-cost constrained, or who can only have access to nutritious foods from complex and fragile urban food systems. Without social support, i.e., sustainable transport systems, fair employment, and housing, the viability criteria required to create equitable food systems, implementing dietary changes ideologically is not appropriate in a number of policy spaces (Rothwell et al., 2020; Weishaupt et al., 2020; Martin-Ortega et al., 2022).

Cellular agriculture, as an alternative form of meat production, has been hypothesized to offer significant farmland and greenhouse gas emission savings relative to conventional meat production. The aim of this section is to review and critique the evidence for and against the carbon and land use savings of cellular agriculture in the field of sustainable intensification and livestock-landscape interactions (Helliwell & Burton, 2021; Bapat et al., 2022; Smith et al., 2022; El Wali et al., 2024).

As highlighted in the previous section, livestock is a significant contributor to greenhouse gas emissions globally. Its impacts occur from land clearing for grazing and crops to feed livestock, methane produced during digestion, and nitrous oxide released in manure management. The large number of livestock, particularly ruminant species, indicates that technological fixes alone are not likely to be sufficient to reduce the greenhouse gas emissions of the sector. Carbon reduction is likely to require a combination of avoided deforestation and carbon sequestration in soils, trees, or alternative land-use systems, alongside the exploration of dietary shifts for key consuming populations (Uwizeye et al., 2020; Zhang et al., 2022; Xu et al., 2021).

6. ECONOMIC VIABILITY OF CELLULAR AGRICULTURE

Similarly, while cellular agriculture clean meat or dairy products look like meat or from a chemical composition viewpoint is similar to dairy products, what our markets—and markets worldwide—will think of these products and these labels is as yet uncertain. Clarification in some political form is necessary to ensure that cellular agriculture firms are not permanently at a disadvantage if their products are labeled cellular, if the labeling of their products is otherwise equivalent to those tabulated for conventional meat and other animal byproducts. Likewise, potential tax stations or subsidies based on the greenhouse gas production footprint are possible means by which the government can assist or stymie the growth of cellular agriculture (Chiles et al., 2021; Soice & Johnston, 2021; Jahir et al., 2023).

Our discussion has primarily taken the perspective of the broad ecology of a cellular agriculture supported economy. How cellular agriculture will fit into the established framework of agriculture, automation technologies, biotechnology, and consumer preferences, while open, is not necessarily a given that we should take for granted. It may be that cellular agriculture products offer valuable complements if not outright economic substitutes, and that the jobs and income potential of cellular agriculture reduce social division associated with market access and division of income between investment and human capital returns. In this review, we have only begun to challenge some of the implications of rethinking fully the economic relationships associated with cellular agriculture-based consumer goods (Chiles et al., 2021;Nyyssölä et al., 2022; Kumar et al., 2022).

6.1. Cost Efficiency

This is so because the tax applies the marginal cost rule, respects the cost-efficiency principle, and eliminates the need for all other regulations and control measures. This is basically the ideal world of Coase and Horn, where the assignment of property rights to polluters should remove the very cause of the problems. Since this ideal world does not exist, taxes might also serve as second-best options, assuming that the costs of information which governments are able to exploit are sufficiently low. However, while in the case of economic instruments the principle does not prescribe a precise instrument but only states that the marginal cost of achieving a given reduction in pollution will be uniform across sources, other regulations may need to be carefully designed (Agrawal & Wildasin, 2020; Hogan, 2022; Johannesen, 2022; Boadway & Flatters, 2023).

Formulated formally, the basic cost-efficiency concept assigns to each level of environmental accomplishment its minimal cost. Every dollar spent to protect or improve the environment under the equilibrium that would prevail in a hypothetical world of zero transaction costs would be spent exactly where it would reduce environmental damages by $1. Cost-efficiency has the following main implications: high environmental targets may prove too costly, environmental regulations will lower value-added, and when consumers buy fewer products due to higher prices, the regulations may adversely affect the economy as a whole. The basic concept can then be used as a benchmark for evaluating actual environmental policies. While the net benefits of a policy falling short of the cost-efficiency benchmark are negative, the net benefits increase to zero under the introduction of pollution taxes (Sun et al., 2020; Hu et al., 2021; Raihan & Tuspekova, 2022; Sun et al., 2023; Yang & Khan, 2021).

6.2. Market Potential

Analysis by the Good Food Institute suggests that cellular agriculture is less costly to scale than some terrestrial plants, with less than $1000 of capital equipment needed to produce a metric ton of ingredient. In total, only around $2000 of equipment would theoretically be needed to produce over 10 million tons of ingredients, equating to over 10% of the global protein market. However, the price elasticity of demand for cellular agriculture products is uncertain. For example, a cube steak produced using cellular agriculture methods has already been produced to a cost of $100,000, and Finless Foods have been quoted as producing a 20,000 kg of cellular agriculture fish for $20,000. Scaling of cellular agriculture will also depend on the role of regulation plays, public and private research, and also demand management from the state (Fujita et al., 2023).

7. REGULATORY AND ETHICAL CONSIDERATIONS

Throughout history, food innovation has periodically been faced with ethical and economic dilemmas, from the cultivation of barley, the development of GMOs, and to a particular extent, its acceptability towards fish farming. It remains to be seen what specific ethical and economic objections are raised in regard to cellular agriculture but, for its acceptance in the long term, these questions must be dealt with openly and transparently, encouraging an open and unbiased dialogue in the public arena. The technology itself, paving the way for a very diverse set of products, would benefit greatly from discussions with a wide range of participants in societal decision-making and an important contribution to solving the challenges of the future, such as food insecurity, the saturation of environmental limits, and the inability to control major communicable diseases (Bryant & Barnett, 2020; Kenigsberg & Zivotofsky, 2020; Chiles et al., 2021; Guan et al., 2021; Siddiqui et al., 2022; Hassoun et al., 2022).

As the field of cellular agriculture grows and draws interest from a broader, more diverse audience, consistent global regulation may be necessary to protect consumer safety, ensure optimal production conditions, and provide regulatory clarity for cellular ag entrepreneurs and researchers. Furthermore, unique jurisdictional attitudes towards genetic engineering could require additional legal considerations which the traditional food and biotechnology industries are not responsible for. Some cellular agriculture research topics have raised philosophical and religious questions associated with the engineering of animal cells. Such discussions will likely be necessary to respond to these concerns and to ascertain whether or not society will accept certain cellular agriculture products. These questions, however, are not trivial and it is uncertain to what extent we, or future generations, will have the ability not only to commercialize cellular agriculture products, but the ability to experiment with existing cellular agricultural models and offering solutions to these questions (Bryant & Barnett, 2020; Hadi & Brightwell, 2021; Eibl et al., 2021; Chiles et al., 2021; Ong et al., 2021).

7.1. Regulatory Frameworks

In 2018, the European Union began to consider the regulation of insect-based cultured meat, much of which was provided to meet the demand for protein and the result of the environmental and health risks in the livestock industry. In 2019, China was more clear-cut on the regulation of cultured meat and recognized the significance of cultured meat as a substitute for traditional animal husbandry. The country began to specifically regulate and standardize the innovation and business activities of the artificial meat industry to blaze a trail to promote the cultivation of meat in our country (Bryant & Barnett, 2020; Hadi & Brightwell, 2021; Lähteenmäki-Uutela et al., 2021; Żuk-Gołaszewska et al., 2022; Mancini & Antonioli, 2022; Mancini & Antonioli, 2022).

According to German law, "Novel Foods" are defined as those that were not used for human consumption in the EU before the enforcement of the Novel Food legislation. Their definition includes food and food ingredients which consist of, contain, or are produced from novel food, or food and food ingredients containing additives or have a new production process applied to them. These must be assessed before they are placed on the market (Sokołowski, 2020; Zarbà et al., 2020; EFSA, 2020).

The scale of government intervention necessary in cellular agriculture is still under debate. Some scholars argue that cellular agriculture is just another method of traditional agriculture and should be regulated as such with no need for a specific regulatory structure. Others propose a new regulatory structure specifically designed for cellular agricultural products because products can be different from both animal products and GM foods. They argue that products from cellular agriculture should have their own set of scientific, industrial, and legal considerations which are unique to the way the products are produced (Rischer et al., 2020; Chiles et al., 2021; Nyyssölä et al., 2022);

Moving from a system of animal agriculture towards cellular agriculture requires wide-reaching regulatory considerations. Regulation should play a number of roles: preventing consumer misunderstanding through accurate labeling; setting necessary safety standards; ensuring security of research outcomes without impeding innovation; coordinating thermostatical effects of the industry with broader policy goals; ensuring fair competition with existing industry stakeholders. Partly because cellular agriculture does not fit neatly into an existing business sector, clear and inclusive regulatory frameworks will be essential (Fox *et al.*, 2021; Touzdjian *et al.*, 2022).

7.2. Consumer Acceptance

The early adopter of cultured meat is therefore the vegan, eater of processed foods, receptor that the entire world will soon be eating at McDonald's cultured meat, or any combination of these three. Their involvement, and that of other non-vegans with facets of the development of the clean meat industry is one way in which non-vegan early adopters can become influential. These early adopters can then influence others. If people are seen to eat these foods, other people are more likely to also eat those foods. If banks are seen to invest money in the development of these products, others are more likely to also co-invest their money (Percival & Percival, 2022; Plec, 2024).

As discussed in previous sections, early consumer acceptance towards cellular agricultural products is vital to mitigate climate change, especially as these products are more environmentally friendly compared to conventionally produced animal products. As such, the success of cellular agricultural products will be dependent on who adopts these products, and how rapidly they are adopted. Early adoption has been explained using Rogers' theory of Technology Adoption. According to this theory, an early adopter is an individual who "agrees to try out a new innovation at some risk". In regard to cultured meat, and other cellular agricultural produced products, it is important to note who is likely to adopt these products in their early stages. According to Guenther, the early adopter is likely to be someone interested in new and progressive technologies, and willing to take risks (Rubio,et al., 2020; Rischer etal., 2020; Hadi and Brightwell, 2021).

8. CHALLENGES AND FUTURE DIRECTIONS

In response to the challenges highlighted, there are several areas of requisite further research. End-to-end animal agricultural cost modeling is a necessity to identify the cost of production of animal-component free culture media animal products desired by consumers. Cellular agriculture technology must continue to develop to lower the cost of production use of a liquid bioreactor platform, which may circumvent the requirement of culture media for cellular agriculture animal product cost modeling. Furthermore, future cellular agriculture regulatory approval would seem to require a partnership between the cellular agriculture industry and the regulatory community. Indeed, it may be necessary to establish a new agency with domain expertise to govern, license, and regulate the production of cellular agriculture animal products, which fundamentally differ from traditional plant-based or cellular agriculture products (Xu *et al.*, 2021; Morrone *et al.*, 2022; Liu *et al.*, 2023; Mysko *et al.*, 2024).

Commensurate with its benefits, cellular agricultural technology is still in its relative infancy. There remain several primary challenges that must be surmounted to enable mainstream adoption of cellular agriculture as the dominant source of animal products. One of the most significant challenges remaining in the production of cellular agriculture is the development of more affordable animal-component free culture media. This media currently constitutes the major limitation for the production of cellular agriculture animal products. In the absence of regulatory and labeling guidelines, it would seem that market demand for cellular agriculture animal products has dropped significantly in the wake of market entry of plant-based competitors, as patented technologies pivot to a direct-to-consumer or B-to-B model with markedly higher return on investment (Ho *et al.*, 2021; (Hubalek et al., 2022; Stout *et al.*, 2022).

8.1. Technological Barriers

In addition to complex lipids, differentiation from stem cells has a number of additional constraints to minimal media formulations. These formulations also require cholesterol, fatty acids, and dimerized fatty acids. Such media composition also contains a defined blend of glucose and inorganic salts that can support the metabolic function of the cell - glucose as a final electron carrier and the salts as cofactors necessary for enzymatic conversion. Nutrient supplementation implies an additional food source for the ex vivo cells, model synaptic systems available for study, to support both stem cell maintenance and differentiation supporting ligand, factor, or antibody supplementation. Furthermore, the rate of mass transport across the stem cell must be matched by the growth rate of the culture. One approach to address this problem is by providing effective mass transport that a sizable, yet unrealistic, cell culture cannot be adequately sustained, nor do these dimensions support efficient and scalable bioprocess unit operations (Clémot *et al.*, 2020; Bispo *et al.*, 2022; Mannully *et al.*, 2022).

Minimally processed bovine serum has a collection of lipid classes, which support lipid biosynthesis in the cell culture. Conditioned basal media serve a similar function, supporting the effective differentiation of stem cells by providing a collection of biologically essential lipids. More recently, there have been continued explorations into culturing cells in chemically defined media that is entirely free from the addition of animal-derived proteins. The significant challenges with media supplementation or replacement with either semi-purified lipids or defined media constituents are sometimes overlooked as a leading challenge of complex lipid biosynthesis. The list of complex lipids necessary for mammalian cell cultivation is both extensive and expensive, making this parameter the key technological barrier in the way of a fully cultivated, serum-free hamburger (Fabris *et al.*, 2020; Hadi and Brightwell, 2021; Bomkamp *et al.*, 2022; Godfray, etal., 2024).

8.2. Scaling Up Production

Building these companies is expensive, but the ability to adopt existing scaling programs and technologies makes them serious industry competitors. An additional and necessary industry component is the manufacture of facility construction materials. In contrast to many agricultural products, the scale of the cell agriculture industry could outstrip the material growth rates of structural building materials. Another challenge is that all manufacturers will not be favored at equal costs. As the primary technology for cultivating cell lines, vertical farming has a unique land and building overhead. Its energy costs of operation are also high, posing resource or access constraints on their widespread use. But other less climate-friendly food producing methods are used in many parts of the world at a large scale. Even in the most advanced economies, large-scale animal raising systems might find transition paths that were less onerous, were environmental externalities better understood. With these realities in mind, money is purposely being spent to perfect and expand a novel industry that could be a net climate benefit for transforming how food is made and sold (Novellino *et al.*, 2021; Zhao *et al.*, 2022; Armstrong, 2023; Jamil, etal., 2023; Wang *et al.*, 2023).

To produce a novel food system using only animal products would be a herculean task. Right now, since it is a brand-new industry, building small production facilities and undertaking novel research is the basis of the art. None of the companies involved have reached full-scale production using cellular agriculture. But scale-up is a natural trajectory for growth-minded entrepreneurs. Because of the technology's roots in the biotechnology and pharmaceutical industries, with time, experience, and economies of scale, significant progress can be made in improving production technologies. As companies actively bring products to market, many raise substantial equity. By licensing their intellectual property, their market success can make established biotechnology companies powerful entry points for the cellular agriculture industry (Risner *et al.*, 2020; Sinke *et al.*, 2023).

9. CONCLUSION AND RECOMMENDATIONS

In this chapter, the likely changes and the new challenges arising from cellular agriculture have been outlined, and a counterfactual thinking on what the food system can expect from the initiation of cellular agriculture has been conjectured. Both the organizational and stakeholders' implications of cellular agriculture have been detailed, presented, and discussed. The future research questions and implications for the main stakeholders, policy shapers, and the masses have been delineated and laid out. A comprehensive conceptual understanding of these issues, which has not yet appeared in the scholarly literature, might establish a valuable platform for further empirical and operational progress. This chapter contributes significantly to the debate on future policy arrangements and system-theoretical framework, mainly involving the application of cellular agriculture to humans, and were important, to some degree, to nonhumans.

For future cellular agriculture development to proceed efficiently, it is imperative that the policy management of cellular agriculture innovation and transition of cellular agriculture are well addressed at the forefront of general discussions. The main obstacles to development are the policy, regulatory, social resistance, consumer behavioral issues, and the cost performance and the inferior tasting issues, among others. Cellular agriculture would ultimately become a pseudo-artificial products industry, and that would alter the social, economic, cultural, moral, legal, and ethical structure of the society, at the individual, organization, or overall levels. The new set of animals provided through cellular agriculture may necessitate a new system of thinking, new business models, new production and marketing system, and a new regulatory and condemnatory course of actions from the statism-established elite establishment or the public.

REFERENCES

Afrifa, G. A., Tingbani, I., Yamoah, F., & Appiah, G. (2020). Innovation input, governance and climate change: Evidence from emerging countries. *Technological Forecasting and Social Change*, 161, 120256. 10.1016/j.techfore.2020.120256

Agrawal, D. R., & Wildasin, D. E. (2020). Technology and tax systems. *Journal of Public Economics*, 185, 104082. 10.1016/j.jpubeco.2019.104082

Agrimonti, C., Lauro, M., & Visioli, G. (2020). Smart agriculture for food quality: Facing climate change in the 21st century. *Critical Reviews in Food Science and Nutrition*, 61(6), 971–981. 10.1080/10408398.2020.174955532270688

Alsarhan, L. M., Alayyar, A. S., Alqahtani, N. B., & Khdary, N. H. (2021). Circular Carbon Economy (CCE): A Way to Invest CO2 and Protect the Environment, a Review. *Sustainability (Basel)*, 13(21), 11625. 10.3390/su132111625

Armstrong, R. (2023). Towards the microbial home: An overview of developments in next-generation sustainable architecture. *Microbial Biotechnology*, 16(6), 1112–1130. 10.1111/1751-7915.1425637070748

Arndt, C., Hristov, A. N., Price, W. J., McClelland, S. C., Pelaez, A. M., Cueva, S. F., Oh, J., Dijkstra, J., Bannink, A., Bayat, A. R., Crompton, L. A., Eugène, M. A., Enahoro, D., Kebreab, E., Kreuzer, M., McGee, M., Martin, C., Newbold, C. J., Reynolds, C. K., & Yu, Z. (2022). Full adoption of the most effective strategies to mitigate methane emissions by ruminants can help meet the 1.5 °C target by 2030 but not 2050. *Proceedings of the National Academy of Sciences of the United States of America*, 119(20), e2111294119–e2111294119. 10.1073/pnas.211129411935537050

Bapat, S., Koranne, V., Shakelly, N., Huang, A., Sealy, M. P., Sutherland, J. W., Rajurkar, K. P., & Malshe, A. P. (2022). Cellular Agriculture: An Outlook on Smart and Resilient Food Agriculture Manufacturing. *Smart and Sustainable Manufacturing Systems*, 6(1), 1–11. 10.1520/SSMS20210020

Barthelmie, R. J. (2022). Impact of Dietary Meat and Animal Products on GHG Footprints: The UK and the US. *Climate (Basel)*, 10(3), 43. 10.3390/cli10030043

Beillouin, D., Schauberger, B., Bastos, A., Ciais, P. & Makowski, D. (2020). Impact of extreme weather conditions on European crop production in 2018. *Philosophical Transactions of the Royal Society of London. Series B, Biological Sciences*, 375(1810), 20190510. 10.1098/rstb.2019.0510

Bhattacharyya, S. S., Leite, F. F. G. D., Adeyemi, M. A., Sarker, A. J., Cambareri, G. S., Faverin, C., Tieri, M. P., Castillo-Zacarías, C., Melchor-Martínez, E. M., Iqbal, H. M. N., & Parra-Saldívar, R. (2021). A paradigm shift to CO2 sequestration to manage global warming – With the emphasis on developing countries. *The Science of the Total Environment*, 790, 148169. 10.1016/j.scitotenv.2021.14816934380249

Bispo, D. S. C., Michálková, L., Correia, M., Jesus, C. S. H., Duarte, I. F., Goodfellow, B. J., Oliveira, M. B., Mano, J. F., & Gil, A. M. (2022). Endo- and Exometabolome Crosstalk in Mesenchymal Stem Cells Undergoing Osteogenic Differentiation. *Cells*, 11(8), 1257. 10.3390/cells1108125735455937

Boadway, R., & Flatters, F. (2023). The Taxation of Natural Resources: Principles and Policy Issues. In *Taxing Choices for Managing Natural Resources, the Environment, and Global Climate Change* (pp. 17–81). Springer International Publishing. 10.1007/978-3-031-22606-9_2

Bomkamp, C., Skaalure, S. C., Fernando, G. F., Ben-Arye, T., Swartz, E. W., & Specht, E. A. (2022). Scaffolding Biomaterials for 3D Cultivated Meat: Prospects and Challenges. *Advanced Science (Weinheim, Baden-Wurttemberg, Germany)*, 9(3), e2102908–e2102908. 10.1002/advs.20210290834786874

Bryant, C., & Barnett, J. (2020). Consumer Acceptance of Cultured Meat: An Updated Review (2018–2020). *Applied Sciences (Basel, Switzerland)*, 10(15), 5201. 10.3390/app10155201

Caputo, S., Schoen, V., Specht, K., Grard, B., Blythe, C., Cohen, N., Fox-Kämper, R., Hawes, J., Newell, J., & Poniży, L. (2021). Applying the food-energy-water nexus approach to urban agriculture: From FEW to FEWP (Food-Energy-Water-People). *Urban Forestry &amp. Urban Forestry & Urban Greening*, 58, 126934. 10.1016/j.ufug.2020.126934

Chen, L., Guttieres, D., Koenigsberg, A., Barone, P. W., Sinskey, A. J., & Springs, S. L. (2022). Large-scale cultured meat production: Trends, challenges and promising biomanufacturing technologies. *Biomaterials*, 280, 121274. 10.1016/j.biomaterials.2021.12127434871881

Chen, W., Jafarzadeh, S., Thakur, M., Ólafsdóttir, G., Mehta, S., Bogason, S., & Holden, N. M. (2021). Environmental impacts of animal-based food supply chains with market characteristics. *The Science of the Total Environment*, 783, 147077. 10.1016/j.scitotenv.2021.14707734088125

Chiles, R. M., Broad, G., Gagnon, M., Negowetti, N., Glenna, L., Griffin, M. A. M., Tami-Barrera, L., Baker, S., & Beck, K. (2021). Democratizing ownership and participation in the 4th Industrial Revolution: Challenges and opportunities in cellular agriculture. *Agriculture and Human Values*, 38(4), 943–961. 10.1007/s10460-021-10237-734456466

Chiriacò, M. V., & Valentini, R. (2021). A land-based approach for climate change mitigation in the livestock sector. *Journal of Cleaner Production*, 283, 124622. 10.1016/j.jclepro.2020.124622

Clarke, B., Otto, F., Stuart-Smith, R., & Harrington, L. (2022). Extreme weather impacts of climate change: An attribution perspective. *Environmental Research: Climate*, 1(1), 12001. 10.1088/2752-5295/ac6e7d

Clémot, M., Sênos Demarco, R., & Jones, D. L. (2020). Lipid Mediated Regulation of Adult Stem Cell Behavior. *Frontiers in Cell and Developmental Biology*, 8, 115. 10.3389/fcell.2020.0011532185173

Cosens, B., Ruhl, J. B., Soininen, N., Gunderson, L., Belinskij, A., Blenckner, T., Camacho, A. E., Chaffin, B. C., Craig, R. K., Doremus, H., Glicksman, R., Heiskanen, A.-S., Larson, R., & Similä, J. (2021). Governing complexity: Integrating science, governance, and law to manage accelerating change in the globalized commons. *Proceedings of the National Academy of Sciences of the United States of America*, 118(36), e2102798118. 10.1073/pnas.210279811834475210

Crippa, M., Solazzo, E., Guizzardi, D., Monforti-Ferrario, F., Tubiello, F. N., & Leip, A. (2021). Food systems are responsible for a third of global anthropogenic GHG emissions. *Nature Food*, 2(3), 198–209. 10.1038/s43016-021-00225-937117443

Cronin, J., Zabel, F., Dessens, O., & Anandarajah, G. (2020). Land suitability for energy crops under scenarios of climate change and land-use. *Global Change Biology. Bioenergy*, 12(8), 648–665. 10.1111/gcbb.12697

Dale, A., Robinson, J., King, L., Burch, S., Newell, R., Shaw, A., & Jost, F. (2019). Meeting the climate change challenge: Local government climate action in British Columbia, Canada. *Climate Policy*, 20(7), 866–880. 10.1080/14693062.2019.1651244

Del Buono, D. (2021). Can biostimulants be used to mitigate the effect of anthropogenic climate change on agriculture? It is time to respond. *The Science of the Total Environment*, 751, 141763. 10.1016/j.scitotenv.2020.14176332889471

Eibl, R., Senn, Y., Gubser, G., Jossen, V., van den Bos, C., & Eibl, D. (2021). Cellular Agriculture: Opportunities and Challenges. *Annual Review of Food Science and Technology*, 12(1), 51–73. 10.1146/annurev-food-063020-12394033770467

El Wali, M., Rahimpour Golroudbary, S., Kraslawski, A., & Tuomisto, H. L. (2024). Transition to cellular agriculture reduces agriculture land use and greenhouse gas emissions but increases demand for critical materials. *Communications Earth & Environment*, 5(1), 61. 10.1038/s43247-024-01227-8

Engler, N., & Krarti, M. (2021). Review of energy efficiency in controlled environment agriculture. *Renewable & Sustainable Energy Reviews*, 141, 110786. 10.1016/j.rser.2021.110786

Ercili-Cura, D., & Barth, D. (2021). *Cellular Agriculture: Lab Grown Foods* (Vol. 8). American Chemical Society.

Errickson, F., Kuruc, K., & McFadden, J. (2021). Animal-based foods have high social and climate costs. *Nature Food*, 2(4), 274–281. 10.1038/s43016-021-00265-137118477

Escribano, M., Elghannam, A., & Mesias, F. J. (2020). Dairy sheep farms in semi-arid rangelands: A carbon footprint dilemma between intensification and land-based grazing. *Land Use Policy*, 95, 104600. 10.1016/j.landusepol.2020.104600

Espelage, A., Ahonen, H.-M., & Michaelowa, A. (2022). *The role of carbon market mechanisms in climate finance*. Edward Elgar Publishing. 10.4337/9781784715656.00023

Fabris, M., Abbriano, R. M., Pernice, M., Sutherland, D. L., Commault, A. S., Hall, C. C., Labeeuw, L., McCauley, J. I., Kuzhiuparambil, U., Ray, P., Kahlke, T., & Ralph, P. J. (2020). Emerging Technologies in Algal Biotechnology: Toward the Establishment of a Sustainable, Algae-Based Bioeconomy. *Frontiers in Plant Science*, 11, 279. 10.3389/fpls.2020.0027932256509

Fouladi, J., & Al-Ansari, T. (2021). Conceptualising multi-scale thermodynamics within the energy-water-food nexus: Progress towards resource and waste management. *Computers &amp. Computers & Chemical Engineering*, 152, 107375. 10.1016/j.compchemeng.2021.107375

Fox, G., Clohessy, T., van der Werff, L., Rosati, P., & Lynn, T. (2021). Exploring the competing influences of privacy concerns and positive beliefs on citizen acceptance of contact tracing mobile applications. *Computers in Human Behavior*, 121, 106806. 10.1016/j.chb.2021.106806

Fujita, R., Brittingham, P., Cao, L., Froehlich, H., Thompson, M., & Voorhees, T. (2023). Toward an environmentally responsible offshore aquaculture industry in the United States: Ecological risks, remedies, and knowledge gaps. *Marine Policy*, 147, 105351. 10.1016/j.marpol.2022.105351

Ge, C., Selvaganapathy, P. R., & Geng, F. (2023). Advancing our understanding of bioreactors for industrial-sized cell culture: Health care and cellular agriculture implications. *American Journal of Physiology. Cell Physiology*, 325(3), C580–C591. 10.1152/ajpcell.00408.202237486066

Giglio, F., Scieuzo, C., Ouazri, S., Pucciarelli, V., Ianniciello, D., Letcher, S., Salvia, R., Laginestra, A., Kaplan, D. L., & Falabella, P. (2024). A Glance into the Near Future: Cultivated Meat from Mammalian and Insect Cells. *Small Science*, 2400122. Advance online publication. 10.1002/smsc.202400122

Glaros, A., Marquis, S., Major, C., Quarshie, P., Ashton, L., Green, A. G., Kc, K. B., Newman, L., Newell, R., Yada, R. Y., & Fraser, E. D. G. (2022). Horizon scanning and review of the impact of five food and food production models for the global food system in 2050. *Trends in Food Science &amp. Trends in Food Science & Technology*, 119, 550–564. 10.1016/j.tifs.2021.11.013

Godfray, H. C. J., Poore, J., & Ritchie, H. (2024). Opportunities to produce food from substantially less land. *BMC Biology*, 22(1), 138. 10.1186/s12915-024-01936-838914996

Goulart, F. F., Chappell, M. J., Mertens, F., & Soares-Filho, B. (2023). Sparing or expanding? The effects of agricultural yields on farm expansion and deforestation in the tropics. *Biodiversity and Conservation*, 32(3), 1089–1104. 10.1007/s10531-022-02540-4

Green, J. F. (2021). Does carbon pricing reduce emissions? A review of ex-post analyses. *Environmental Research Letters*, 16(4), 43004. 10.1088/1748-9326/abdae9

Guan, X., Lei, Q., Yan, Q., Li, X., Zhou, J., Du, G., & Chen, J. (2021). Trends and ideas in technology, regulation and public acceptance of cultured meat. *Future Foods : a Dedicated Journal for Sustainability in Food Science*, 3, 100032. 10.1016/j.fufo.2021.100032

Guerrero-Pineda, C., Iacona, G. D., Mair, L., Hawkins, F., Siikamäki, J., Miller, D., & Gerber, L. R. (2022). An investment strategy to address biodiversity loss from agricultural expansion. *Nature Sustainability*, 5(7), 610–618. 10.1038/s41893-022-00871-2

Hadi, J., & Brightwell, G. (2021). Safety of Alternative Proteins: Technological, Environmental and Regulatory Aspects of Cultured Meat, Plant-Based Meat, Insect Protein and Single-Cell Protein. *Foods*, 10(6), 1226. 10.3390/foods1006122634071292

Hassoun, A., Cropotova, J., Trif, M., Rusu, A. V., Bobiş, O., Nayik, G. A., Jagdale, Y. D., Saeed, F., Afzaal, M., Mostashari, P., Khaneghah, A. M., & Regenstein, J. M. (2022). Consumer acceptance of new food trends resulting from the fourth industrial revolution technologies: A narrative review of literature and future perspectives. *Frontiers in Nutrition*, 9, 972154. 10.3389/fnut.2022.97215436034919

Helliwell, R., & Burton, R. J. F. (2021). The promised land? Exploring the future visions and narrative silences of cellular agriculture in news and industry media. *Journal of Rural Studies*, 84, 180–191. 10.1016/j.jrurstud.2021.04.002

Ho, Y. Y., Lu, H. K., Lim, Z. F. S., Lim, H. W., Ho, Y. S., & Ng, S. K. (2021). Applications and analysis of hydrolysates in animal cell culture. *Bioresources and Bioprocessing*, 8(1), 93. 10.1186/s40643-021-00443-w34603939

Hogan, W. W. (2022). Electricity Market Design and Zero-Marginal Cost Generation. *Current Sustainable/Renewable. Energy Reports*, 9(1), 15–26. 10.1007/s40518-021-00200-9

Home, P. D., & Mehta, R. (2021). Insulin therapy development beyond 100 years. *The Lancet Diabetes &amp. Endocrinology*, 9(10), 695–707. 10.1016/S2213-8587(21)00182-034480874

Hu, D., Jiao, J., Tang, Y., Han, X., & Sun, H. (2021). The effect of global value chain position on green technology innovation efficiency: From the perspective of environmental regulation. *Ecological Indicators*, 121, 107195. 10.1016/j.ecolind.2020.107195

Hubalek, S., Post, M. J., & Moutsatsou, P. (2022). Towards resource-efficient and cost-efficient cultured meat. *Current Opinion in Food Science*, 47, 100885. 10.1016/j.cofs.2022.100885

Jahir, N. R., Ramakrishna, S., Abdullah, A. A. A., & Vigneswari, S. (2023). Cultured meat in cellular agriculture: Advantages, applications and challenges. *Food Bioscience*, 53, 102614. 10.1016/j.fbio.2023.102614

Jamil, U., Bonnington, A., & Pearce, J. M. (2023). The Agrivoltaic Potential of Canada. *Sustainability (Basel)*, 15(4), 3228. 10.3390/su15043228

Johannesen, N. (2022). The global minimum tax. *Journal of Public Economics*, 212, 104709. 10.1016/j.jpubeco.2022.104709

Kaufman, D., McKay, N., Routson, C., Erb, M., Dätwyler, C., Sommer, P. S., Heiri, O., & Davis, B. (2020). Holocene global mean surface temperature, a multi-method reconstruction approach. *Scientific Data*, 7(1), 201. 10.1038/s41597-020-0530-732606396

Kenigsberg, J. A., & Zivotofsky, A. Z. (2020). A Jewish Religious Perspective on Cellular Agriculture. *Frontiers in Sustainable Food Systems*, 3, 128. Advance online publication. 10.3389/fsufs.2019.00128

Klerkx, L., & Rose, D. (2020). Dealing with the game-changing technologies of Agriculture 4.0: How do we manage diversity and responsibility in food system transition pathways? *Global Food Security*, 24, 100347. 10.1016/j.gfs.2019.100347

Kumar, P., Mehta, N., Abubakar, A. A., Verma, A. K., Kaka, U., Sharma, N., Sazili, A. Q., Pateiro, M., Kumar, M., & Lorenzo, J. M. (2022). Potential Alternatives of Animal Proteins for Sustainability in the Food Sector. *Food Reviews International*, 39(8), 5703–5728. 10.1080/87559129.2022.2094403

Lähteenmäki-Uutela, A., Rahikainen, M., Lonkila, A., & Yang, B. (2021). Alternative proteins and EU food law. *Food Control*, 130, 108336. 10.1016/j.foodcont.2021.108336

Lee, S.-H., & Yoon, K.-H. (2021). A Century of Progress in Diabetes Care with Insulin: A History of Innovations and Foundation for the Future. *Diabetes & Metabolism Journal*, 45(5), 629–640. 10.4093/dmj.2021.016334610718

Lee, S. Y., Lee, D. Y., Mariano, E. J., Yun, S. H., Lee, J., Park, J., Choi, Y., Han, D., Kim, J. S., Joo, S.-T., & Hur, S. J. (2023). Study on the current research trends and future agenda in animal products: An Asian perspective. *Journal of Animal Science and Technology*, 65(6), 1124–1150. 10.5187/jast.2023.e12138616880

Liu, J., Xiao, D., Liu, Y., & Huang, Y. (2023). A Pig Mass Estimation Model Based on Deep Learning without Constraint. *Animals (Basel)*, 13(8), 1376. 10.3390/ani13081376 37106939

Liu, Y., Dong, X., Wang, B., Tian, R., Li, J., Liu, L., Du, G., & Chen, J. (2021). Food synthetic biology-driven protein supply transition: From animal-derived production to microbial fermentation. *Chinese Journal of Chemical Engineering*, 30, 29–36. 10.1016/j.cjche.2020.11.014

Logan, T. J. (2020). Sustainable Agriculture and Water Quality. In *Sustainable Agricultural Systems* (pp. 582–613). CRC Press. 10.1201/9781003070474-40

Mancini, M. C., & Antonioli, F. (2022). The future of cultured meat between sustainability expectations and socio-economic challenges. In *Future Foods* (pp. 331–350). Elsevier. 10.1016/B978-0-323-91001-9.00024-4

Mannully, C. T., Bruck-Haimson, R., Zacharia, A., Orih, P., Shehadeh, A., Saidemberg, D., Kogan, N. M., Alfandary, S., Serruya, R., Dagan, A., Petit, I., & Moussaieff, A. (2022). Lipid desaturation regulates the balance between self-renewal and differentiation in mouse blastocyst-derived stem cells. *Cell Death & Disease*, 13(12), 1027. 10.1038/s41419-022-05263-036477438

Martin-Ortega, J., Rothwell, S. A., Anderson, A., Okumah, M., Lyon, C., Sherry, E., Johnston, C., Withers, P. J. A., & Doody, D. G. (2022). Are stakeholders ready to transform phosphorus use in food systems? A transdisciplinary study in a livestock intensive system. *Environmental Science & Policy*, 131, 177–187. 10.1016/j.envsci.2022.01.01135505912

Mathieu, C., Martens, P.-J., & Vangoitsenhoven, R. (2021). One hundred years of insulin therapy. *Nature Reviews. Endocrinology*, 17(12), 715–725. 10.1038/s41574-021-00542-w34404937

Meissner, H., Blignaut, J., Smith, H. & Du Toit, L. (2023). *The broad-based eco-economic impact of beef and dairy production: A global review.* https://doi.org/10.4314/sajas.v53i2.11

Mendly-Zambo, Z., Powell, L. J., & Newman, L. L. (2021). Dairy 3.0: Cellular agriculture and the future of milk. *Food, Culture &amp. Food, Culture, & Society*, 24(5), 675–693. 10.1080/15528014.2021.1888411

Morice, C. P., Kennedy, J. J., Rayner, N. A., Winn, J. P., Hogan, E., Killick, R. E., Dunn, R. J. H., Osborn, T. J., Jones, P. D., & Simpson, I. R. (2021). An Updated Assessment of Near-Surface Temperature Change From 1850: The HadCRUT5 Data Set. *Journal of Geophysical Research. Atmospheres*, 126(3), e2019JD032361. Advance online publication. 10.1029/2019JD032361

Morrone, S., Dimauro, C., Gambella, F., & Cappai, M. G. (2022). Industry 4.0 and Precision Livestock Farming (PLF): An up to Date Overview across Animal Productions. *Sensors (Basel)*, 22(12), 4319. 10.3390/s2212431935746102

Munteanu, C., Mireşan, V., Răducu, C., Ihuţ, A., Uiuiu, P., Pop, D., Neacşu, A., Cenariu, M., & Groza, I. (2021). Can Cultured Meat Be an Alternative to Farm Animal Production for a Sustainable and Healthier Lifestyle? *Frontiers in Nutrition*, 8, 749298. 10.3389/fnut.2021.74929834671633

Mysko, L., Minviel, J.-J., Veysset, P., & Veissier, I. (2024). How to concurrently achieve economic, environmental, and animal welfare performances in French suckler cattle farms. *Agricultural Systems*, 218, 103956. 10.1016/j.agsy.2024.103956

Naser, M. M., & Pearce, P. (2022). Evolution of the International Climate Change Policy and Processes: UNFCCC to Paris Agreement. In *Oxford Research Encyclopedia of Environmental Science*. Oxford University Press. 10.1093/acrefore/9780199389414.013.422

Newman, L., Fraser, E., Newell, R., Bowness, E., Newman, K., & Glaros, A. (2023). Cellular agriculture and the sustainable development goals. In *Genomics and the Global Bioeconomy* (pp. 3–23). Elsevier. 10.1016/B978-0-323-91601-1.00010-9

Newman, L., Newell, R., Mendly-Zambo, Z., & Powell, L. (2021). Bioengineering, telecoupling, and alternative dairy: Agricultural land use futures in the Anthropocene. *The Geographical Journal*, 188(3), 342–357. 10.1111/geoj.12392

Novellino, A., Brown, T. J., Bide, T., Th c Anh, N. T., Petavratzi, E., & Kresse, C. (2021). Using Satellite Data to Analyse Raw Material Consumption in Hanoi, Vietnam. *Remote Sensing (Basel)*, 13(3), 334. 10.3390/rs13030334

Nyyssölä, A., Suhonen, A., Ritala, A., & Oksman-Caldentey, K.-M. (2022). The role of single cell protein in cellular agriculture. *Current Opinion in Biotechnology*, 75, 102686. 10.1016/j.copbio.2022.10268635093677

O'Neill, E. N., Cosenza, Z. A., Baar, K., & Block, D. E. (2020). Considerations for the development of cost-effective cell culture media for cultivated meat production. *Comprehensive Reviews in Food Science and Food Safety*, 20(1), 686–709. 10.1111/1541-4337.1267833325139

Ong, K. J., Johnston, J., Datar, I., Sewalt, V., Holmes, D., & Shatkin, J. A. (2021). Food safety considerations and research priorities for the cultured meat and seafood industry. *Comprehensive Reviews in Food Science and Food Safety*, 20(6), 5421–5448. 10.1111/1541-4337.1285334633147

Ong, S., Choudhury, D., & Naing, M. W. (2020). Cell-based meat: Current ambiguities with nomenclature. *Trends in Food Science &Trends in Food Science & Technology*, 102, 223–231. 10.1016/j.tifs.2020.02.010

Penuelas, J., Coello, F., & Sardans, J. (2023). A better use of fertilizers is needed for global food security and environmental sustainability. *Agriculture &Agriculture & Food Security*, 12(1), 5. Advance online publication. 10.1186/s40066-023-00409-5

Percival, B., & Percival, F. (n.d.). Reinventing the Wheel. In *Milk, Microbes, and the Fight for Real Cheese*. University of California Press. https://doi.org/doi:10.1525/9780520964464

Plec, E. (2024). The Animal People. In *Animal Activism On and Off Screen* (pp. 65–88). Sydney University Press. 10.2307/jj.16110774.7

Post, M. J., Levenberg, S., Kaplan, D. L., Genovese, N., Fu, J., Bryant, C. J., Negowetti, N., Verzijden, K., & Moutsatsou, P. (2020). Scientific, sustainability and regulatory challenges of cultured meat. *Nature Food*, 1(7), 403–415. 10.1038/s43016-020-0112-z

Potopová, V., Musiolková, M., Gaviria, J. A., Trnka, M., Havlík, P., Boere, E., Trifan, T., Muntean, N., & Chawdhery, M. R. A. (2023). Water Consumption by Livestock Systems from 2002–2020 and Predictions for 2030–2050 under Climate Changes in the Czech Republic. *Agriculture*, 13(7), 1291. 10.3390/agriculture13071291

Raihan, A., & Tuspekova, A. (2022). Dynamic impacts of economic growth, energy use, urbanization, tourism, agricultural value-added, and forested area on carbon dioxide emissions in Brazil. *Journal of Environmental Studies and Sciences*, 12(4), 794–814. 10.1007/s13412-022-00782-w

Reiss, J., Robertson, S., & Suzuki, M. (2021). Cell Sources for Cultivated Meat: Applications and Considerations throughout the Production Workflow. *International Journal of Molecular Sciences*, 22(14), 7513. 10.3390/ijms2214751334299132

Ren, Z., Dong, Y., Lin, D., Zhang, L., Fan, Y., & Xia, X. (2022). Managing energy-water-carbon-food nexus for cleaner agricultural greenhouse production: A control system approach. *The Science of the Total Environment*, 848, 157756. 10.1016/j.scitotenv.2022.15775635926594

Rischer, H., Szilvay, G. R., & Oksman-Caldentey, K.-M. (2020). Cellular agriculture — Industrial biotechnology for food and materials. *Current Opinion in Biotechnology*, 61, 128–134. 10.1016/j.copbio.2019.12.00331926477

Risner, D., Li, F., Fell, J. S., Pace, S. A., Siegel, J. B., Tagkopoulos, I., & Spang, E. S. (2020). Preliminary Techno-Economic Assessment of Animal Cell-Based Meat. *Foods*, 10(1), 3. 10.3390/foods1001000333374916

Rodríguez Escobar, M. I., Cadena, E., Nhu, T. T., Cooreman-Algoed, M., De Smet, S., & Dewulf, J. (2021). Analysis of the Cultured Meat Production System in Function of Its Environmental Footprint: Current Status, Gaps and Recommendations. *Foods*, 10(12), 2941. 10.3390/foods1012294134945492

Roe, S., Streck, C., Beach, R., Busch, J., Chapman, M., Daioglou, V., Deppermann, A., Doelman, J., Emmet-Booth, J., Engelmann, J., Fricko, O., Frischmann, C., Funk, J., Grassi, G., Griscom, B., Havlik, P., Hanssen, S., Humpenöder, F., Landholm, D., & Lawrence, D. (2021). Land-based measures to mitigate climate change: Potential and feasibility by country. *Global Change Biology*, 27(23), 6025–6058. 10.1111/gcb.1587334636101

Rothwell, S. A., Doody, D. G., Johnston, C., Forber, K. J., Cencic, O., Rechberger, H., & Withers, P. J. A. (2020). Phosphorus stocks and flows in an intensive livestock dominated food system. *Resources, Conservation and Recycling*, 163, 105065. 10.1016/j.resconrec.2020.10506533273754

Rubio, N. R., Xiang, N., & Kaplan, D. L. (2020). Plant-based and cell-based approaches to meat production. *Nature Communications*, 11(1), 6276. 10.1038/s41467-020-20061-y33293564

Siddiqui, S. A., Zannou, O., Karim, I., Kasmiati, , Awad, N. M. H., Gołaszewski, J., Heinz, V., & Smetana, S. (2022). Avoiding Food Neophobia and Increasing Consumer Acceptance of New Food Trends—A Decade of Research. *Sustainability (Basel)*, 14(16), 10391. 10.3390/su141610391

Sinke, P., Swartz, E., Sanctorum, H., van der Giesen, C., & Odegard, I. (2023). Ex-ante life cycle assessment of commercial-scale cultivated meat production in 2030. *The International Journal of Life Cycle Assessment*, 28(3), 234–254. 10.1007/s11367-022-02128-8

Smith, D. J., Helmy, M., Lindley, N. D., & Selvarajoo, K. (2022). The transformation of our food system using cellular agriculture: What lies ahead and who will lead it? *Trends in Food Science &amp. Trends in Food Science & Technology*, 127, 368–376. 10.1016/j.tifs.2022.04.015

Soice, E., & Johnston, J. (2021). How Cellular Agriculture Systems Can Promote Food Security. *Frontiers in Sustainable Food Systems*, 5, 753996. Advance online publication. 10.3389/fsufs.2021.753996

Sokołowski, Ł. M. (2020). *The placing of novel foods on the EU market in the light of new EU regulations*. Adam Mickiewicz University Press. 10.14746/amup.9788323236153

Stavi, I., Paschalidou, A., Kyriazopoulos, A. P., Halbac-Cotoara-Zamfir, R., Siad, S. M., Suska-Malawska, M., Savic, D., Roque de Pinho, J., Thalheimer, L., Williams, D. S., Hashimshony-Yaffe, N., van der Geest, K., Cordovil, C. M. S., & Ficko, A. (2021). Multidimensional Food Security Nexus in Drylands under the Slow Onset Effects of Climate Change. *Land (Basel)*, 10(12), 1350. 10.3390/land10121350

Stephens, N., & Ellis, M. (2020). Cellular agriculture in the UK: A review. *Wellcome Open Research*, 5, 12. 10.12688/wellcomeopenres.15685.132090174

Stout, A. J., Mirliani, A. B., Rittenberg, M. L., Shub, M., White, E. C., Yuen, J. S. K.Jr, & Kaplan, D. L. (2022). Simple and effective serum-free medium for sustained expansion of bovine satellite cells for cell cultured meat. *Communications Biology*, 5(1), 466. 10.1038/s42003-022-03423-835654948

Sun, C., Zhan, Y., & Du, G. (2020). Can value-added tax incentives of new energy industry increase firm's profitability? Evidence from financial data of China's listed companies. *Energy Economics*, 86, 104654. 10.1016/j.eneco.2019.104654

Sun, C., Zhan, Y., & Gao, X. (2023). Does environmental regulation increase domestic value-added in exports? An empirical study of cleaner production standards in China. *World Development*, 163, 106154. 10.1016/j.worlddev.2022.106154

Thisted, E. V., & Thisted, R. V. (2019). The diffusion of carbon taxes and emission trading schemes: The emerging norm of carbon pricing. *Environmental Politics*, 29(5), 804–824. 10.1080/09644016.2019.1661155

Touzdjian Pinheiro Kohlrausch Távora, F., de Assis Dos Santos Diniz, F., de Moraes Rêgo-Machado, C., Chagas Freitas, N., Barbosa Monteiro Arraes, F., Chumbinho de Andrade, E., Furtado, L. L., Osiro, K. O., Lima de Sousa, N., Cardoso, T. B., Márcia Mertz Henning, L., Abrão de Oliveira Molinari, P., Feingold, S. E., Hunter, W. B., Fátima Grossi de Sá, M., Kobayashi, A. K., Lima Nepomuceno, A., Santiago, T. R., & Correa Molinari, H. B. (2022). CRISPR/Cas- and Topical RNAi-Based Technologies for Crop Management and Improvement: Reviewing the Risk Assessment and Challenges Towards a More Sustainable Agriculture. *Frontiers in Bioengineering and Biotechnology*, 10, 913728. 10.3389/fbioe.2022.91372835837551

Treich, N. (2021). Cultured Meat: Promises and Challenges. *Environmental and Resource Economics*, 79(1), 33–61. 10.1007/s10640-021-00551-333758465

Tuomisto, H. L., Allan, S. J., & Ellis, M. J. (2022). Prospective life cycle assessment of a bioprocess design for cultured meat production in hollow fiber bioreactors. *The Science of the Total Environment*, 851, 158051. 10.1016/j.scitotenv.2022.15805135985596

Turck, D., Castenmiller, J., De Henauw, S., Hirsch-Ernst, K. I., Kearney, J., Maciuk, A., Mangelsdorf, I., McArdle, H. J., Naska, A., Pelaez, C., Pentieva, K., Siani, A., Thies, F., Tsabouri, S., Vinceti, M., Cubadda, F., Frenzel, T., Heinonen, M., Marchelli, R., & Knutsen, H. K.EFSA Panel on Nutrition. (2020). Safety of the extension of use of plant sterol esters as a novel food pursuant to Regulation (EU) 2015/2283. *EFSA Journal.EFSA Journal*, 18(6), e06135–e06135. 10.2903/j.efsa.2020.6135

Twine, R. (2021). Emissions from Animal Agriculture—16.5% Is the New Minimum Figure. *Sustainability (Basel)*, 13(11), 6276. 10.3390/su13116276

Tyagi, J., Ahmad, S., & Malik, M. (2022). Nitrogenous fertilizers: Impact on environment sustainability, mitigation strategies, and challenges. *International Journal of Environmental Science and Technology*, 19(11), 11649–11672. 10.1007/s13762-022-04027-9

Uwizeye, A., de Boer, I. J. M., Opio, C. I., Schulte, R. P. O., Falcucci, A., Tempio, G., Teillard, F., Casu, F., Rulli, M., Galloway, J. N., Leip, A., Erisman, J. W., Robinson, T. P., Steinfeld, H., & Gerber, P. J. (2020). Nitrogen emissions along global livestock supply chains. *Nature Food*, 1(7), 437–446. 10.1038/s43016-020-0113-y

Venmans, F., Ellis, J., & Nachtigall, D. (2020). Carbon pricing and competitiveness: Are they at odds? *Climate Policy*, 20(9), 1070–1091. 10.1080/14693062.2020.1805291

Walsh, J. E., Ballinger, T. J., Euskirchen, E. S., Hanna, E., Mård, J., Overland, J. E., Tangen, H., & Vihma, T. (2020). Extreme weather and climate events in northern areas: A review. *Earth-Science Reviews*, 209, 103324. 10.1016/j.earscirev.2020.103324

Wang, S., Hausfather, Z., Davis, S., Lloyd, J., Olson, E. B., Liebermann, L., Núñez-Mujica, G. D., & McBride, J. (2023). Future demand for electricity generation materials under different climate mitigation scenarios. *Joule*, 7(2), 309–332. 10.1016/j.joule.2023.01.001

Weishaupt, A., Ekardt, F., Garske, B., Stubenrauch, J., & Wieding, J. (2020). Land Use, Livestock, Quantity Governance, and Economic Instruments—Sustainability Beyond Big Livestock Herds and Fossil Fuels. *Sustainability (Basel)*, 12(5), 2053. 10.3390/su12052053

Xu, B., Wang, W., Guo, L., Chen, G., Wang, Y., Zhang, W., & Li, Y. (2021). Evaluation of Deep Learning for Automatic Multi-View Face Detection in Cattle. *Agriculture*, 11(11), 1062. 10.3390/agriculture11111062

Xu, X., Sharma, P., Shu, S., Lin, T.-S., Ciais, P., Tubiello, F. N., Smith, P., Campbell, N., & Jain, A. K. (2021). Global greenhouse gas emissions from animal-based foods are twice those of plant-based foods. *Nature Food*, 2(9), 724–732. 10.1038/s43016-021-00358-x37117472

Yang, X., & Khan, I. (2021). Dynamics among economic growth, urbanization, and environmental sustainability in IEA countries: The role of industry value-added. *Environmental Science and Pollution Research International*, 29(3), 4116–4127. 10.1007/s11356-021-16000-z34402019

Zarbà, C., Chinnici, G., & D'Amico, M. (2020). Novel Food: The Impact of Innovation on the Paths of the Traditional Food Chain. *Sustainability (Basel)*, 12(2), 555. 10.3390/su12020555

Zeraatpisheh, M., Bakhshandeh, E., Hosseini, M., & Alavi, S. M. (2020). Assessing the effects of deforestation and intensive agriculture on the soil quality through digital soil mapping. *Geoderma*, 363, 114139. 10.1016/j.geoderma.2019.114139

Zerbe, S. (2022). Global Land-Use Development Trends: Traditional Cultural Landscapes Under Threat. In *Landscape Series* (pp. 129–199). Springer International Publishing. 10.1007/978-3-030-95572-4_4

Zhang, G., Zhao, X., Li, X., Du, G., Zhou, J., & Chen, J. (2020). Challenges and possibilities for bio-manufacturing cultured meat. *Trends in Food Science &amp. Trends in Food Science & Technology*, 97, 443–450. 10.1016/j.tifs.2020.01.026

Zhang, H., Li, Z., Dai, C., Wang, P., Fan, S., Yu, B., & Qu, Y. (2021). Antibacterial properties and mechanism of selenium nanoparticles synthesized by Providencia sp. DCX. *Environmental Research*, 194, 110630. 10.1016/j.envres.2020.11063033345899

Zhang, L., Tian, H., Shi, H., Pan, S., Chang, J., Dangal, S. R. S., Qin, X., Wang, S., Tubiello, F. N., Canadell, J. G., & Jackson, R. B. (2022). A 130-year global inventory of methane emissions from livestock: Trends, patterns, and drivers. *Global Change Biology*, 28(17), 5142–5158. 10.1111/gcb.1628035642457

Zhao, J. R., Zheng, R., Tang, J., Sun, H. J., & Wang, J. (2022). A mini-review on building insulation materials from perspective of plastic pollution: Current issues and natural fibres as a possible solution. *Journal of Hazardous Materials*, 438, 129449. 10.1016/j.jhazmat.2022.12944935792430

Żuk-Gołaszewska, K., Gałęcki, R., Obremski, K., Smetana, S., Figiel, S., & Gołaszewski, J. (2022). Edible Insect Farming in the Context of the EU Regulations and Marketing-An Overview. *Insects*, 13(5), 446. 10.3390/insects1305044635621781

Chapter 12
Regulatory Landscape and Public Perception

Peter Yu
APAC Society for Cellular Agriculture, Singapore

Calisa Lim
APAC Society for Cellular Agriculture, Singapore

ABSTRACT

The current food system faces global disruptions from climate change and population growth, destabilising the state of food security in many countries. In this new global context, cultivated meat (CM) presents an innovative solution to provide a sustainable source of protein. There are two essential aspects of the CM industry: development of clear and transparent regulatory frameworks and positive consumer perception. Regulatory frameworks allow companies to demonstrate food safety, a pre-market requirement that most countries mandate before commercialisation. Post-commercialisation success will be measured by how well the CM products are received by the public. This chapter explores both topics by providing a comprehensive overview of the current state of each area, with further insights on key drivers that facilitate progressive development, such as international coordination and harmonised labelling guidelines. The chapter concludes with a combined outlook of the CM industry going forward.

REGULATORY LANDSCAPE AND PUBLIC PERCEPTION

The current food system—in particular the meat and dairy livestock sector—feeds a large part of the world population and has been a source of social and economic growth for developed and emerging economies (United Nations Environment Pro-

DOI: 10.4018/979-8-3693-4115-5.ch012

gramme, 2023). Today the food system faces global disruptions, destabilising the state of food security in many countries (Mbow et al., 2019). These include climate change, leading to an increase in zoonotic diseases and loss of biodiversity, exacerbated by income and population growth. International organisations including the United Nations have called for consumers to reduce their meat consumption to limit their annual carbon footprint (United Nations, n.d.). For example, it has been mentioned that adopting a vegan diet can reduce 2.1 tonnes of an individual's annual carbon footprint. Yet, consumers are not willing to give up eating meat due to factors including enjoyment of the taste profile, cultural traditions and identities, and the desire to maintain their status quo in a meat-eating environment (May & Kumar, 2023).

In this new global context, the Intergovernmental Panel on Climate Change (IPCC) has named cultivated meat (CM) as a transformative solution to the world's food system that "can promise substantial reductions in direct greenhouse gas emissions from food production" (Intergovernmental Panel on Climate Change, 2022, p. 113).

Since the reveal of the first burger made with cultivated beef over a decade ago (Jha, 2013), more than 170 CM companies have been established (Good Food Institute, 2024). The selection of cultivated protein produced by the companies ranges from beef, chicken, seafood, and pork to more specialised entries such as foie gras and caviar. Transitioned from small-scale research and development to global commercialisation, many companies have, or will soon possess, the capacity to go to market. Demonstrating the safety of such products and understanding consumer perceptions are now primary considerations for the CM industry.

CM companies should pay attention to the intricate relationship between the industry, regulators, media, and consumers to effectively bring their products to market, as seen in Figure 1.

Figure 1. A Simplified Interplay Between the CM Industry, Regulators, Media, and Consumers That May Influence the Perception and Acceptance of CM As a Source of Safe, Sustainable, and Delicious Food Note. Authors' own elaboration.

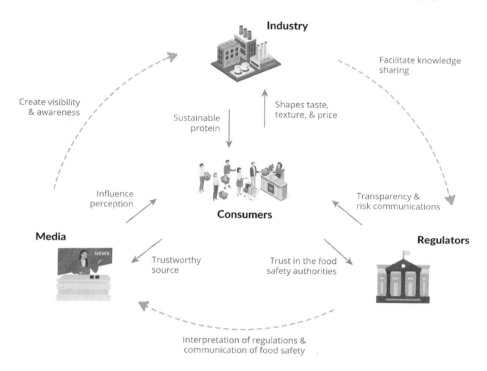

There are understandably various uncertainties regulators may have regarding CM. These may include questions about the safety and manufacturing aspects of the CM production process (Food and Agriculture Organisation of the United Nations & World Health Organisation, 2023). The industry must engage with food safety authorities to facilitate continuous knowledge sharing, as a clear and transparent regulatory regime can ensure consumer safety and trust in the approved CM product. Allowing tasting sessions for yet-to-be-approved CM products helps companies tailor their CM products to consumers' preferences, thus de-risking the adoption of CM products when they enter the market. The first part of the chapter provides an overview of the current regulatory frameworks and activities carried out by the industry and regulators to support the commercialisation of CM.

Due to the general public's unfamiliarity with CM, information disseminated through media sources can greatly influence consumers' perception and acceptance of CM products (Chriki et al., 2020). Media can also create visibility and awareness for the industry. Therefore, aside from the conventional variables that could affect consumer acceptance, care should be taken in the language and imagery used to convey an informed and balanced perspective on CM. The second part of the chapter provides an overview of market drivers that may influence public perception; this is followed by an overview of contingencies of the future, such as labelling and messaging.

REGULATORY LANDSCAPE

An increasing number of policymakers globally have signaled their support for CM. Yet, commercialisation of these products has only been achieved in three countries—Singapore in 2020 and 2024, Israel in 2023, and the United States in 2023. Considering the rapid pace of industry development, many national authorities have either established or are developing regulatory frameworks and infrastructure required for the market entry of CM. From an industry perspective, cross-border regulatory coordination will ensure the efficiency and interoperability of regulatory processes, which will in turn drive growth and commercialisation efforts.

This section provides an overview of key regional progress in Asia Pacific, Europe, and the Americas on CM regulations that allow companies to demonstrate food safety for human consumption. This is followed by an outline of activities carried out by national food safety authorities and industry to create a supportive environment for the commercialisation of CM.

REGIONAL REGULATORY DEVELOPMENTS

Asia Pacific

Australia and New Zealand. Food Standards Australia New Zealand (FSANZ) is the agency that draws up the Food Standards Code for the regulation of new food technology in Australia and New Zealand. However, enforcement of the code lies with the government of each participating jurisdiction (New Zealand and each respective Australian State and territory) (Food Standards Australia New Zealand, 2023b).

Although there are no permissions or requirements in the code specific to CM, the agency noted that it will be regulated by the existing novel food standards 1.5.1 of the code and shall require pre-market approval (Food Standards Australia New Zealand, 2023c). Depending on its composition, a specific CM product may require regulatory compliance across different standards of the code. These could include compliance with the standards for novel foods, processing aids, food additives, foods produced using gene technology, vitamins and minerals, labelling that indicates the true nature of the food, the definition of CM, and food safety requirements.

In December 2023, FSANZ commenced its hazard and risk assessment of a cultured quail product as a food ingredient (Food Standards Australia New Zealand, 2023a). The assessment evaluated the chemical, nutritional, microbiological, and dietary exposure levels and provided a detailed examination of the applicant's production process. As FSANZ has deemed the application as a major procedure, the process includes two rounds of public consultation (Food Standards Australia New Zealand, 2024a). The first call of submission was completed in February 2024, while the second call of submission is expected to start in late May 2024. Subsequently, FSANZ will propose an amendment to the code and prepare a draft food regulation measure for the Ministerial Council (Australian Government, 2024). If no reviews are requested, the anticipated gazettal and final approval of the cultivated quail product is expected in December 2024.

Singapore. The food safety authority governing the manufacture, sales, import, and export of CM in Singapore is the Singapore Food Agency (SFA).

As a small island state, Singapore imports more than 90% of its food sources (Singapore Food Agency, 2023c). Novel foods such as CM are highlighted as a potential food source that could strengthen the country's food security. This led to the SFA's introduction of the novel food regulatory framework in 2019, accompanied by a guidance document titled *Requirements for the Safety Assessment of Novel Foods and Novel Food Ingredients* (Singapore Food Agency, 2024a). Outlining the required food safety information for pre-market approval, the document has since undergone five iterations based on industry and expert feedback (Singapore Food Agency, 2023b). To date, it is recognised as one of the most comprehensive resources globally for CM regulations.

Acknowledging that the science for producing CM is still at an early stage, the SFA does not prescribe a specific dossier format on the safety assessment submitted (Singapore Food Agency, 2023b). The SFA may accept safety assessment reports conducted by overseas regulatory agencies in Australia, Canada, New Zealand, Japan, the European Union (EU), and the United States, provided that the assessments are carried out in accordance with the list of reference documents requested by the SFA (Food and Agriculture Organisation of the United Nations & World Health Organisation, 2023). Due to the novelty of CM products, the SFA requires companies to inform and seek the SFA's approval if changes made to the manufacturing process could affect the validity of the original safety assessment submitted. While the SFA allows for sensory evaluation, companies must submit supporting evidence to prove that the one-time consumption of the pre-approved CM product will be safe (Singapore Food Agency, 2023b). A Novel Food Safety Expert Working Group was also established in March 2020 to provide the SFA with scientific advice on the submitted safety assessments (Singapore Food Agency, 2021). The group is comprised of experts in food science, food toxicology, bioinformatics, nutrition, epidemiology, public health, genetics, carcinogenicity, metabolomics, fermentation technology, microbiology, and pharmacology.

Singapore's support for the CM industry expands beyond regulations. Figure 2 provides an overview of initiatives available for novel food companies. The strong ecosystem and SFA's pro-activeness led to a series of successes for the industry.

Figure 2. Singapore Has Introduced a Catalogue of Government Initiatives Ranging From Regulatory Support to Guidance on Infrastructure to Build the Agri-Food Industry's Capability and Capacity

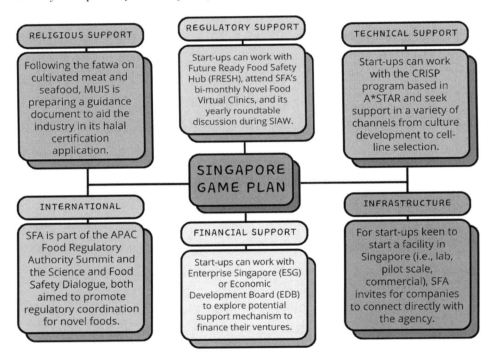

In November 2020, Singapore became the first country in the world to grant regulatory approval for the commercialisation of cultivated chicken bites, with subsequent approval of its new format in November 2021 (Eat Just, 2023).

In July 2021, the SFA granted a contract development and manufacturing organisation (CDMO) the world's first licence for the manufacturing of CM (Esco Aster, 2021).

In November 2021, the SFA issued a world-first regulatory approval for the use of serum-free media in the production of CM (Eat Just, 2023).

In April 2024, a second company received regulatory approval for their cultivated quail to be used as a food ingredient (Tan, 2024). This allows flexibility for the company to develop all kinds of quail-derived products, including whole cuts, without the need for further SFA approvals.

Israel. The food safety authority governing the manufacture, sales, and import of CM in Israel is the National Food Control Service (FCS) within the Ministry of Health (MOH).

CM is considered a novel food and is regulated under the Novel Food Directive 2006 (004-08) (Ministry of Health, 2024b). The FCS is required to issue pre-market approvals and a food manufacturing licence prior to the sale of CM within the country (Food and Agriculture Organisation of the United Nations & World Health Organisation, 2023). Yet, the existing novel food framework does not detail the provisions required to evaluate the safety of CM. The Israeli MOH thus hosted the Food and Agriculture Organisation of the United Nations (FAO) in September 2022 to discuss the safety aspects, regulatory processes, and consumer acceptance of CM products (Food and Agriculture Organisation of the United Nations, 2022). The Food Risk Management Department under the FCS further initiated a two-year pilot submission programme to determine the additional safety criteria (Ministry of Health, 2024a). Working with four companies, the department was able to draft preliminary assessment guidelines.

A successful case study arising from the pilot programme was the regulatory approval of Aleph Farms' cultivated beef product in December 2023 (Aleph Farms, 2024). This approval grants Aleph Farms the permission to commericalise their product in Israel. Issued in the form of a "No Questions" letter, the Director of the Department noted that the cultivated beef product had undergone a comprehensive assessment of critical factors, including "toxicology, allergenic potential, nutritional composition, the microbiological and chemical safety of the new food and all aspects of its manufacturing process, from initial isolation of the cells to processing and packing of the final food product" (Ministry of Health, 2024a, para. 3). Aleph Farms is awaiting final instructions from the FCS on consumer labelling and marketing, alongside a final inspection of its pilot production facility in central Israel. Once the requirements are fulfilled, Aleph Farms will offer exclusive tasting experiences in selected restaurants (Mridul, 2024).

EUROPE

European Union

The European Food Safety Authority (EFSA) is the scientific agency of the EU that provides independent scientific advice on the safety of CM. The European Commission and the EU member states are key stakeholders to authorise the labelling and sale of CM on the EU market.

In 2018, the Commission implemented Regulation (EU) 2015/2283 on Novel Foods (EUR-Lex, 2015). The regulatory framework mandates the EFSA to perform safety assessment on a novel food upon the request of the Commission. Based on the EFSA's scientific opinion that the novel food is found safe, the Commission will launch a legislative process to put forward a draft implementing act to the Standing Committee on Plants, Animals, Food and Feed (European Commission, n.d.). Post approval, the product will be updated into the union's list of authorised novel foods and can be sold in the EU market.

The Commission previously remarked on the potential benefits of CM as a "promising and innovative solution to help achieve the objectives of the farm to fork strategy for fair, safe, healthy and environmentally-friendly food systems" (European Commission, 2022, Topic Description section, para. 2). Growing interest in the industry led to a two-day scientific colloquium organised by the EFSA in May 2023 to understand the production and safety aspects of CM (European Food Safety Authority, 2023). Soon after, the EFSA noted its plans to increase communication activities and scientific discussions around the topic (European Food Safety Authority, 2024b) and introduced the provisions to assess CM products as part of the mandate from the Commission in June 2023 (European Food Safety Authority, 2024a). Based on the preliminary draft, the proposed assessment criteria include compositional data, production process, toxicological information, exposure, nutritional information, and potential allergenicity. The finalised guidance document is expected in June 2024.

The EU novel food framework does not specify pre-market sensory evaluation (EUR-Lex, 2015). Recognising the benefits of pre-market sensory evaluation on food innovation, the Dutch government introduced a pilot programme to permit pre-market approval tasting sessions for CM in the Netherlands (Government of the Netherlands, 2023). A Code of Practice document was issued in July 2023 to guide companies in their application. Subjected to an independent third-party committee sanctioned by the Dutch government, companies may apply for a year-long licence to hold up to 10 tasting sessions in a controlled environment.

United Kingdom of Great Britain. Following Brexit, the procedure for novel food application has been directed to the Food Standards Agency (FSA) (Food Standards Agency, 2024). The FSA uses the same evaluation criteria as stipulated under Regulation (EU) 2015/2283 on Novel Foods to determine the safety of CM. Pre-market tasting sessions are allowed as part of a research process to develop CM products (Food Standards Agency, n.d.).

AMERICAS

Brazil

The Food General Office (Gerência-Geral de Alimentos [GGALI]) at the Brazilian National Agency of Sanitary Surveillance (Agência Nacional de Vigilância Sanitária [Anvisa]) and the Animal Products Inspection Department within the Ministry of Agriculture, Livestock, and Food Supply (Ministério da Agricultura, Pecuária e Abastecimento [MAPA]) are the key regulatory agencies governing CM products.

Brazil is the world's largest beef and chicken supplier (Mano, 2024). The country exported a record of 2.01 million metric tonnes of fresh beef in 2023 and commands 37% of the world's total chicken sales. Despite its vested interest in the conventional meat production system, Brazilian regulators and companies showed great optimism toward CM as a novel food source and have taken steps to support the industry (Good Food Institute Brazil, 2022).

In December 2022, the GGALI completed a comprehensive report acknowledging the gaps and inconsistencies of the existing framework to regulate novel food and food ingredients (Agência Nacional de Vigilância Sanitária, 2022). This led to the publication of Resolution RDC No. 839/2023, which came into force on 16 March 2024 (Agência Nacional de Vigilância Sanitária, 2023). The new regulatory framework updates the definition of novel food and food ingredients and outlines the specific requirements needed for novel food companies to obtain a pre-market approval. The framework further allows for companies to seek Anvisa's verification on whether their food or food ingredient falls within the novel food definition (Agência Nacional de Vigilância Sanitária, n.d.). With the new regulation, MAPA can start its legislative process for product registration (Good Food Institute, 2024). Good Food Institute Brazil is currently working with MAPA to develop the required regulations. These include instructions for labelling, inspections of manufacturing facilities, and the identity and quality standards of the products.

United States of America. In the United States, the U.S. Food and Drug Administration (FDA) and the U.S. Department of Agriculture's Food Safety and Inspective Service (USDA-FSIS) overseas the manufacture, sales, import, and export for CM products.

In 2019, a formal agreement was made by the FDA and USDA-FSIS to establish a joint regulatory framework to oversee CM products derived from livestock, poultry, and siluriformes fish (Food and Drug Administration, 2019). The FDA's approach to regulating CM products involves a thorough pre-market consultation and inspections of records and facilities. The FDA reviews how the cells are isolated, how the cell bank is formed, the substances used in the culture process, and the harvest material before it leaves the sealed vessel such as a bioreactor (Food and Drug Administration, 2023b). This process allows applicants to work with the FDA on a product-by-product basis to ensure that the production process complies with the Federal Food, Drug, and Cosmetic Act. Subsequently, oversight is transferred to the USDA-FSIS for their evaluation of the conventional processing and production, packaging, and labelling. Applicants can apply for a grant of inspection to verify that their processing facilities meet the requirements set out by the agency. Notable here is that the FDA has the sole jurisdiction over cultivated seafood products (except siluriformes fish) and thus oversees the entire process.

To date, the joint effort resulted in two cultivated chicken products approved for manufacture and sale in 2023. Complied in an inventory page, the FDA disclosed the documents related to the pre-market consultation (Food and Drug Administration, 2023a). These include the FDA's scientific memo that describes the agency's evaluation of the data received, the FDA's response letter to the company, an additional correspondence, and the company's own safety assessment of their product. In February 2024, the FDA noted its intention to develop a draft guidance document on the pre-market consultation process (Food and Drug Administration, 2024). The draft document is expected to be published by the end of December 2024.

INTERNATIONAL REGULATORY COOPERATION

FAO, WHO, and Codex

In April 2023, two United Nations agencies—the FAO and the World Health Organisation (WHO)—jointly published a comprehensive report on the food safety aspects of CM (Food and Agriculture Organisation of the United Nations & World Health Organisation, 2023). The report brought together 24 leading experts from 15 countries to identify potential hazards associated with CM production and share appropriate food safety recommendations. The document also included analyses around terminology, regulations, and food safety communication. With an eye towards national food safety authorities, the report called upon UN Member States to start developing regulatory frameworks for CM.

Understanding that national food safety authorities are at different developmental phases when it comes to regulating CM, the FAO regularly engages a broad subset of stakeholders. For example, the FAO introduced a series of roundtable discussion meetings with selected regulatory bodies (Israel in 2022, China in 2023, and Canada in 2024). The roundtable discussion invites industry players to showcase their products and helps regulators understand the specific risks or hazards that may arise in the production process (Food and Agriculture Organisation of the United Nations, n.d.). The FAO also facilities an informal technical working group comprised of regulatory experts, jurisdictions, and agencies from 34 countries to exchange information on the food safety aspects of CM (Food and Agriculture Organisation of the United Nations, 2023). This ongoing initiative started in 2021, with WHO joining the group in 2023.

The FAO and WHO further extends its support for the industry via the Codex Alimentarius Commission. Through the codex, UN Member States work together to develop international food standards, guidelines, and codes of practice aimed at ensuring safe foods for consumers (Codex Alimentarius Commission, n.d.). While these resources are made as recommendations, it was highlighted that "Codex standards serve in many cases as a basis for national legislation" (Codex Alimentarius Commission, n.d., para. 7). At the 44th Codex session, the FAO and WHO submitted an agenda noting that it is only a matter of time before food biotechnologies such as CM become mainstream and urged the Codex Executive Committee to address and facilitate the development of appropriate guidance at the international level (Codex Alimentarius Commission, 2021). The Codex Executive Committee has since encouraged UN Member States to submit proposals relating to new food sources and production systems such as alternative proteins through existing Codex mechanisms (Codex Alimentarius Commission, 2023). It has also invited members and observer organisations to identify possible issues that requires new codex mechanisms.

This global progress bodes well for positive recognition of the emerging industry and the development of an integrated ecosystem that supports end-to-end growth across value chains.

INTERNATIONAL COLLABORATION

Few countries have experience processing food safety applications for CM products, and some may have questions about the existing safety assessment methodologies that can prevent or mitigate potential hazards. Many countries may also lack the time, resources, and infrastructure needed to evaluate the safety of novel foods. Thus, collaboration and knowledge sharing among regulators can better equip food safety authorities with the technical skills required.

An example would be the SFA's annual roundtable programme on novel food regulations. Started in 2019, the roundtable is an international platform for regulators, the industry, academics, and key stakeholders in the novel food ecosystem to share best practices on safety and regulatory aspects of novel foods (Singapore Food Agency, 2023a). The latest roundtable in 2023 focused on the strategies by which safety assessments and regulatory approvals can be harmonised and streamlined at an international level (Singapore Food Agency, 2024b). A strategy highlighted to achieve regulatory convergence includes the establishment of common lists or databases to reduce duplicative assessments for both the industry and regulators. Concluding the programme, the SFA noted that international standards for CM products would help to assure food safety while facilitating fair international trade practices.

At the regional level, regulators are also working in partnership to facilitate information exchange and harmonisation of novel food regulations. Partnerships include the Asia-Pacific Food Regulatory Authority Summit headed by South Korea's Ministry of Food and Drug Safety (Ministry of Food and Drug Safety, 2023) and the Science and Food Safety Dialogue established by FSANZ (Food Standards Australia New Zealand, 2024b).

INDUSTRY-DRIVEN INITIATIVES

APAC Regulatory Coordination Forum

In 2023, the Asia-Pacific Society for Cellular Agriculture (APAC-SCA) and the Good Food Institute Asia Pacific (GFI APAC) launched an industry-led regulatory forum to establish a mechanism for continuous and systematic cross-border dialogue between CM companies, industry associations, think tanks, governmental agencies, and regulators in different jurisdictions (APAC Regulatory Coordination Forum, 2023a).

Bringing together 15 key stakeholders across 10 countries, the forum focuses on all aspects of CM products fit for human consumption and is intended to complement the activities carried out by food safety authorities (APAC Regulatory Coordination Forum, 2023b). This includes the mutual recognition of coordinated regulatory frameworks in Asia Pacific (i.e., alignment on criteria for safety testing, labelling, and inspections). The forum hopes to streamline the review processes for companies that want to simultaneously enter multiple markets, which in turn minimises trade barriers and cost to consumers. From an industry perspective, coordination further brings forth efficiency in application and transparency, encouraging growth and commercialisation efforts.

Current work at the forum includes discussion around cell line development and cell culture media components. White papers are projected to be published in November 2024 to inform regulators of the challenges and opportunities available for regulatory convergence. Future workflow could include religious rulings and standards (i.e., halal and kosher) alongside standardised regulatory approaches towards hybrid product definitions and novel cell cultivation technologies (Asia-Pacific Society for Cellular Agriculture & Good Food Institute Asia Pacific, 2023).

JOINT PARTNERSHIPS

In an effort to support the fast-growing industry, the APAC-SCA entered into a tripartite alliance with the U.S.-based Association for Meat, Poultry, and Seafood Innovation (AMPS-Innovation) and Cellular Agriculture Europe (CAE) (Global Cellular Agriculture Alliance, 2023a). The Global Cellular Agriculture Alliance (GCAA) represents over 30 CM companies with the aim of identifying regional synergies, advocating for regulatory cooperation, and communicating accurate information about the potential of cellular agriculture. At the COP28 UAE, the GCAA hosted a series of dialogues to showcase the role of protein diversification as a key element of an inclusive transition towards sustainable, resilient, and equitable food systems (Global Cellular Agriculture Alliance, 2023b).

SUMMARY OF REGULATORY LANDSCAPE

Despite growing interest from food safety authorities worldwide in the potential of CM as a complementary source of protein, many stakeholders may have uncertainty about its implications for public health. Other barriers to acceptance, like consumer perception and religious rulings, could also delay the introduction of CM products in the market. Thus, collaborative approaches, such as the APAC Regulatory Coordination Forum and roundtable series organised by the SFA and the FAO, are necessary to keep regulators abreast of the latest changes in the industry. This will result in better-informed decision-making and the creation of evidence-based policies that support the growth and acceptance of CM products as a sustainable and safe food source.

PUBLIC PERCEPTION

Transparent and coordinated regulatory frameworks play an essential role in the succession of stages for the CM industry. Yet, public perception is paramount for its long-term viability. To effectively bring CM products to the market, companies need to understand the key market drivers affecting consumer perception post-consumption.

It is generally difficult to shift consumer perceptions on meat (May & Kumar, 2023). However, the CM industry introduces a unique pathway for the consumer to food products that could be indistinguishable from conventional meat ("A conversation about", 2023). The change in consumer perception may therefore be less pronounced compared to consumer perception of other novel or foreign foods. That being said, CM is produced by growing cells in-vitro with limited animal interaction. Apart from the conventional variables that could affect consumer acceptance, effective consumer education strategies should consider other contributors such as food safety of novel production technologies and perceived benefits of CM to increase acceptance levels (McNamara, n.d.).

This section begins with an overview of market drivers that may influence public perception, as seen in Figure 3; this is followed by an overview of the contingencies of the future that can drive consumer understanding, such as labelling and messaging.

Figure 3. Consumer Adoption Represented Through a Generalised Bell-Shape Curve That May Shift Left or Right Based on the Development and Representation of the Key Market Drivers

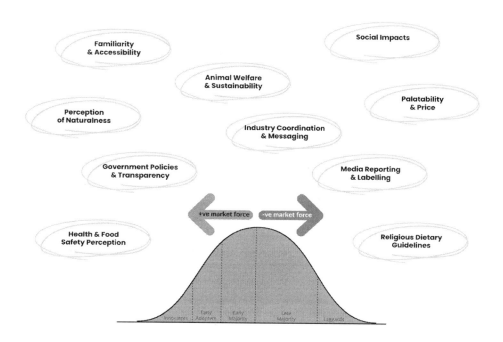

KEY DRIVERS OF PUBLIC PERCEPTION

Palatability and Price

Determinants such as price, texture, and taste play a significant role in consumer perception of food (Hidayat & Haryanti, 2020). Reasonably, CM is not an outlier. These factors are among the highest priorities for the majority of potential consumers surveyed in the United States (Wilks & Phillips, 2017) and South Korea (Asia-Pacific Society for Cellular Agriculture, 2023).

There is little public information on palatability due to limited CM products accessible in the market (To et al., 2024). Company-specific formulation of the final product, protein profile, type of intended cut or meat product, potential processing aids and additives, and intended way of cooking are all factors contributing to palatability. Consumer awareness of CM and subjective consumer mindset could also be a contributor to the outcome of taste judgment (Rolland et al., 2020). For commercially available CM in Singapore, it was shown that post-consumption perception of cultivated chicken was significantly higher than pre-consumption (Chong et al., 2024). This may be indicative of a positive reception to taste and corroborates the idea that CM is a strong candidate not only as a complementary source of protein but also as a food option on par with conventional cuisines.

It is predicted that CM may reach price parity or be lower than that of conventional meat products by 2030 (Brennan et al., 2021). These figures are generalised and based on trends and predictions surrounding the progression of key cost-determinants such as scale-up, research and development efficiency, and final product compositions. However, with a predicted pathway towards price parity, CM will become more accessible, which may have a synergistic effect on familiarity and long-term adoption.

HEALTH & SAFETY

CM is produced in a controlled environment, significantly reducing exposure to contaminants, diseases, or other harmful substances (Datar & Betti, 2010). Jurisdictions with regulatory pathways for CM further mandate robust safety demonstrations throughout the entire production chain. Yet, with limited accessibility and familiarity, consumers may be wary of the long-term effects of consumption (Nielsen et al., 2009; Verbeke et al., 2015). For example, health and safety were reported to be among the top three priorities for Japanese consumers in a 2023 consumer report (Asia-Pacific Society for Cellular Agriculture, 2024). Noted within the report is the importance of effective communication between consumers, regulatory entities, and the frameworks to which the safety of CM is assessed. Public trust in domestic regulations will be important to mitigate safety concerns, while institutional transparency and proactive communication may facilitate increased trust (Rosenfeld & Tomiyama, 2023).

SOCIAL IMPACT & RELIGION

Some consumer groups may have concerns about how the CM industry could impact the conventional livestock sector (Wilks & Phillips, 2017). With the introduction of most novel industries, there will inevitably be social changes. However, COVID-19 highlighted the reality that many countries need to build a robust contingency plan for food security (Kakaei et al., 2022). The rapid increase in global populations, reduction in arable land space, effects of climate change, supply chain disruptions, and severity of zoonotic disease outbreaks demand an urgent need to find solutions. CM acts not as a competitor to conventional livestock products but as a complementary source of protein for the foreseeable future. In addition, CM production will involve skilled, specialised labour and will not compete in that capacity with farmers (Stephens et al., 2018). Nevertheless, policies to support both conventional farmers and novel food producers may be necessary to ensure a food-secure future while mitigating social concerns that may arise (Tomiyama et al., 2020).

Eating habits and consumers' purchasing behaviours are greatly influenced by religion (Santovito et al., 2023). With 85% of the world's population identifying with a religion (Hackett et al., 2012), religious rulings, guidelines, and certifications are essential to assuring religious consumers that CM products are permissible for consumption (Tamimah et al., 2018). To date, developments have been made for both kosher and halal foods. Israel's Chief Rabbi declared Aleph Farms' cultivated beef product as kosher in 2023, noting that the production process falls into the category of *parve*—foods not classified as meat or dairy (Aleph Farms, 2023). The MUIS issued a fatwa on CM stating "assuming the right conditions are fulfilled, CM is generally halal or permissible to be consumed by Muslims" (Majlis Ugama Islam Singapura, 2024, para. 1). These indications could imply further and increased acceptance among followers of either kosher or halal diets.

NATURALNESS

Food novelty warrants special attention when it comes to the topic of naturalness. What constitutes natural is subjective. One might classify it as a "gut feeling" that undoubtedly has a powerful effect on the minds of future consumers (Hocquette et al., 2015). Even if the perceived benefits are higher, many novel technologies face similar obstacles when it comes to consumer acceptance (Siegrist et al., 2018).

A negative perception of naturalness could stem from a lack of understanding of conventional industrialised farming practices or CM production (Tomiyama et al., 2020). From a consumer perspective, highlighting the unnaturalness of conventional livestock practices increased the positive perception of CM to a greater extent than arguing for the concept of naturalness of CM directly (Schaefer & Savulescu, 2014). Showcasing CM technology rather than focusing on the similarities between CM and conventional meat may further reduce consumer perception of CM (Fidder & Graça, 2023). Thus, an effective method to overcome consumers' fear of unnatural products could include drawing parallels to conventional meat, explaining the purpose of CM, and adding context to how industrialised farming has progressed to date.

ANIMAL WELFARE & SUSTAINABILITY

While conventional food production is necessary today to meet the growing demand of the global population, it is well-documented that it comes at a cost. Humans consume, on a global average, 44 kilograms of meat per person per year (Godfray et al., 2018). Food production is a significant contributor to greenhouse emissions, the majority of which is attributed to animal farming (Xu et al., 2021). Industrial animal production faces a multitude of other concerns such as antibiotic resistance, zoonotic disease outbreaks, and ocean acidification (Higuita et al., 2023). CM is a good candidate to mitigate many of these concerns and has a strong potential for sustainable resource efficiency in the context of land and water utilisation, climate change emissions, and pollution (Good Food Institute, 2018). Notable here is that more studies are needed once CM fully materialises at scale to bring theoretical estimations into reality.

CM production significantly reduces the involvement of animals and obviates the necessity for animal slaughter (Tomiyama et al., 2020). Tailored to the growing demand for animal welfare (Freedonia Group, 2017), CM is touted as a good alternative to conventional meat for this segment of consumers (Siegrist et al., 2018). It may be important for companies to showcase information regarding any substitutes for animal-derived components utilised during the production of CM to address this consumer segment. Alternatively, companies can showcase a contingency plan to minimise the use of animal-derived components.

LABELLING

The topic of CM labelling is increasingly important due to its framing and anchoring effect on consumers (Malerich & Bryant, 2022). Without much sensory experience or direct understanding, the consumer is left with a conceptual frame of mind based on the association of CM labelling to that of its emotions and context (Bryant & Barnett, 2019). Labelling may therefore play a powerful role in the phase of consumer adoption. This is exemplified through studies conducted on negative labelling and messaging of genetically modified (GM) foods and their anchoring effect on the consumer surpassing the underlying tenets of potential benefits and technological possibilities (Mohorčich & Reese, 2019). Notable here is that although CM is not the same or a predecessor to GM, the lessons learned from the latter can be comparable as many of the underlying consumer conceptualisations are directly applicable to both industries.

A problem of labelling is in the multitude of terminologies being utilised today. Labels such as cultivated, cultured, or cell-based (or terms of similar nature) are commonly utilised by the industry (Food and Agriculture Organisation of the United Nations & World Health Organisation, 2023). In scientific literature, CM descriptors include synthetic, in-vitro, or artificial, whereas, in media and public resources, not-so-descriptive labels such as franken, clean, or lab-grown are sometimes used (Chriki et al., 2020). Consistency in the choice of terminology, and cross-sector coordination may assist in promoting consumer transparency, perception, and adoption of CM (Hallman et al., 2023). In addition, companies filing for CM approvals must also meet regulatory clearance for the labelling relevant to the governing jurisdiction. For example, the SFA mandates the utilisation of cultured, cell-based, or other suitable qualified terms indicative of the true nature of CM (Singapore Food Agency, 2023b). Hence, while certain terminologies are viable as accurate descriptors for CM, communication and education regarding non-descriptive or inaccurate labels could be effective towards convergence (Cellular Agriculture Australia, 2023a).

FUTURE OUTLOOK AND A COORDINATED PERSPECTIVE FOR CONSUMER ENGAGEMENT

Industry Coordination on Terminology and Messaging

Industry coordination will continue to play an important role in fostering a productive environment for consumer perception. Developments here may assist in driving efforts for consumer transparency and bring forth effective means of communication. Some efforts in the industry include:

In 2023, the Cellular Agriculture Australia (CAA) launched a series of consensus-driven and informative public guides on the topic of "key terms & language" about cellular agriculture (Cellular Agriculture Australia, 2023b). This is an important stepping stone towards regional coordination and awareness especially towards media or other public-facing entities.

In 2022, the APAC-SCA and GFI APAC coordinated a regional memorandum of understanding, bringing together over 30 industry stakeholders as signatories for "cultivated" as the preferred descriptor of CM (Good Food Institute Asia Pacific, 2022). This work culminated from previous research conducted by the GFI APAC identifying cultivated as the preferential choice among key industry players (Friedrich, 2021).

Since 2019, the SFA has held annual roundtables on novel food regulations at the Singapore International Agri-food Week (Singapore Food Agency, 2023a). Regulator-led platforms are an important asset for international coordination as dialogue on selected topics is coordinated between industry participants and other stakeholders.

As with regulatory coordination, work by industry groups and government entities serves a centralised role in streamlining communication efforts directed at the consumer. Going forward, industry coordination could ultimately manifest through industry standards that further specify voluntary requirements regarding labelling and terminology. Furthermore, consumer-driven communication platforms such as Communication Leadership in Future Food (Communication Leadership in Future Foods, n.d.), may facilitate better coordination of messaging towards consumers.

FAMILIARITY, ACCESSIBILITY, AND MEDIA PORTRAYAL

Limited consumer accessibility to CM products has constrained studies to populations without direct sensory experience. Familiarity with food is a driver for consumer preference, and hence conventional food products may, at this stage, enjoy an added benefit over novel foods (Nacef et al., 2019). However, as studies progress from hypothetical pre-consumption drivers of acceptance to post-consumption consumer behaviour, the trend towards increased familiarity may manifest through rising consumer preference. This view was corroborated in a first-of-its-kind study in Singapore conducted on a commercially available CM product (Chong et al., 2024). Here, post-consumption acceptance was shown to be significantly higher than that of pre-consumption, supporting the correlation between familiarity and consumer perception. With additional CM products coming to the market, consumer preference may increase.

The first public tasting for CM happened over a decade ago in London (Jha, 2013). Since then, sensory evaluation for pre-market approved products has been held in various countries including Australia, China, France, Israel, the Netherlands, Singapore, Switzerland, and the United States (Good Food Institute, 2022, 2023, 2024; Meatable, 2024). The availability of tasting sessions for yet-to-be-approved CM products helps companies tailor their CM products to consumers' preferences, thus de-risking the adoption of CM products when they enter the market. Therefore, it is important for countries to implement consumer testing regimes that bring forth accessibility and familiarity to CM products during a pre-market approval phase.

In addition to familiarity and accessibility, information disseminated through media sources can greatly influence consumers' perception and acceptance of CM products (Chriki et al., 2020). Although media coverage has substantially increased since 2010, there is still a continuous divergence when it comes to descriptors utilised for CM (Food and Agriculture Organisation of the United Nations & World Health Organisation, 2023). Terminologies such as fake or artificial meat do not accurately describe CM and may further exacerbate any negative perception of misinformed viewpoints. It will be important to reverberate accurate language and messaging through multi-channel informational sources to facilitate consumer understanding and perception. For example, using imagery depicting test tubes, petri dishes, or other science-related concepts to describe CM could exacerbate negative perceptions (Cellular Agriculture Australia, 2024). News media partners can access an imagery guide put together by industry partners that features real photos of CM products (Good Food Institute, 2020).

SUMMARY OF PUBLIC PERCEPTION

As CM products become more accessible to the general consumer, the ultimate success of the industry lies with the consumer. CM products are unique in their capability of producing real meat with minimal animal interaction. Similar to conventional food, factors such as palatability, price, health, sustainability, safety, and animal welfare play a significant role in the way consumers choose their next meal. Additional factors such as the perception of naturalness, social impact, and religious guidance may further elicit emotional responses towards CM. Going forward, industry coordination and effective messaging will play an important part in shaping an environment that facilitates accurate, transparent, and clear information to the consumer. With the future in mind and the realisation of the tenets supporting CM production, there is tremendous potential for global consumer success.

CHAPTER CONCLUSION: REGULATORY LANDSCAPE AND PUBLIC PERCEPTION

The chapter has outlined the important aspects of CM regulations and public perception. As the CM industry is still in its nascency, the contingency forward will be highly dependent on the progression of both areas. To commercialise, companies need agency-procured regulatory frameworks to showcase that CM products are safe for consumption. To become a market success, there is a need for coordinated messaging driving positive consumer perceptions.

From a global regulatory standpoint, many food safety authorities have recently begun to develop or enforce regulatory frameworks specific to CM. To date, regulatory approvals have been issued in countries such as Singapore, the United States, and Israel, with more likely to follow. As developing an efficient framework involves a great amount of time and resources for agencies, it will be paramount to strive for international coordination and cross-border information sharing. Transparency in regulatory development and open communication can foster an environment that increases not only consumer trust for CM but also frameworks that are interoperable between different jurisdictions. Such development would benefit the industry and accelerate the process for global commercialisation.

With more CM products introduced to the market, the focus now shifts towards consumers attitudes and behaviour. Similar to conventional food, certain attributes such as palatability and price are reported to be among the highest of priority for potential CM consumers in some markets. Religious labelling and guidelines are also important to assure consumers that CM meets the dietary requirements for that religion. Furthermore, a plethora of other factors such as the perception of naturalness, familiarity, animal welfare, sustainability, health and safety, and social impact contribute to shaping the perception of CM. Notable here is that going forward, it will be important to corroborate pre-consumption research results with that of post-consumption behaviour. In addition, as consumers' familiarity with CM products is heavily influenced by media portrayal and messaging, utilising industry standard terminology and imagery will be fundamental to conveying an informed and balanced perspective on CM.

To realise the full potential of the industry, multi-stakeholder collaboration will be a necessity. As a nascent industry with tremendous potential for success, there are also significant challenges ahead. A belief in a shared responsibility among all stakeholders gives rise to a collaborative industry that will facilitate regulatory coordination, increase unity behind strong CM messaging, and ultimately drive a positive image of CM for the consumers. This becomes especially important as the work done today will be long-lasting into the future.

REFERENCES

A conversation about cultivated meat. (2023). *Nature Communications, 14*(1). 10.1038/s41467-023-43984-8

Agência Nacional de Vigilância Sanitária. (2022). *Relatório de AIR sobre a modernização do marco regulatório, fluxos e procedimentos para novos alimentos e novos ingredientes.* https://www.gov.br/anvisa/pt-br/assuntos/regulamentacao/air/analises-de-impacto-regulatorio/2023/arquivos-relatorios-de-air-2023-2/relatorio_de_air_novos_ alimentos___final.pdf

Agência Nacional de Vigilância Sanitária. (2023). *Resolução da diretoria colegiada – RDC Nº 839, de 14 de Dezembro de 2023.* https://antigo.anvisa.gov.br/documents/10181/6582266/RDC_839_2023_.pdf/a064b871-55dd-44b9-ab40-16ca7672497d Agência Nacional de Vigilância Sanitária. (n.d.). *4144 – Consulta sobre a classificação de um novo alimento e novo ingredient.* https://consultas.anvisa.gov.br/#/consultadeassuntos

Aleph Farms. (2023). Chief Rabbi of Israel affirms Aleph Farms' cultivated steak is kosher. *Aleph Farms.* https://aleph-farms.com/journals/kosher-ruling/

Aleph Farms. (2024). Aleph Farms granted world's first regulatory approval for cultivated beef. *Aleph Farms.* https://aleph-farms.com/journals/aleph-farms-granted-worlds-first-regulatory-approval-for-cultivated-beef/

APAC Regulatory Coordination Forum. (2023a). About. *APAC Regulatory Coordination Forum.* https://www.cellagforum.info

APAC Regulatory Coordination Forum. (2023b). New cellular agriculture forum aims to accelerate regulatory approvals across APAC. *APAC Regulatory Coordination Forum.* https://www.cellagforum.info/post/new-cellular-agriculture-forum-aims-to-accelerate-regulatory-approvals-across-apac

Asia-Pacific Society for Cellular Agriculture, & Good Food Institute Asia Pacific. (2023). *Memorandum of understanding.* https://687315f7-d318-48d8-a759-cb77fa9836e9.usrfiles.com/ugd/687315_a30dfed3b7cd477282f1da795d12d05c.pdf

Asia-Pacific Society for Cellular Agriculture. (2023). South Korean consumers willing to try cultivated meat and seafood products. *Asia-Pacific Society for Cellular Agriculture.* https://www.apac-sca.org/post/south-korean-consumers-willing-to-try-cultivated-meat-and-seafood-products

Asia-Pacific Society for Cellular Agriculture. (2024). Safety of cultivated meat and seafood ranks highest for Japanese consumers. *Asia-Pacific Society for Cellular Agriculture*. https://www.apac-sca.org/post/safety-of-cultivated-meat-and-seafood-ranks-highest-for-japanese-consumers

Australian Government. (2024, May 2). https://www.foodregulation.gov.au/activities-committees/food-ministers-meeting

Brennan, T., Katz, J., Quint, Y., & Spencer, B. (2021, June 16). *Cultivated meat: Out of the lab, into the frying pan*. McKinsey & Company. https://www.mckinsey.com/industries/agriculture/our-insights/cultivated-meat-out-of-the-lab-into-the-frying-pan# Bryant, C., & Barnett, J. (2019). What's in a name? Consumer perceptions of in vitro meat under different names. *Appetite, 137*, 104–113. 10.1016/j.appet.2019.02.021

Cellular Agriculture Australia. (2023a, December 11). What's next for Vow on the path to regulatory approval in Australia? *Cellular Agriculture Australia*. https://www.cellularagricultureaustralia.org/publications/whats-next-for-vow-on-the-path-to-regulatory-approval-in-australia

Cellular Agriculture Australia. (2023b). Key terms & language. *Cellular Agriculture Australia*. https://www.cellularagricultureaustralia.org/resources/key-terms

Cellular Agriculture Australia. (2024). Media guide how to talk about cellular agriculture effectively. *Cellular Agriculture Australia*. https://assets-global.website-files.com/641400f02d9306cbd4fb3f94/6614cde01226b58d37bee299_Cellular%20Agriculture%20Australia%20-%20Media%20Guide%20V1.0%20(1).pdf

Chong, M., Leung, A. K., & Fernandez, T. M. (2024). On-site sensory experience boosts acceptance of cultivated chicken. *Future Foods : a Dedicated Journal for Sustainability in Food Science*, 100326, 100326. Advance online publication. 10.1016/j.fufo.2024.100326

Chriki, S., Ellies-Oury, M., Fournier, D., Liu, J., & Hocquette, J. (2020). Analysis of scientific and press articles related to cultured meat for a better understanding of its perception. *Frontiers in Psychology*, 11, 1845. Advance online publication. 10.3389/fpsyg.2020.0184532982823

Codex Alimentarius Commission. (2021). New food sources and production systems: Need for codex attention and guidance? *Codex Alimentarius Commission*. https://www.fao.org/fao-who-codexalimentarius/sh-proxy/en/?lnk=1&url=https%253A%252F%252Fworkspace.fao.org%252Fsites%252Fcodex%252FMeetings%252FCX-701-44%252FWorking%2BDocuments%252Fcac44_15.Add.1e.pdf

Codex Alimentarius Commission. (2023). Report of the 45th session of the codex alimentarius commission. *Codex Alimentarius Commission.* https://www.fao.org/fao-who-codexalimentarius/sh-proxy/en/?lnk=1&url=https%253A%252F%252Fworkspace.fao.org%252Fsites%252Fcodex%252FMeetings%252FCX-701-45%252FFinal%2BReport%2BCAC45%252FCompiled%2BREP22_CAC.pdf

Codex Alimentarius Commission. (n.d.). About codex alimentarius. *Codex Alimentarius Commission.* https://www.fao.org/fao-who-codexalimentarius/about-codex/en/#:~:text=the%20Codex%20Alimentarius-,Purpose%20of%20the%20Codex%20Alimentarius,practices%20in%20the%20food%20trade

Communication Leadership in Future Foods. (n.d.). Home. *Future Foods.* https://futurefoods.info

Datar, I., & Betti, M. (2010). Possibilities for an in vitro meat production system. *Innovative Food Science and Emerging Technologies. Innovative Food Science & Emerging Technologies*, 11(1), 13–22. 10.1016/j.ifset.2009.10.007

Eat Just. (2023). GOOD meat receives approval to commercialize serum-free media. *Eat Just.* https://www.goodmeat.co/all-news/good-meat-receives-approval-to-commercialize-serum-free-media

Esco Aster. (2021). Esco Aster receives food processing license to manufacture cell-based cultivated meat from Singapore authorities. *Esco Aster.* https://escoaster.com/news/Esco-Aster-Receives-Food-Processing-License-to-Manufacture-Cell-Based-Cultivated-Meat-from-Singapore-Authorities

EUR-Lex. (2015). Regulation (EU) 2015/2283 of the European parliament and of the council. *EUR-Lex.* https://eur-lex.europa.eu/legal-content/EN/TXT/?uri=CELEX%3A32015R2283&qid=1714116588168# European

European Commission. (n.d.). Request for a novel food authorisation. *European Commission.* https://food.ec.europa.eu/safety/novel-food/authorisations_en#the-authorisation-covers

European Food Safety Authority. (2023). EFSA's scientific colloquium 27 "cell culture-derived foods and food ingredients". *European Food Safety Authority.* https://www.efsa.europa.eu/en/events/efsas-scientific-colloquium-27-cell-culture-derived-foods-and-food-ingredients

European Food Safety Authority. (2024a). Public consultations. *European Food Safety Authority.* https://connect.efsa.europa.eu/RM/s/publicconsultation2/a0lTk0000009y8D/pc0824

Afonso, A. L., Gelbmann, W., Germini, A., Fernández, E. N., Parrino, L., Precup, G., & Ververis, E.European Food Safety Authority. (2024b). EFSA scientific colloquium 27: Cell culture-derived foods and food ingredients. *EFSA Supporting Publications*, 21(3). Advance online publication. 10.2903/sp.efsa.2024.EN-8664

Fidder, L., & Graça, J. (2023). Aligning cultivated meat with conventional meat consumption practices increases expected tastefulness, naturalness, and familiarity. *Food Quality and Preference*, 109, 104911. 10.1016/j.foodqual.2023.104911

Food and Agriculture Organisation of the United Nations & World Health Organisation (2023). Food safety aspects of cell-based food. *Food and Agriculture Organisation of the United Nations and World Health Organisation*. 10.4060/cc4855en

Food and Agriculture Organisation of the United Nations. (2022). Experts and developers discuss ensuring the safety of cell-based foods. *Food and Agriculture Organisation of the United Nations*. https://www.fao.org/food-safety/news/news-details /fr/c/1604145/

Food and Agriculture Organisation of the United Nations. (2023). Food safety aspects of cell-based food: Global perspectives. *Food and Agriculture Organisation of the United Nations*. https://afz.zootechnie.fr/images/pdf/food_and_feed _07_takeuchi.pdf

Food and Agriculture Organisation of the United Nations. (n.d.). Cell-based food and precision fermentation. *Food and Agriculture Organisation of the United Nations*. https://www.fao.org/food-safety/scientific-advice/crosscutting-and-emerging -issues/cell-based-food/en/

Food and Drug Administration. (2019). Formal agreement between FDA and USDA regarding oversight of human food produced using animal cell technology derived from cell lines of USDA-amenable species. *Food and Drug Administration*. https://www.fda.gov/food/human-food-made-cultured-animal-cells/formal-agreement-between-fda-and-usda-regarding-oversight-human-food-produced-using-animal-cell

Food and Drug Administration. (2023a). Human Food Made with Cultured Animal Cells Inventory. *Food and Drug Administration*. https://www.cfsanappsexternal.fda.gov/scripts/fdcc/?set=AnimalCellCultureFoods

Food and Drug Administration. (2023b). Human food made with cultured animal cells. *Food and Drug Administration*. https://www.fda.gov/food/food-ingredients -packaging/human-food-made-cultured-animal-cells

Food and Drug Administration. (2024). Foods program guidance under development. *Food and Drug Administration*. https://www.fda.gov/food/guidance-documents-regulatory-information-topic-food-and-dietary-supplements/foods-program-guidance-under-development

Food Standards Agency. (2024). Novel foods authorisation guidance. *Food Standards Agency*. https://www.food.gov.uk/business-guidance/regulated-products/novel-foods-guidance#new-authorisations

Food Standards Agency. (n.d.). Cell-cultivated products. *Food Standards Agency*. https://www.food.gov.uk/business-guidance/cell-cultivated-products

Food Standards Australia New Zealand. (2023a). A1269 cultured quail as a novel food. *Food Standards Australia New Zealand*. https://consultations.foodstandards.gov.au/sas/a1269-cultured-quail-as-a-novel-food/

Food Standards Australia New Zealand. (2023b). What we do (and don't do). *Food Standards Australia New Zealand*. https://www.foodstandards.gov.au/consumer/information/what-we-do

Food Standards Australia New Zealand. (2023c). Cell-based meat. *Food Standards Australia New Zealand*. https://www.foodstandards.gov.au/consumer/safety/Cell-based-meat

Food Standards Australia New Zealand. (2024a). Food standards work plan – Proposed standards development and variations to standards for applications and proposals. *Food Standards Australia New Zealand*. https://www.foodstandards.gov.au/sites/default/files/2024-04/Work%20Plan%20April%2024.pdf

Food Standards Australia New Zealand. (2024b). Annual report 2022-23. *Food Standards Australia New Zealand*. https://www.foodstandards.gov.au/publications/annual-report-2022-23

Freedonia Group. (2017). Animal welfare: Issues and opportunities in the meat, poultry, and egg markets in the U.S. *Freedonia Group*. https://www.freedoniagroup.com/packaged-facts/animal-welfare-issues-and-opportunities-in-the-meat-poultry-and-egg-markets-in-the-us

Friedrich, B. (2021). Cultivated meat: A growing nomenclature consensus. *Good Food Institute*. https://gfi.org/blog/cultivated-meat-a-growing-nomenclature-consensus/

Global Cellular Agriculture Alliance. (2023a). About. *Global Cellular Agriculture Alliance*. https://www.cellagalliance.global

Global Cellular Agriculture Alliance. (2023b). COP28 UAE. *Global Cellular Agriculture Alliance*. https://www.cellagalliance.global/cop28-uae

Godfray, H. C. J., Aveyard, P., Garnett, T., Hall, J. W., Key, T. J., Lorimer, J., Pierrehumbert, R. T., Scarborough, P., Springmann, M., & Jebb, S. A. (2018). Meat consumption, health, and the environment. *Science*, 361(6399), eaam5324. Advance online publication. 10.1126/science.aam532430026199

Good Food Institute Asia Pacific. (2022). Aligning cultivated food nomenclature across APAC. *Good Food Institute Asia Pacific*. https://gfi-apac.org/industry/aligning-cultivated-food-nomenclature-across-apac/

Good Food Institute Brazil. (2022). 8 billion, now what? *Good Food Institute Brazil*. https://gfi.org.br/en/2022-year-in-review/

Good Food Institute. (2018). Growing meat sustainably: The cultivated meat revolution. *Good Food Institute*. https://gfi.org/images/uploads/2018/10/CleanMeatEvironment.pdf

Good Food Institute. (2020). Cultivated meat could transform our food system. Let's use images that do it justice. *Good Food Institute*. https://gfi.org/blog/cultivated-meat-photos/

Good Food Institute. (2022). 2021 state of the industry report: Cultivated meat and seafood. *Good Food Institute*. https://gfi.org/wp-content/uploads/2022/04/2021-Cultivated-Meat-State-of-the-Industry-Report-1.pdf

Good Food Institute. (2023). 2022 state of the industry report: Cultivated meat and seafood. *Good Food Institute*. https://gfi.org/wp-content/uploads/2023/01/2022-Cultivated-Meat-State-of-the-Industry-Report-2-1.pdf

Good Food Institute. (2024). 2023 state of the industry report: Cultivated meat and seafood. *Good Food Institute*. https://gfi.org/wp-content/uploads/2024/04/State-of-the-Industry-report_Cultivated_2023.pdf

Government of the Netherlands. (2023). Report code of practice safely conducting tastings cultivated foods prior to EU approval. *Government of the Netherlands*. https://open.overheid.nl/documenten/39127f7e-b18b-4ddf-95a7-0be5ff660aed/file

Hackett, C., Grim, B. J., Lugo, L., Cooperman, A., Esparza Ochoa, J. C., Gao, C., Connor, P., Shi, A. F., & Kuriakose, N. (2012). The global religious landscape: A report on the size and distribution of the world's major religious groups as of 2010. *Pew Research Center's Forum on Religion & Public Life*. https://assets.pewresearch.org/wp-content/uploads/sites/11/2014/01/global-religion-full.pdf

Hallman, W. K., Hallman, W. K.II, & Hallman, E. E. (2023). Cell-based, cell-cultured, cell-cultivated, cultured, or cultivated. What is the best name for meat, poultry, and seafood made directly from the cells of animals? *NPJ Science of Food*, 7(1), 62. Advance online publication. 10.1038/s41538-023-00234-x38057390

Hidayat, R., & Haryanti, I. (2020). The effect of price and taste on the purchase. *Almana/Almana. Jurnal Manajemen Dan Bisnis*, 4(2), 244–252. 10.36555/almana.v4i2.1403

Higuita, N. I. A., LaRocque, R. C., & McGushin, A. (2023). Climate change, industrial animal agriculture, and the role of physicians – Time to act. *The Journal of Climate Change and Health*, 13, 100260. 10.1016/j.joclim.2023.100260

Hocquette, A., Lambert, C., Sinquin, C., Peterolff, L., Wagner, Z., Bonny, S., Lebert, A., & Hocquette, J. (2015). Educated consumers don't believe artificial meat is the solution to the problems with the meat industry. *Journal of Integrative Agriculture. Journal of Integrative Agriculture*, 14(2), 273–284. 10.1016/S2095-3119(14)60886-8

Intergovernmental Panel on Climate Change. (2022). *Climate change 2022: Mitigation of climate change. Contribution of working group III to the sixth assessment report of the intergovernmental panel on climate change.* Cambridge University Press. https://www.ipcc.ch/report/ar6/wg3/downloads/report/IPCC_AR6_WGIII_FullReport.pdf

Jha, A. (2013, August). First hamburger made from lab-grown meat to be served at press conference. *The Guardian.* https://www.theguardian.com/science/2013/aug/05/first-hamburger-lab-grown-meat-press-conference

Kakaei, H., Nourmoradi, H., Bakhtiyari, S., Jalilian, M., & Mirzaei, A. (2022). Effect of COVID-19 on food security, hunger, and food crisis. In M. H. Dehgani, R. R. Karri, & S. Roy (Eds.), *Covid-19 and the sustainable development goals* (pp. 3–29). Elsevier eBooks. 10.1016/B978-0-323-91307-2.00005-5

Majlis Ugama Islam Singapura. (2024). Fatwa on cultivated meat. *Majlis Ugama Islam Singapura.* https://www.muis.gov.sg/Media/Media-Releases/2024/2/3-Feb-24-Fatwa-on-Cultivated-Meat

Malerich, M., & Bryant, C. (2022). Nomenclature of cell-cultivated meat & seafood products. *NPJ Science of Food*, 6(1), 56. Advance online publication. 10.1038/s41538-022-00172-036496502

Mano, A. (2024, March). Brazil slaughtering records make it global meat king, IBGE data shows. *Reuters.* https://www.reuters.com/markets/commodities/brazil-slaughtering-records-make-it-global-meat-king-ibge-data-shows-2024-03-14/

May, J., & Kumar, V. (2023). Harnessing moral psychology to reduce meat consumption. *Journal of the American Philosophical Association*, 9(2), 367–387. https://www.cambridge.org/core/journals/journal-of-the-american-philosophical-association/article/harnessing-moral-psychology-to-reduce-meat-consumption/C3DD02CA582919B2BC7F25B126CD9DD5. 10.1017/apa.2022.2

Mbow, C., Rosenzweig, C., Barioni, L. G., Benton, T. G., Herrero, M., Krishnapillai, M., Liwenga, E., Pradhan, P., Rivera-Ferre, M. G., Sapkota, T., Tubiello, F. N., & Xu, Y. (2019). Food security. In *Climate change and land: An IPCC special report on climate change, desertification, land degradation, sustainable land management, food security, and greenhouse gas fluxes in terrestrial ecosystems* (pp. 437–550). https://www.ipcc.ch/site/assets/uploads/sites/4/2022/11/SRCCL_Chapter_5.pdf

McNamara, E. (n.d.). Consumer education on the food safety of cultivated meat. *Good Food Institute*. https://gfi.org/solutions/consumer-education-food-safety-cultivated-meat/

Meatable (2024). Press release: First EU tasting. *Meatable*. https://meatable.com/news-room/

Milford, A. B., Le Mouël, C., Bodirsky, B. L., & Rolinski, S. (2019). Drivers of meat consumption. *Appetite*, 141, 104313. 10.1016/j.appet.2019.06.00531195058

Ministry of Food and Drug Safety. (2023). [Press Release] Republic of Korea elected as inaugural chair of APFRAS. *Ministry of Food and Drug Safety*. https://www.mfds.go.kr/eng/brd/m_61/view.do?seq=149

Ministry of Health. (2024a). First in the world: The Ministry of Health has approved cattle-based cultivated meat. *Ministry of Health*. https://www.gov.il/en/pages/17012024-02

Ministry of Health. (2024b). New food in Israel. *Ministry of Health*. https://www.gov.il/he/pages/novel-food

Mohorčich, J., & Reese, J. (2019). Cell-cultured meat: Lessons from GMO adoption and resistance. *Appetite*, 143, 104408. 10.1016/j.appet.2019.10440831449883

Mridul, A. (2024, January). Aleph Farms: Israel awards the world's first regulatory approval for cultivated beef. *Green Queen*. https://www.greenqueen.com.hk/aleph-farms-israel-cultured-meat-regulatory-approval-beef/

Nacef, M., Lelièvre-Desmas, M., Symoneaux, R., Jombart, L., Flahaut, C., & Chollet, S. (2019). Consumers' expectation and liking for cheese: Can familiarity effects resulting from regional differences be highlighted within a country? *Food Quality and Preference*, 72, 188–197. 10.1016/j.foodqual.2018.10.004

Nielsen, H. B., Sonne, A., Grunert, K. G., Bánáti, D., Pollák-Tóth, A., Lakner, Z., Olsen, N. V., Žontar, T. P., & Peterman, M. (2009). Consumer perception of the use of high-pressure processing and pulsed electric field technologies in food production. *Appetite*, 52(1), 115–126. 10.1016/j.appet.2008.09.01018845196

Rolland, N. C. M., Markus, C. R., & Post, M. J. (2020). The effect of information content on acceptance of cultured meat in a tasting context. *PLoS One*, 15(4), e0231176. 10.1371/journal.pone.023117632298291

Rosenfeld, D. L., & Tomiyama, A. J. (2023). Toward consumer acceptance of cultured meat. *Trends in Cognitive Sciences*, 27(8), 689–691. 10.1016/j.tics.2023.05.00237246026

Santovito, S., Campo, R., Rosato, P., & Khuc, L. D. (2023). Impact of faith on food marketing and consumer behaviour: A review. *British Food Journal*, 125(13), 462–481. 10.1108/BFJ-02-2023-0112

Schaefer, G. O., & Savulescu, J. (2014). The ethics of producing in vitro meat. *Journal of Applied Philosophy*, 31(2), 188–202. 10.1111/japp.1205625954058

Siegrist, M., Sütterlin, B., & Hartmann, C. (2018). Perceived naturalness and evoked disgust influence acceptance of cultured meat. *Meat Science*, 139, 213–219. 10.1016/j.meatsci.2018.02.00729459297

Singapore Food Agency. (2021). A growing culture of safe, sustainable meat. *Singapore Food Agency*. https://www.sfa.gov.sg/food-for-thought/article/detail/a-growing-culture-of-safe-sustainable-meat

Singapore Food Agency. (2023a). The codex alimentarius commission and Singapore's involvement. *Singapore Food Agency*. https://www.sfa.gov.sg/food-information/singapore's-national-codex-office/the-codex-alimentarius-commission-and-singapore-s-involvement

Singapore Food Agency. (2023b). Requirements for the safety assessment of novel foods and novel food ingredients. *Singapore Food Agency*. https://www.sfa.gov.sg/docs/default-source/food-information/requirements-for-the-safety-assessment-of-novel-foods-and-novel-food-ingredients.pdf

Singapore Food Agency. (2023c). Our Singapore food story. *Singapore Food Agency*. https://www.sfa.gov.sg/food-farming/sgfoodstory/our-singapore-food-story

Singapore Food Agency. (2024a). Novel food. *Singapore Food Agency*. https://www.sfa.gov.sg/food-information/novel-food

Singapore Food Agency. (2024b). Roundtable on novel food regulation 2023. *Singapore Food Agency*. https://www.sfa.gov.sg/docs/default-source/default-document-library/roundtable-on-novel-food-regulations-2023_post-event-summary-and-full-report.pdf

Stephens, N., Di Silvio, L., Dunsford, I., Ellis, M., Glencross, A., & Sexton, A. (2018). Bringing cultured meat to market: Technical, socio-political, and regulatory challenges in cellular agriculture. *Trends in Food Science & Technology*, 78, 155–166. 10.1016/j.tifs.2018.04.01030100674

Tamimah, T., Herianingrum, S., Ratih, I. S., Rofi'ah, K., & Kulsum, U. (2018). Halalan thayyiban: The key of successful halal food industry development. *Ulumuna: Jurnal Studi Keislaman/Ulumuna, 4*(2), 171–186. 10.36420/ju.v4i2.3501

Tan, C. (2024, April). Singapore approves lab-grown quail for consumption. *The Straits Times*. https://www.straitstimes.com/singapore/s-pore-approves-lab-grown-quail-for-consumption-here

To, K. V., Comer, C. C., O'Keefe, S. F., & Lahne, J. (2024). A taste of cell-cultured meat: A scoping review. *Frontiers in Nutrition*, 11, 1332765. Advance online publication. 10.3389/fnut.2024.133276538321991

Tomiyama, A. J., Kawecki, N. S., Rosenfeld, D. L., Jay, J. A., Rajagopal, D., & Rowat, A. C. (2020). Bridging the gap between the science of cultured meat and public perceptions. *Trends in Food Science & Technology*, 104, 144–152. 10.1016/j.tifs.2020.07.019

United Nations Environment Programme. (2023). What's cooking? An assessment of the potential impacts of selected novel alternatives to conventional animal products. *Frontiers*. Advance online publication. 10.59117/20.500.11822/44236

United Nations. (n.d.). Your guide to climate action: Food. *United Nations*. https://www.un.org/en/actnow/food

Verbeke, W., Marcu, A., Rutsaert, P., Gaspar, R., Seibt, B., Fletcher, D., & Barnett, J. (2015). 'Would you eat cultured meat?': Consumers' reactions and attitude formation in Belgium, Portugal and the United Kingdom. *Meat Science*, 102, 49–58. 10.1016/j.meatsci.2014.11.01325541372

Wilks, M., & Phillips, C. J. C. (2017). Attitudes to in vitro meat: A survey of potential consumers in the United States. *PLoS One*, 12(2), e0171904. 10.1371/journal.pone.017190428207878

Xu, X., Sharma, P., Shu, S., Lin, T., Ciais, P., Tubiello, F. N., Smith, P., Campbell, N., & Jain, A. K. (2021). Global greenhouse gas emissions from animal-based foods are twice those of plant-based foods. *Nature Food*, 2(9), 724–732. 10.1038/s43016-021-00358-x37117472

ADDITIONAL READING

Food and Agriculture Organisation of the United Nations & World Health Organisation. (2023). Food safety aspects of cell-based food. 10.4060/cc4855en

Hartmann, C., & Siegrist, M. (2023). Consumer attitudes to cultured meat products: Improving understanding and acceptance. In Post, M., Connon, C., & Bryant, C. (Eds.), *Advances in cultured meat technology* (pp. 341–353). Burleigh Dodds Science Publishing. 10.19103/AS.2023.0130.15

Ong, K. J., Case, F., & Shatkin, J. A. (2024). Food safety of fermented proteins and cultivated meat and seafood. In Fraser, E. D. G., Kaplan, D. L., Newman, L., & Yada, R. Y. (Eds.), *Cellular agriculture technology, society, sustainability, and science* (1st ed., pp. 77–94). Elsevier. 10.1016/B978-0-443-18767-4.00033-0

Yeung, J., Tan, Y. Q., Chan, S. H., Chng, K. R., Yeo, C., Poh, J. L., Low, T. Y., & Harn, J. S. (2023). Creating a regulatory framework for cultured meat products: Singapore. In M. Post, C. Connon, & C. Bryant (Eds.), *Advances in cultured meat technology* (pp. 43–80). Burleigh Dodds Science Publishing.

KEY TERMS AND DEFINITIONS

Cultivated: A prefix that could be utilised to describe a variety of cultivated products produced by companies ranging from beef, chicken, seafood, and pork to more specialised entries such as foie gras and caviar.

Cultivated Meat: A generalised term to describe meat or seafood derived from growing cells in vitro.

Labelling: How cultivated meat products are described and portrayed to the consumer.

Naturalness: Pertains to a feeling about food in the context of its originality, to which higher adulteration generates a negative perception.

Novel Foods: A subset of food or food development practices without a history of consumption or utilisation for the last 25 years or stipulated by each country's food code. These include cultivated meat, precision fermentation, biomass fermentation, and certain insects.

Regulatory Coordination: Regulators work together internationally and with other stakeholders to offer support and interoperability of frameworks. Regulatory Frameworks: The food safety protocols and infrastructure established to allow companies to demonstrate food safety of their products.

APPENDIX

Table 1.

AMPS-Innovation	U.S.-Based Association for Meat, Poultry, and Seafood Innovation
Anvisa	Agência Nacional de Vigilância Sanitária
APAC-SCA	Asia-Pacific Society for Cellular Agriculture
CAA	Cellular Agriculture Australia
CAE	Cellular Agriculture Europe
EFSA	European Food Safety Authority
FAO	Food and Agriculture Organisation of the United Nations
FCS	National Food Control Service
GCAA	Global Cellular Agriculture Alliance
GFI APAC	Good Food Institute Asia Pacific
GGALI	Gerência-Geral de Alimentos
MAPA	Ministério da Agricultura, Pecuária e Abastecimento
MUIS	Majlis Ugama Islam Singapura
SFA	Singapore Food Agency
U.S FDA	Unites States Federal Drug Administration
USDA	United States Department of Agriculture
WHO	World Health Organisation

Compilation of References

A conversation about cultivated meat. (2023). *Nature Communications, 14*(1). 10.1038/s41467-023-43984-8

Abitbol, T., Rivkin, A., Cao, Y., Nevo, Y., Abraham, E., Ben-Shalom, T., & Shoseyov, O. (2016). Nanocellulose, a tiny fiber with huge applications. *Current Opinion in Biotechnology*, 39, 76–88. 10.1016/j.copbio.2016.01.00226930621

Adams, C. J., Crary, A., & Gruen, L. (2023). The empty promises of cultured meat. The good it promises, the harm it does: Critical essays on effective altruism, 149.

Afazeli, H., Jafari, A., Rafiee, S., & Nosrati, M. (2014). An investigation of biogas production potential from livestock and slaughterhouse wastes. *Renewable & Sustainable Energy Reviews*, 34, 380–386. 10.1016/j.rser.2014.03.016

Afonso, A. L., Gelbmann, W., Germini, A., Fernández, E. N., Parrino, L., Precup, G., & Ververis, E.European Food Safety Authority. (2024b). EFSA scientific colloquium 27: Cell culture-derived foods and food ingredients. *EFSA Supporting Publications*, 21(3). Advance online publication. 10.2903/sp.efsa.2024.EN-8664

Afrifa, G. A., Tingbani, I., Yamoah, F., & Appiah, G. (2020). Innovation input, governance and climate change: Evidence from emerging countries. *Technological Forecasting and Social Change*, 161, 120256. 10.1016/j.techfore.2020.120256

Agência Nacional de Vigilância Sanitária. (2022). *Relatório de AIR sobre a modernização do marco regulatório, fluxos e procedimentos para novos alimentos e novos ingredientes*. https://www.gov.br/anvisa/pt-br/assuntos/regulamentacao/air/analises-de-impacto-regulatorio/2023/arquivos-relatorios-de-air-2023-2/relatorio_de_air_novos_ alimentos___final.pdf

Agência Nacional de Vigilância Sanitária. (2023). *Resolução da diretoria colegiada – RDC Nº 839, de 14 de Dezembro de 2023.* https://antigo.anvisa.gov.br/documents/10181/6582266/RDC_839_2023_.pdf/a064b871-55dd-44b9-ab40-16ca7672497d Agência Nacional de Vigilância Sanitária. (n.d.). *4144 – Consulta sobre a classificação de um novo alimento e novo ingredient.* https://consultas.anvisa.gov.br/#/consultadeassuntos

Agrawal, D. R., & Wildasin, D. E. (2020). Technology and tax systems. *Journal of Public Economics*, 185, 104082. 10.1016/j.jpubeco.2019.104082

Agrimonti, C., Lauro, M., & Visioli, G. (2020). Smart agriculture for food quality: Facing climate change in the 21st century. *Critical Reviews in Food Science and Nutrition*, 61(6), 971–981. 10.1080/10408398.2020.174955532270688

Alcorta, A., Porta, A., Tárrega, A., Alvarez, M. D., & Vaquero, M. P. (2021). Foods for plant-based diets: Challenges and innovations. *Foods*, 10(2), 293. 10.3390/foods1002029333535684

Aleph Farms. (2023). Chief Rabbi of Israel affirms Aleph Farms' cultivated steak is kosher. *Aleph Farms.* https://aleph-farms.com/journals/kosher-ruling/

Aleph Farms. (2024). Aleph Farms granted world's first regulatory approval for cultivated beef. *Aleph Farms.* https://aleph-farms.com/journals/aleph-farms-granted-worlds-first-regulatory-approval-for-cultivated-beef/

Alexander, P., Brown, C., Arneth, A., Dias, C., Finnigan, J., Moran, D., & Rounsevell, M. D. A. (2017). Could consumption of insects, cultured meat or imitation meat reduce global agricultural land use? *Global Food Security*, 15, 22–32. 10.1016/j.gfs.2017.04.001

Ali, A., & Ahmed, S. (2018). Recent advances in edible polymer-based hydrogels as a sustainable alternative to conventional polymers. *Journal of Agricultural and Food Chemistry*, 2018(66), 6940–6967. 10.1021/acs.jafc.8b0105229878765

Allan, G. J., Lecca, P., McGregor, P. G., & Swales, J. K. (2014). The economic impacts of marine energy developments: A case study from Scotland. *Marine Policy*, 43, 122–131. 10.1016/j.marpol.2013.05.003

Allan, S. J., De Bank, P. A., & Ellis, M. J. (2019). Bioprocess design considerations for cultured meat production with a focus on the expansion bioreactor. *Frontiers in Sustainable Food Systems*, 3, 44. 10.3389/fsufs.2019.00044

Allan, S. J., Ellis, M. J., & De Bank, P. A. (2021). Decellularized grass as a sustainable scaffold for skeletal muscle tissue engineering. *Journal of Biomedical Materials Research. Part A*, 109(12), 2471–2482. 10.1002/jbm.a.3724134057281

Alonso, E. B. (2015). The impact of culture, religion and traditional knowledge on food and nutrition security in developing countries. Retrieved from https://ageconsearch.umn.edu/record/285169/files/30_briones.pdf

Alonso, M. E., González-Montaña, J. R., & Lomillos, J. M. (2020). Consumers' concerns and perceptions of farm animal welfare. *Animals (Basel)*, 10(3), 385. 10.3390/ani1003038532120935

Alsarhan, L. M., Alayyar, A. S., Alqahtani, N. B., & Khdary, N. H. (2021). Circular Carbon Economy (CCE): A Way to Invest CO2 and Protect the Environment, a Review. *Sustainability (Basel)*, 13(21), 11625. 10.3390/su132111625

Amit, M., Carpenter, M. K., Inokuma, M. S., Chiu, C. P., Harris, C. P., Waknitz, M. A., Itskovitz-Eldor, J., & Thomson, J. A. (2000, November). Clonally Derived Human Embryonic Stem Cell Lines Maintain Pluripotency and Proliferative Potential for Prolonged Periods of Culture. *Developmental Biology*, 227(2), 271–278. 10.1006/dbio.2000.991211071754

Ammar, H., Abidi, S., Ayed, M., Moujahed, N., deHaro Martí, M. E., Chahine, M., & Hechlef, H. (2020). Estimation of Tunisian greenhouse gas emissions from different livestock species. *Agriculture*, 10(11), 562. 10.3390/agriculture10110562

Anderson, E. N. (2014). Everyone eats: Understanding food and culture. NYU Press. Retrieved from https://books.google.com/books?hl=en&lr=&id=OKEUCgAAQBAJ&oi=fnd&pg=PP9&dq=Food+is+a+fundamental+component+of+cultural+identity,+and+conventional+meals+often+incorporate+specialized+ways+of+animal+rearing,+slaughtering,+and+processing.+&ots=bJwX_OqW0X&sig=BaqSQGblYkeIAMTXHx25cJ-p48E

Andronie, M., Lăzăroiu, G., Iatagan, M., Hurloiu, I., & Dijmărescu, I. (2021). Sustainable cyber-physical production systems in big data-driven smart urban economy: A systematic literature review. *Sustainability (Basel)*, 13(2), 751. 10.3390/su13020751

Anomaly, J., Browning, H., Fleischman, D., & Veit, W. (2023). Flesh without blood: The public health benefits of lab-grown meat. *Journal of Bioethical Inquiry*, 21(1), 167–175. 10.1007/s11673-023-10254-737656382

Anthis, J. R., Berenstein, P., Beurrier, C., Biehl, C., Birdie, P., Burley, J., . . . d'Origny, G. (2023). Modern meat: the next generation of meat from cells. Retrieved from https://ora.ox.ac.uk/objects/uuid:6dd61d4b-781a-4ccb-bcd6-9029728348fe/files/sfx719n99b

APAC Regulatory Coordination Forum. (2023a). About. *APAC Regulatory Coordination Forum*. https://www.cellagforum.info

APAC Regulatory Coordination Forum. (2023b). New cellular agriculture forum aims to accelerate regulatory approvals across APAC. *APAC Regulatory Coordination Forum.* https://www.cellagforum.info/post/new-cellular-agriculture-forum-aims-to-accelerate-regulatory-approvals-across-apac

Arab, W., Rauf, S., Al-Harbi, O., & Hauser, C. A. (2018). Novel ultrashort self-assembling peptide bioinks for 3D culture of muscle myoblast cells. *International Journal of Bioprinting,* 4(2). Advance online publication. 10.18063/ijb.v4i1.12933102913

Armstrong, R. (2023). Towards the microbial home: An overview of developments in next-generation sustainable architecture. *Microbial Biotechnology,* 16(6), 1112–1130. 10.1111/1751-7915.1425637070748

Arndt, C., Hristov, A. N., Price, W. J., McClelland, S. C., Pelaez, A. M., Cueva, S. F., Oh, J., Dijkstra, J., Bannink, A., Bayat, A. R., Crompton, L. A., Eugène, M. A., Enahoro, D., Kebreab, E., Kreuzer, M., McGee, M., Martin, C., Newbold, C. J., Reynolds, C. K., & Yu, Z. (2022). Full adoption of the most effective strategies to mitigate methane emissions by ruminants can help meet the 1.5 °C target by 2030 but not 2050. *Proceedings of the National Academy of Sciences of the United States of America,* 119(20), e2111294119–e2111294119. 10.1073/pnas.211129411935537050

Arora, S., Kataria, P., Nautiyal, M., Tuteja, I., Sharma, V., Ahmad, F., Haque, S., Shahwan, M., Capanoglu, E., Vashishth, R., & Gupta, A. K. (2023). Comprehensive Review on the Role of Plant Protein As a Possible Meat Analogue: Framing the Future of Meat. *ACS Omega,* 8(26), 23305–23319. 10.1021/acsomega.3c0137337426217

Arshad, M. S., Javed, M., Sohaib, M., Saeed, F., Imran, A., & Amjad, Z. (2017). Tissue engineering approaches to develop cultured meat from cells: A mini review. *Cogent Food & Agriculture,* 3(1), 1320814. 10.1080/23311932.2017.1320814

Aryal, J. P., Manchanda, N., & Sonobe, T. (2022). Expectations for household food security in the coming decades: A global scenario. In *Future foods* (pp. 107–131). Academic Press.

Asia-Pacific Society for Cellular Agriculture, & Good Food Institute Asia Pacific. (2023). *Memorandum of understanding.* https://687315f7-d318-48d8-a759-cb77fa9836e9.usrfiles.com/ugd/687315_a30dfed3b7cd477282f1da795d12d05c.pdf

Asia-Pacific Society for Cellular Agriculture. (2023). South Korean consumers willing to try cultivated meat and seafood products. *Asia-Pacific Society for Cellular Agriculture.* https://www.apac-sca.org/post/south-korean-consumers-willing-to-try-cultivated-meat-and-seafood-products

Asia-Pacific Society for Cellular Agriculture. (2024). Safety of cultivated meat and seafood ranks highest for Japanese consumers. *Asia-Pacific Society for Cellular Agriculture*. https://www.apac-sca.org/post/safety-of-cultivated-meat-and-seafood-ranks-highest-for-japanese-consumers

Asioli, D., Bazzani, C., & Nayga, R.Jr. (2021). Are consumers willing to pay for in-vitro meat? an investigation of naming effects. *Journal of Agricultural Economics*, 73(2), 356–375. 10.1111/1477-9552.12467

Auer, J. A., Goodship, A., Arnoczky, S., Pearce, S., Price, J., Claes, L., von Rechenberg, B., Hofmann-Amtenbrinck, M., Schneider, E., Müller-Terpitz, R., Thiele, F., Rippe, K.-P., & Grainger, D. W. (2007). Refining animal models in fracture research: Seeking consensus in optimising both animal welfare and scientific validity for appropriate biomedical use. *BMC Musculoskeletal Disorders*, 8(1), 1–13. 10.1186/1471-2474-8-7217678534

Australian Government. (2024, May 2). https://www.foodregulation.gov.au/activities-committees/food-ministers-meeting

Azhar, A., Zeyaullah, M., Bhunia, S., Kacham, S., Patil, G., Muzammil, K., Khan, M. S., & Sharma, S. (2023). Cell-based meat: The molecular aspect. *Frontiers in Food Science and Technology*, 3, 1126455. 10.3389/frfst.2023.1126455

Bach, A., Stern-Straeter, J., Beier, J., Bannasch, H., & Stark, G. (2003, October). Engineering of muscle tissue. *Clinics in Plastic Surgery*, 30(4), 589–599. 10.1016/S0094-1298(03)00077-414621307

Baden, M. Y., Liu, G., Satija, A., Li, Y., Sun, Q., Fung, T. T., Rimm, E. B., Willett, W. C., Hu, F. B., & Bhupathiraju, S. N. (2019). Changes in Plant-Based Diet Quality and Total and Cause-Specific Mortality. *Circulation*, 140(12), 979–991. 10.1161/CIRCULATIONAHA.119.041014

Bahar, N. H., Lo, M., Sanjaya, M., Van Vianen, J., Alexander, P., Ickowitz, A., & Sunderland, T. (2020). Meeting the food security challenge for nine billion people in 2050: What impact on forests? *Global Environmental Change*, 62, 102056. 10.1016/j.gloenvcha.2020.102056

Bailey, R. L. (2020). Current regulatory guidelines and resources to support research of dietary supplements in the United States. *Critical Reviews in Food Science and Nutrition*, 60(2), 298–309. 10.1080/10408398.2018.152436430421981

Bai, Y., Herforth, A., & Masters, W. A. (2022). Global variation in the cost of a nutrient-adequate diet by population group: An observational study. *The Lancet. Planetary Health*, 6(1), e19–e28. 10.1016/S2542-5196(21)00285-034998455

Bakhsh, A., Lee, E.-Y., Ncho, C. M., Kim, C.-J., Son, Y.-M., Hwang, Y.-H., & Joo, S.-T. (2022). Quality characteristics of meat analogs through the incorporation of textured vegetable protein: A systematic review. *Foods*, 11(9), 1242. 10.3390/foods11091242 35563965

Bala, M. M., Strzeszynski, L., & Topor-Madry, R. (2017). Mass media interventions for smoking cessation in adults. *Cochrane Database of Systematic Reviews*, 11(11), CD004704. Advance online publication. 10.1002/14651858.CD004704.pub4 29159862

Bapat, S., Koranne, V., Shakelly, N., Huang, A., Sealy, M. P., Sutherland, J. W., Rajurkar, K. P., & Malshe, A. P. (2022). Cellular agriculture: An outlook on smart and resilient food agriculture manufacturing. *Smart and Sustainable Manufacturing Systems*, 6(1), 1–11. 10.1520/SSMS20210020

Barbosa, W., Correia, P., Vieira, J., Leal, I., Rodrigues, L., Nery, T., Barbosa, J., & Soares, M. (2023). Trends and Technological Challenges of 3D Bioprinting in Cultured Meat: Technological Prospection. *Applied Sciences (Basel, Switzerland)*, 13(22), 12158. 10.3390/app132212158

Barbulova, A., Apone, F., & Colucci, G. (2014, April 22). Plant Cell Cultures as Source of Cosmetic Active Ingredients. *Cosmetics*, 1(2), 94–104. 10.3390/cosmetics1020094

Bar-Nur, O., Gerli, M. F., Di Stefano, B., Almada, A. E., Galvin, A., Coffey, A., Huebner, A. J., Feige, P., Verheul, C., Cheung, P., Payzin-Dogru, D., Paisant, S., Anselmo, A., Sadreyev, R. I., Ott, H. C., Tajbakhsh, S., Rudnicki, M. A., Wagers, A. J., & Hochedlinger, K. (2018). Direct reprogramming of mouse fibroblasts into functional skeletal muscle progenitors. *Stem Cell Reports*, 10(5), 1505–1521. 10.1016/j.stemcr.2018.04.009 29742392

Bar-On, Y. M., Phillips, R., & Milo, R. (2018, May 21). The biomass distribution on Earth. *Proceedings of the National Academy of Sciences of the United States of America*, 115(25), 6506–6511. 10.1073/pnas.1711842115 29784790

Barrett, C. B. (2021). Overcoming global food security challenges through science and solidarity. *American Journal of Agricultural Economics*, 103(2), 422–447. 10.1111/ajae.12160

Bar-Shai, N., Sharabani-Yosef, O., Zollmann, M., Lesman, A., & Golberg, A. (2021). Seaweed cellulose scaffolds derived from green macroalgae for tissue engineering. *Scientific Reports*, 11(1), 11843. 10.1038/s41598-021-90903-2 34088909

Barthelmie, R. J. (2022). Impact of Dietary Meat and Animal Products on GHG Footprints: The UK and the US. *Climate (Basel)*, 10(3), 43. 10.3390/cli10030043

Bartholet, J. (2011). Inside the meat lab. *Scientific American*, 304(6), 64–69. 10.1038/scientificamerican0611-6421608405

Bartnikowski, M., Dargaville, T. R., Ivanovski, S., & Hutmacher, D. W. (2019). Degradation mechanisms of polycaprolactone in the context of chemistry, geometry and environment. *Progress in Polymer Science*, 96, 1–20. 10.1016/j.progpolymsci.2019.05.004

Batra, S. (1981). Cows and cow-slaughter in India: Religious, political and social aspects. Retrieved from https://repub.eur.nl/pub/34927/

Bayvel, A. D., & Cross, N. (2010). Animal welfare: A complex domestic and international public-policy issue—who are the key players? *Journal of Veterinary Medical Education*, 37(1), 3–12. 10.3138/jvme.37.1.320378871

Beauchamp, T. L., & DeGrazia, D. (2019). *Principles of animal research ethics*. Oxford University Press. 10.1017/S0962728600009052

Begley, C. G., & Ioannidis, J. P. (2015). Reproducibility in science: Improving the standard for basic and preclinical research. *Circulation Research*, 116(1), 116–126. 10.1161/CIRCRESAHA.114.30381925552691

Behm, K., Nappa, M., Aro, N., Welman, A., Ledgard, S., Suomalainen, M., & Hill, J. (2022). Comparison of carbon footprint and water scarcity footprint of milk protein produced by cellular agriculture and the dairy industry. *The International Journal of Life Cycle Assessment*, 27(8), 1017–1034. 10.1007/s11367-022-02087-0

Beillouin, D., Schauberger, B., Bastos, A., Ciais, P. & Makowski, D. (2020). Impact of extreme weather conditions on European crop production in 2018. *Philosophical Transactions of the Royal Society of London. Series B, Biological Sciences*, 375(1810), 20190510. 10.1098/rstb.2019.0510

Bellani, C. F., Ajeian, J., Duffy, L., Miotto, M., Groenewegen, L., & Connon, C. J. (2020). Scale-Up Technologies for the Manufacture of Adherent Cells. *Frontiers in Nutrition*, 7, 575146. 10.3389/fnut.2020.57514633251241

Ben-Arye, T., & Levenberg, S. (2019). Tissue engineering for clean meat production. *Frontiers in Sustainable Food Systems*, 3, 46. 10.3389/fsufs.2019.00046

Ben-Arye, T., Shandalov, Y., Ben-Shaul, S., Landau, S., Zagury, Y., Ianovici, I., Lavon, N., & Levenberg, S. (2020). Textured soy protein scaffolds enable the generation of three-dimensional bovine skeletal muscle tissue for cell-based meat. *Nature Food*, 1(4), 210–220. 10.1038/s43016-020-0046-5

Benjaminson, M., Gilchriest, J., & Lorenz, M. (2002, December). In vitro edible muscle protein production system (mpps): Stage 1, fish. *Acta Astronautica*, 51(12), 879–889. 10.1016/S0094-5765(02)00033-412416526

Benny, A., Pandi, K., & Upadhyay, R. (2022). Techniques, challenges and future prospects for cell-based meat. *Food Science and Biotechnology*, 31(10), 1225–1242. 10.1007/s10068-022-01136-6

Bhat, Z. F., Bhat, H., & Pathak, V. (2013). Prospects for In Vitro Cultured Meat - A Future Harvest. In Principles of Tissue Engineering: Fourth Edition (pp. 1663–1683). Elsevier Inc. 10.1016/B978-0-12-398358-9.00079-3

Bhattacharyya, S. S., Leite, F. F. G. D., Adeyemi, M. A., Sarker, A. J., Cambareri, G. S., Faverin, C., Tieri, M. P., Castillo-Zacarías, C., Melchor-Martínez, E. M., Iqbal, H. M. N., & Parra-Saldívar, R. (2021). A paradigm shift to CO2 sequestration to manage global warming – With the emphasis on developing countries. *The Science of the Total Environment*, 790, 148169. 10.1016/j.scitotenv.2021.14816934380249

Bhat, Z. F., & Bhat, H. (2011). Animal-free meat biofabrication. *American Journal of Food Technology*, 6(6), 441–459. 10.3923/ajft.2011.441.459

Bhat, Z. F., & Fayaz, H. (2011). Prospectus of cultured meat—Advancing meat alternatives. *Journal of Food Science and Technology*, 48(2), 125–140. 10.1007/s13197-010-0198-7

Bhat, Z. F., Kumar, S., & Bhat, H. F. (2015, May 5). In vitro meat: A future animal-free harvest. *Critical Reviews in Food Science and Nutrition*, 57(4), 782–789. 10.1080/10408398.2014.92489925942290

Bhat, Z. F., Kumar, S., & Fayaz, H. (2015). In vitro meat production: Challenges and benefits over conventional meat production. *Journal of Integrative Agriculture*, 14(2), 241–248. 10.1016/S2095-3119(14)60887-X

Bhat, Z. F., Morton, J. D., Mason, S. L., Bekhit, A. E. A., & Bhat, H. F. (2019). Technological, Regulatory, and Ethical Aspects of In Vitro Meat: A Future Slaughter-Free Harvest. *Comprehensive Reviews in Food Science and Food Safety*, 18(4), 1192–1208. 10.1111/1541-4337.12473

Bhuyan, B. P., Tomar, R., & Cherif, A. R. (2022). A systematic review of knowledge representation techniques in smart agriculture (Urban). *Sustainability (Basel)*, 14(22), 15249. 10.3390/su142215249

Bian, W., & Bursac, N. (2009). Engineered skeletal muscle tissue networks with controllable architecture. *Biomaterials*, 30(7), 1401–1412. 10.1016/j.biomaterials.2008.11.015

Bierbaum, R., Leonard, S. A., Rejeski, D., Whaley, C., Barra, R. O., & Libre, C. (2020). Novel entities and technologies: Environmental benefits and risks. *Environmental Science & Policy*, 105, 134–143. 10.1016/j.envsci.2019.11.002

Bin Rahman, A. R., & Zhang, J. (2023). Trends in rice research: 2030 and beyond. *Food and Energy Security*, 12(2), e390. 10.1002/fes3.390

Birhanu, M. Y., Alemayehu, T., Bruno, J. E., Kebede, F. G., Sonaiya, E. B., Goromela, E. H., & Dessie, T. (2021). Technical efficiency of traditional village chicken production in Africa: Entry points for sustainable transformation and improved livelihood. *Sustainability (Basel)*, 13(15), 8539. 10.3390/su13158539

Bishop, Mostafa, Pakvasa, Luu, Lee, Wolf, Ameer, He, & Reid. (2017). 3-D bioprinting technologies in tissue engineering and regenerative medicine: current and future trends. Genes Dis 4:185–195. https://doi.org/.2017.10.00210.1016/j.gendis

Bispo, D. S. C., Michálková, L., Correia, M., Jesus, C. S. H., Duarte, I. F., Goodfellow, B. J., Oliveira, M. B., Mano, J. F., & Gil, A. M. (2022). Endo- and Exometabolome Crosstalk in Mesenchymal Stem Cells Undergoing Osteogenic Differentiation. *Cells*, 11(8), 1257. 10.3390/cells1108125735455937

Biswal, T. (2021). Biopolymers for tissue engineering applications: A review. *Materials Today: Proceedings*, 41, 397–402. 10.1016/j.matpr.2020.09.628

Boadway, R., & Flatters, F. (2023). The Taxation of Natural Resources: Principles and Policy Issues. In *Taxing Choices for Managing Natural Resources, the Environment, and Global Climate Change* (pp. 17–81). Springer International Publishing. 10.1007/978-3-031-22606-9_2

Boden, M., Cagnin, C., Carabias-Hütter, V., Haegemann, K., & Könnölä, T. (2010). Facing the future: time for the EU to meet global challenges. Retrieved from https://digitalcollection.zhaw.ch/handle/11475/6040

Bodiou, V., Moutsatsou, P., & Post, M. J. (2020). Microcarriers for upscaling cultured meat production. *Frontiers in Nutrition*, 7, 10. 10.3389/fnut.2020.0001032154261

Bodiou, V., Moutsatsou, P., & Post, M. J. (2020, February 20). Microcarriers for Upscaling Cultured Meat Production. *Frontiers in Nutrition*, 7.32154261

Bomkamp, C., Carter, M., Cohen, M., Gertner, D., Ignaszewski, E., Murray, S., O'Donnell, M., Pierce, B., Swartz, E., & Voss, S. (2022a). Cultivated meat state of the industry report. https://gfi.org/resource/cultivated-meat-eggs-and-dairy-state-of-the-industry-report/

Bomkamp, C., Skaalure, S. C., Fernando, G. F., Ben-Arye, T., Swartz, E. W., & Specht, E. A. (2022). Scaffolding Biomaterials for 3D Cultivated Meat: Prospects and Challenges. *Advanced Science (Weinheim, Baden-Wurttemberg, Germany)*, 9(3), e2102908. 10.1002/advs.20210290834786874

Bostock, J., McAndrew, B., Richards, R., Jauncey, K., Telfer, T., Lorenzen, K., Little, D., Ross, L., Handisyde, N., Gatward, I., & Corner, R. (2010). Aquaculture: Global status and trends. In Philosophical Transactions of the Royal Society B: Biological Sciences (Vol. 365, Issue 1554, pp. 2897–2912). Royal Society. 10.1098/rstb.2010.0170

Bouvard, V., Loomis, D., Guyton, K. Z., Grosse, Y., El Ghissassi, F., Benbrahim-Tallaa, L., Guha, N., Mattock, H., & Straif, K. (2015): Carcinogenicity of consumption of red and processed meat. Lancet Oncol., 16, 1599.

Brennan, L., & Owende, P. **(2010).** Biofuels from microalgae-A review of technologies for production, processing, and extractions of biofuels and co-products. In Renewable and Sustainable Energy Reviews (Vol. 14, Issue 2, pp. 557–577). https://doi.org/10.1016/j.rser.2009.10.009

Brennan, T., Katz, J., Quint, Y., & Spencer, B. (2021). Cultivated meat: Out of the lab, into the frying pan. McKinsey & Company.

Brennan, T., Katz, J., Quint, Y., & Spencer, B. (2021, June 16). *Cultivated meat: Out of the lab, into the frying pan*. McKinsey & Company. https://www.mckinsey.com/industries/agriculture/our-insights/cultivated-meat-out-of-the-lab-into-the-frying-pan# Bryant, C., & Barnett, J. (2019). What's in a name? Consumer perceptions of in vitro meat under different names. *Appetite, 137*, 104–113. 10.1016/j.appet.2019.02.021

Broom, D. M. (2007). Quality of life means welfare: How is it related to other concepts and assessed? *Animal Welfare (South Mimms, England)*, 16(S1), 45–53. 10.1017/S0962728600031729

Broucke, K., Van Pamel, E., Van Coillie, E., Herman, L., & Van Royen, G. (2023). Cultured meat and challenges ahead: A review on nutritional, technofunctional and sensorial properties, safety and legislation. *Meat Science*, 195, 109006. 10.1016/j.meatsci.2022.10900636274374

Bruckner, G. K. (2009). The role of the World Organisation for Animal Health (OIE) to facilitate the international trade in animals and animal products: Policy and trade issues. *The Onderstepoort Journal of Veterinary Research*, 76(1), 141–146. 10.4102/ojvr.v76i1.7819967940

Brüggenthies, J. B., Fiore, A., Russier, M., Bitsina, C., Brötzmann, J., Kordes, S., Menninger, S., Wolf, A., Conti, E., Eickhoff, J. E., & Murray, P. J. (2022). A cell-based chemical-genetic screen for amino acid stress response inhibitors reveals torins reverse stress kinase GCN2 signaling. *The Journal of Biological Chemistry*, 298(12), 102629. 10.1016/j.jbc.2022.10262936273589

Brunner, D. (2010). Serum-free cell culture: The serum-free media interactive online database. *ALTEX*, 53–62. 10.14573/altex.2010.1.5320390239

Bryant, C. (2020). *Exploring the Nature of Consumer Preferences between Conventional and Cultured Meat*. University of Bath.

Bryant, C. J. (2020). Culture, meat, and cultured meat. *Journal of Animal Science*, 98(8), skaa172. Advance online publication. 10.1093/jas/skaa17232745186

Bryant, C., & Barnett, J. (2018). Consumer acceptance of cultured meat: A systematic review. *Meat Science*, 143, 8–17. 10.1016/j.meatsci.2018.04.00829684844

Bryant, C., & Barnett, J. (2020). Consumer acceptance of cultured meat: An updated review (2018–2020). *Applied Sciences (Basel, Switzerland)*, 10(15), 5201. 10.3390/app10155201

Bryant, C., Szejda, K., Parekh, N., Desphande, V., & Tse, B. (2019). A survey of consumer perceptions of plant-based and clean meat in the USA, India, and China. *Frontiers in Sustainable Food Systems*, 3, 11. Advance online publication. 10.3389/fsufs.2019.00011

Bryant, C., van Nek, L., & Rolland, N. C. M. (2020). European Markets for Cultured Meat: A Comparison of Germany and France. *Foods*, 2020(9), 1152. 10.3390/foods909115232825592

Burattini, S., Ferri, P., Battistelli, M., Curci, R., Luchetti, F., & Falcieri, E. (2004). C2C12 murine myoblasts as a model of skeletal muscle development: Morphofunctional characterization. *European Journal of Histochemistry*, 48(3), 223–234.

Burdock, G. A., Carabin, I. G., & Griffiths, J. C. (2006). The importance of GRAS to the functional food and nutraceutical industries. *Toxicology*, 221(1), 17–27. 10.1016/j.tox.2006.01.01216483705

Burton, A. C., Neilson, E., Moreira, D., Ladle, A., Steenweg, R., Fisher, J. T., Bayne, E., & Boutin, S. (2015). Wildlife camera trapping: A review and recommendations for linking surveys to ecological processes. *Journal of Applied Ecology*, 52(3), 675–685. 10.1111/1365-2664.1243226211047

Burton, R. J. (2019). The Potential Impact of Synthetic Animal Protein on Livestock Production: The New "War against Agriculture"? *Journal of Rural Studies*, 68, 33–45. 10.1016/j.jrurstud.2019.03.002

Butler, M. (2015). Serum and protein free media. Animal Cell Culture, 223-236.

Camidge, R. (2001). The Story of Taxol: Nature and politics in the pursuit of an anti-cancer drug. *BMJ (Clinical Research Ed.)*, 323(7304), 115. 10.1136/bmj.323.7304.115/a

Campuzano, S., & Pelling, A. E. (2019, May 17). Scaffolds for 3D Cell Culture and Cellular Agriculture Applications Derived From Non-animal Sources. *Frontiers in Sustainable Food Systems*, 3, 3. 10.3389/fsufs.2019.00038

Caputo, S., Schoen, V., Specht, K., Grard, B., Blythe, C., Cohen, N., Fox-Kämper, R., Hawes, J., Newell, J., & Poniży, L. (2021). Applying the food-energy-water nexus approach to urban agriculture: From FEW to FEWP (Food-Energy-Water-People). *Urban Forestry &amp. Urban Forestry & Urban Greening*, 58, 126934. 10.1016/j.ufug.2020.126934

Cardoso, J. C., Sheng Gerald, L. T., & Teixeira da Silva, J. A. (2018). Micropropagation in the Twenty-First Century. *Methods in Molecular Biology (Clifton, N.J.)*, 1815, 17–46. 10.1007/978-1-4939-8594-4_229981112

Carletti, E., Motta, A., & Migliaresi, C. (2011). *Scaffolds for tissue engineering and 3D cell culture*. Methods Mol Bio. 10.1007/978-1-60761-984-0_2

Carpentier, A. V., Bonnet, M., Campard, D., & Moisan, A. (2021). Perfusion bioreactors for 3D skin cell culture: State of the art and perspectives. *Journal of Tissue Engineering and Regenerative Medicine*, 15(2), 109–123.

Carrel, A. (1912, May 1). On the permanent life of tissues outside of the organism. *The Journal of Experimental Medicine*, 15(5), 516–528. 10.1084/jem.15.5.51619867545

Carrington, D. (2020). No-kill, lab-grown meat to go on sale for first time. Retrieved on September 18, 2021 from The Guardian. website: https://www.theguardian.com/environment/2020/dec/02/nokill-lab-grown-meat-to-go-on-sale-for-first-time

Cattet, M. R. (2013). Falling through the cracks: Shortcomings in the collaboration between biologists and veterinarians and their consequences for wildlife. *ILAR Journal*, 54(1), 33–40. 10.1093/ilar/ilt01023904530

Catts, O., & Zurr, I. (2002). Growing semi-living sculptures: The tissue culture project. *Leonardo*, 35(4), 365–370. 10.1162/002409402760181123

Cecchinato, A. (2021). The Plant-Based Meat: a study of packaging to communicate a sustainable food product's innovation.

CellMEAT. (2023). Cell MEAT begins approval process for cell-cultured meat from the Ministry of Food and Drug Safety of Korea. Available from: https://www.thecellmeat.com/bbs/board.php?bo_table=gallery&wr_id=44

Cellular Agriculture Australia. (2023a, December 11). What's next for Vow on the path to regulatory approval in Australia? *Cellular Agriculture Australia*. https://www.cellularagricultureaustralia.org/publications/whats-next-for-vow-on-the-path-to-regulatory-approval-in-australia

Cellular Agriculture Australia. (2023b). Key terms & language. *Cellular Agriculture Australia*. https://www.cellularagricultureaustralia.org/resources/key-terms

Cellular Agriculture Australia. (2024). Media guide how to talk about cellular agriculture effectively. *Cellular Agriculture Australia*. https://assets-global.website-files.com/641400f02d9306cbd4fb3f94/6614cde01226b58d37bee299_Cellular%20Agriculture%20Australia%20-%20Media%20Guide%20V1.0%20(1).pdf

Cesarz, Z., & Tamama, K. (2016). Spheroid Culture of Mesenchymal Stem Cells. *Stem Cells International*, 2016(1), 1–11. 10.1155/2016/917635726649054

Chan, D. L.-K., Lim, P.-Y., Sanny, A., Georgiadou, D., Lee, A. P., & Tan, A. H.-M. (2024). Technical, commercial, and regulatory challenges of cellular agriculture for seafood production. *Trends in Food Science & Technology*, 144, 104341. 10.1016/j.tifs.2024.104341

Chandimali, N., Park, E. H., Bak, S. G., Won, Y. S., Lim, H. J., & Lee, S. J. (2024). Not seafood but seafood: A review on cell-based cultured seafood in lieu of conventional seafood. *Food Control*, 162, 110472. 10.1016/j.foodcont.2024.110472

Chandramohan, J., Asoka Chakravarthi, R. P., & Ramasamy, U. (2023). A comprehensive inventory management system for non-instantaneous deteriorating items in supplier- retailer-customer supply chains. *Supply Chain Analytics*, 3, 100015. https://doi.org/https://doi.org/10.1016/j.sca.2023.100015

Chang, K. H., Liao, H. T., & Chen, J. P. (2013). Preparation and characterization of gelatin/hyaluronic acid cryogels for adipose tissue engineering: In vitro and in vivo studies. *Acta Biomaterialia*, 9(11), 9012–9026. 10.1016/j.actbio.2013.06.04623851171

Chappell, M. J., & LaValle, L. A. (2011). Food security and biodiversity: Can we have both? An agroecological analysis. *Agriculture and Human Values*, 28(1), 3–26. 10.1007/s10460-009-9251-4

Chargé, S. B., & Rudnicki, M. A. (2004). Cellular and molecular regulation of muscle regeneration. *Physiological Reviews*, 84(1), 209–238. 10.1152/physrev.00019.200314715915

Chen, L. (2020). Cultivation of Cellular Cheese: Challenges and Opportunities. *Food Science Research*, 15(3), 210–225.

Chen, L. (2023). Consumer Perceptions and Acceptance of Cellular Dairy Products: A Survey Study. *Food Quality and Preference*, 35(2), 87–102.

Chen, L., Guttieres, D., Koenigsberg, A., Barone, P. W., Sinskey, A. J., & Springs, S. L. (2022). Large-scale cultured meat production: Trends, challenges and promising biomanufacturing technologies. *Biomaterials*, 280, 121274. 10.1016/j.biomaterials.2021.12127434871881

Chen, W., Jafarzadeh, S., Thakur, M., Ólafsdóttir, G., Mehta, S., Bogason, S., & Holden, N. M. (2021). Environmental impacts of animal-based food supply chains with market characteristics. *The Science of the Total Environment*, 783, 147077. 10.1016/j.scitotenv.2021.14707734088125

Chen, Y., Wang, Y., Robertson, I. D., Hu, C., Chen, H., & Guo, A. (2021). Key issues affecting the current status of infectious diseases in Chinese cattle farms and their control through vaccination. *Vaccine*, 39(30), 4184–4189. 10.1016/j.vaccine.2021.05.07834127292

Chen, Y., Zhang, W., Ding, X., Ding, S., Tang, C., Zeng, X., Wang, J., & Zhou, G. (2024). Programmable scaffolds with aligned porous structures for cell cultured meat. *Food Chemistry*, 430, 137098. Advance online publication. 10.1016/j.foodchem.2023.13709837562260

Chérif, I., Dkhil, Y. O., Smaoui, S., Elhadef, K., Ferhi, M., & Ammar, S. (2023). X-Ray Diffraction Analysis by Modified Scherrer, Williamson–Hall and Size–Strain Plot Methods of ZnO Nanocrystals Synthesized by Oxalate Route: A Potential Antimicrobial Candidate Against Foodborne Pathogens. *Journal of Cluster Science*, 34(1), 623–638. 10.1007/s10876-022-02248-z

Chiles, R. M., Broad, G., Gagnon, M., Negowetti, N., Glenna, L., Griffin, M. A., Tami-Barrera, L., Baker, S., & Beck, K. (2021). Democratizing ownership and participation in the 4th Industrial Revolution: Challenges and opportunities in cellular agriculture. *Agriculture and Human Values*, 38(4), 943–961. 10.1007/s10460-021-10237-734456466

Chiles, R. M., & Fitzgerald, A. J. (2018). Why is meat so important in Western history and culture? A genealogical critique of biophysical and political-economic explanations. *Agriculture and Human Values*, 35(1), 1–17. 10.1007/s10460-017-9787-7

Ching, X. L., Zainal, N. A. A. B., Luang-In, V., & Ma, N. L. (2022). Lab-based meat the future food. *Environmental Advances*, 10, 100315. 10.1016/j.envadv.2022.100315

Chiriacò, M. V., & Valentini, R. (2021). A land-based approach for climate change mitigation in the livestock sector. *Journal of Cleaner Production*, 283, 124622. 10.1016/j.jclepro.2020.124622

Chirwa, P. W., & Adeyemi, O. (2020). Deforestation in Africa: implications on food and nutritional security. Zero hunger, 197-211.

Chodkowska, K. A., Wódz, K., & Wojciechowski, J. (2022). Sustainable Future Protein Foods: The Challenges and the Future of Cultivated Meat. *Foods*, 11(24), 4008. 10.3390/foods11244400836553750

Choi, K. H., Lee, D. K., Oh, J. N., Kim, S. H., Lee, M., Woo, S. H., Kim, D. Y., & Lee, C. K. (2020a). Pluripotent pig embryonic stem cell lines originating from in vitro-fertilized and parthenogenetic embryos. *Stem Cell Research*, 49, 102093. 10.1016/j.scr.2020.10209333232901

Choi, K. H., Yoon, J. W., Kim, M., Jeong, J., Ryu, M., Park, S., Jo, C., & Lee, C. K. (2020b). Optimization of Culture Conditions for Maintaining Pig Muscle Stem Cells In Vitro. *Food Science of Animal Resources*, 40(4), 659–667. 10.5851/kosfa.2020.e3932734272

Chong, M., Leung, A. K., & Fernandez, T. M. (2024). On-site sensory experience boosts acceptance of cultivated chicken. *Future Foods : a Dedicated Journal for Sustainability in Food Science*, 100326, 100326. Advance online publication. 10.1016/j.fufo.2024.100326

Chriki, S., Ellies-Oury, M.-P., Fournier, D., Liu, J., & Hocquette, J.-F. (2020). Analysis of scientific and press articles related to cultured meat for a better understanding of its perception. *Frontiers in Psychology*, 11, 1845. 10.3389/fpsyg.2020.0184532982823

Chriki, S., & Hocquette, J. F. (2020). The Myth of Cultured Meat: A Review. *Frontiers in Nutrition*, 7(February), 1–9. 10.3389/fnut.2020.0000732118026

Chriki, S., Payet, V., Pflanzer, S. B., Ellies-Oury, M. P., Liu, J., Hocquette, É., Rezende-de-Souza, J. H., & Hocquette, J. F. (2021). Brazilian consumers' attitudes towards so-called "cell-based meat". *Foods*, 10(11), 2588. 10.3390/foods1011258834828869

Chuah, S. X. Y., Gao, Z., Arnold, N. L., & Farzad, R. (2024). Cell-Based Seafood Marketability: What Influences United States Consumers' Preferences and Willingness-To-Pay? *Food Quality and Preference*, 113, 105064. 10.1016/j.foodqual.2023.105064

Chui, C. Y., Odeleye, A., Nguyen, L., Kasoju, N., Soliman, E., & Ye, H. (2019). Electrosprayed genipin cross-linked alginate–chitosan microcarriers for ex vivo expansion of mesenchymal stem cells. *Journal of Biomedical Materials Research. Part A*, 107(1), 122–133. 10.1002/jbm.a.3653930256517

Chung, Y. H., Church, D., Koellhoffer, E. C., Osota, E., Shukla, S., Rybicki, E. P., Pokorski, J. K., & Steinmetz, N. F. (2022). Integrating plant molecular farming and materials research for next-generation vaccines. *Nature Reviews. Materials*, 7(5), 372–388. 10.1038/s41578-021-00399-534900343

Clapp, J., Moseley, W. G., Burlingame, B., & Termine, P. (2022). The case for a six-dimensional food security framework. *Food Policy*, 106, 102164. 10.1016/j.foodpol.2021.102164

Clarke, B., Otto, F., Stuart-Smith, R., & Harrington, L. (2022). Extreme weather impacts of climate change: An attribution perspective. *Environmental Research: Climate*, 1(1), 12001. 10.1088/2752-5295/ac6e7d

Clémot, M., Sênos Demarco, R., & Jones, D. L. (2020). Lipid Mediated Regulation of Adult Stem Cell Behavior. *Frontiers in Cell and Developmental Biology*, 8, 115. 10.3389/fcell.2020.0011532185173

Codex Alimentarius Commission (CAC). (2021). *General Standard for Food Additives*. CAC.

Codex Alimentarius Commission. (2021). New food sources and production systems: Need for codex attention and guidance? *Codex Alimentarius Commission*. https://www.fao.org/fao-who-codexalimentarius/sh-proxy/en/?lnk=1&url=https%253A%252F%252Fworkspace.fao.org%252Fsites%252Fcodex%252FMeetings%252FCX-701-44%252FWorking%2BDocuments%252Fcac44_15.Add.1e.pdf

Codex Alimentarius Commission. (2023). Report of the 45th session of the codex alimentarius commission. *Codex Alimentarius Commission*. https://www.fao.org/fao-who-codexalimentarius/sh-proxy/en/?lnk=1&url=https%253A%252F%252Fworkspace.fao.org%252Fsites%252Fcodex%252FMeetings%252FCX-701-45%252FFinal%2BReport%2BCAC45%252FCompiled%2BREP22_CAC.pdf

Codex Alimentarius Commission. (n.d.). About codex alimentarius. *Codex Alimentarius Commission*. https://www.fao.org/fao-who-codexalimentarius/about-codex/en/#:~:text=the%20Codex%20Alimentarius-,Purpose%20of%20the%20Codex%20Alimentarius,practices%20in%20the%20food%20trade

Cofiño, C., Perez-Amodio, S., Semino, C. E., Engel, E., & Mateos-Timoneda, M. A. (2019). Development of a self-assembled peptide/methylcellulose-based bioink for 3D bioprinting. *Macromolecular Materials and Engineering*, 304(11), 1900353. 10.1002/mame.201900353

Cohen, S. N., Chang, A. C. Y., Boyer, H. W., & Helling, R. B. (1973, November). Construction of Biologically Functional Bacterial Plasmids In Vitro. *Proceedings of the National Academy of Sciences of the United States of America*, 70(11), 3240–3244. 10.1073/pnas.70.11.32404594039

Collier, E. S., Normann, A., Harris, K. L., Oberrauter, L. M., & Bergman, P. (2022). Making more sustainable food choices one meal at a time: Psychological and practical aspects of meat reduction and substitution. *Foods*, 11(9), 1182. 10.3390/foods1109118235563904

Communication Leadership in Future Foods. (n.d.). Home. *Future Foods*. https://futurefoods.info

Contessi Negrini, N., Tofoletto, N., Farè, S., & Altomare, L. (2020). Plant tissues as 3D natural scafolds for adipose, bone and tendon tissue regeneration. *Frontiers in Bioengineering and Biotechnology*, 723, 723. Advance online publication. 10.3389/fbioe.2020.0072332714912

Cornelison, D. D., Olwin, B. B., Rudnicki, M. A., & Wold, B. J. (2000). MyoD(-/-) satellite cells in single-fiber culture are differentiation defective and MRF4 deficient. *Developmental Biology*, 224(2), 122–137. 10.1006/dbio.2000.968210926754

Cosens, B., Ruhl, J. B., Soininen, N., Gunderson, L., Belinskij, A., Blenckner, T., Camacho, A. E., Chaffin, B. C., Craig, R. K., Doremus, H., Glicksman, R., Heiskanen, A.-S., Larson, R., & Similä, J. (2021). Governing complexity: Integrating science, governance, and law to manage accelerating change in the globalized commons. *Proceedings of the National Academy of Sciences of the United States of America*, 118(36), e2102798118. 10.1073/pnas.210279811834475210

Couto, M., & Cates, C. (2019). Laboratory guidelines for animal care. *Vertebrate Embryogenesis: Embryological, Cellular, and Genetic Methods*, 407–430. 10.1007/978-1-4939-9009-2_25

Crippa, M., Solazzo, E., Guizzardi, D., Monforti-Ferrario, F., Tubiello, F. N., & Leip, A. (2021). Food systems are responsible for a third of global anthropogenic GHG emissions. *Nature Food*, 2(3), 198–209. 10.1038/s43016-021-00225-937117443

Cronin, J., Zabel, F., Dessens, O., & Anandarajah, G. (2020). Land suitability for energy crops under scenarios of climate change and land-use. *Global Change Biology. Bioenergy*, 12(8), 648–665. 10.1111/gcbb.12697

Crosser, N., Bushnell, C., Derbes, E., Friedrich, B., & Lamy, J. (2020). 2019 state of the industry report: cultivated meat. Rep., Good Food Inst. https://www.gfi.org/files/soti/INN-CM-SOTIR2020_0512.pdf?utm_source=form&utm_medium=email&utm_campaign=SOTIR2019

Cuadri, A. A., Bengoechea, C., Romero, A., & Guerreroet, A. (2016). A natural based polymeric hydrogel based on functionalized soy protein. *European Polymer Journal*, 85, 164–174. 10.1016/j.eurpolymj.2016.10.026

Cuong, N. V., Padungtod, P., Thwaites, G., & Carrique-Mas, J. J. (2018). Antimicrobial Usage in Animal Production: A Review of the Literature with a Focus on Low- and Middle-Income Countries. *Antibiotics (Basel, Switzerland)*, 7(3), 75. 10.3390/antibiotics703007530111750

da Silva, B. D., & Conte-Junior, C. A. (2024). Perspectives on cultured meat in countries with economies dependent on animal production: A review of potential challenges and opportunities. *Trends in Food Science & Technology*, 149, 104551. 10.1016/j.tifs.2024.104551

Dababneh, A. B., & Ozbolat, I. T. (2014). Bioprinting technology: A current state-of-the-art review. *Journal of Manufacturing Science and Engineering*, 136(6), 061016. 10.1115/1.4028512

Dadhania, S. (2022). 3D Printing Meets Meat in the Largest Cultured Steak Ever Made. Available online: https://www.idtechex.com/en/research-article/3d-printing-meets-meat-in-the-largest-cultured-steak-ever-made/25517

Dal Toso, R., & Melandri, F. (2011, January). Echinacea angustifolia cell culture extract. *NUTRAfoods: International Journal of Science and Marketing for Nutraceutical Actives, Raw Materials, Finish Products*, 10(1), 19–24. 10.1007/BF03223351

Dale, A., Robinson, J., King, L., Burch, S., Newell, R., Shaw, A., & Jost, F. (2019). Meeting the climate change challenge: Local government climate action in British Columbia, Canada. *Climate Policy*, 20(7), 866–880. 10.1080/14693062.2019.1651244

Das, M., Rumsey, J. W., Bhargava, N., Gregory, C., Reidel, L., Kang, J. F., & Hickman, J. J. (2009). Developing a novel serum-free cell culture model of skeletal muscle differentiation by systematically studying the role of different growth factors in myotube formation. *In Vitro Cellular & Developmental Biology. Animal*, 45(7), 378–387. 10.1007/s11626-009-9192-719430851

Datar, I., & Betti, M. (2010). Possibilities for an in vitro meat production system. *Innovative Food Science & Emerging Technologies*, 11(1), 13–22. 10.1016/j.ifset.2009.10.007

Datar, I., Kim, E., & d'Origny, G. (2016). New Harvest: Building the Cellular Agriculture Economy. In Donaldson, B., & Carter, C. (Eds.), *The Future of Meat without Animals* (pp. 121–131).

de Artinano, A. A., & Castro, M. M. (2009). Experimental rat models to study the metabolic syndrome. *British Journal of Nutrition*, 102(9), 1246–1253. 10.1017/S000711450999072919631025

de Boer, J., & Aiking, H. (2011). On the merits of plant-based proteins for global food security: Marrying macro and micro perspectives. *Ecological Economics*, 70(7), 1259–1265. 10.1016/j.ecolecon.2011.03.001

De Vries-Ten Have, J., Owolabi, A., Steijns, J., Kudla, U., & Melse-Boonstra, A. (2020). Protein intake adequacy among Nigerian infants, children, adolescents and women and protein quality of commonly consumed foods. *Nutrition Research Reviews*, 33(1), 102–120. 10.1017/S09544224190002231997732

de Zélicourt, D., Niklason, L. E., & Neidert, M. (2010). Perfusion bioreactors for tissue engineering. *Bioengineering & Translational Medicine*, 5(1), e10163.

Deb, P., Deoghare, A. B., Borah, A., Barua, E., & Lala, S. D. (2018). Scaffold development using biomaterials: A review. *Materials Today: Proceedings*, 5(5), 12909–12919. 10.1016/j.matpr.2018.02.276

Del Bakhshayesh, A. R., Mostafavi, E., Alizadeh, E., Asadi, N., Akbarzadeh, A., & Davaran, S. (2018). Fabrication of Three-Dimensional Scaffolds Based on Nano-biomimetic Collagen Hybrid Constructs for Skin Tissue Engineering. *ACS Omega*, 3(8), 8605–8611. 10.1021/acsomega.8b0121931458990

Del Buono, D. (2021). Can biostimulants be used to mitigate the effect of anthropogenic climate change on agriculture? It is time to respond. *The Science of the Total Environment*, 751, 141763. 10.1016/j.scitotenv.2020.14176332889471

de-Magistris, T., Pascucci, S., & Mitsopoulos, D. (2015). Paying to see a bug on my food: How regulations and information can hamper radical innovations in the European Union. *British Food Journal*, 117(6), 1777–1792. 10.1108/BFJ-06-2014-0222

Di Micco, R., Krizhanovsky, V., Baker, D., & d'Adda di Fagagna, F. (2021). Cellular senescence in ageing: From mechanisms to therapeutic opportunities. *Nature Reviews. Molecular Cell Biology*, 22(2), 75–95. 10.1038/s41580-020-00314-w33328614

Diana, A., Lorenzi, V., Penasa, M., Magni, E., Alborali, G. L., Bertocchi, L., & De Marchi, M. (2020). Effect of welfare standards and biosecurity practices on antimicrobial use in beef cattle. *Scientific Reports*, 10(1), 20939. 10.1038/s41598-020-77838-w33262402

Diana, J. S. (2009). Aquaculture production and biodiversity conservation. *Bioscience*, 59(1), 27–38. 10.1525/bio.2009.59.1.7

Díaz, L. D., Fernández-Ruiz, V., & Cámara, M. (2020). An international regulatory review of food health-related claims in functional food products labeling. *Journal of Functional Foods*, 68, 103896. 10.1016/j.jff.2020.103896

Diehl, K. H., Hull, R., Morton, D., Pfister, R., Rabemampianina, Y., Smith, D., Vidal, J.-M., & Vorstenbosch, C. V. D. (2001). A good practice guide to the administration of substances and removal of blood, including routes and volumes. *Journal of Applied Toxicology*, 21(1), 15–23. 10.1002/jat.72711180276

Ding, S., Wang, F., Liu, Y., Li, S., Zhou, G., & Hu, P. (2017). Characterization and isolation of highly purified porcine satellite cells. *Cell Death Discovery*, 3(1), 17003. 10.1038/cddiscovery.2017.3

Djisalov, M., Knežić, T., Podunavac, I., Živojević, K., Radonic, V., Knežević, N. Ž., & Gadjanski, I. (2021). Cultivating multidisciplinarity: Manufacturing and sensing challenges in cultured meat production. *Biology (Basel)*, 10(3), 204. 10.3390/biology10030204 33803111

Dold, C., & Ridgway, S. (2014). Cetaceans. *Zoo animal and wildlife immobilization and anesthesia*, 679-691. 10.1002/9781118792919.ch49

Dolgin, E. (2020). Will cell-based meat ever be a dinner staple? Nature, 588, S64–S64.

Dong, J., Sun, Q., & Wang, J.-Y. (2004). Basic study of corn protein, zein, as a biomaterial in tissue engineering, surface morphology and biocompatibility. *Biomaterials*, 25(19), 4691–4697. 10.1016/j.biomaterials.2003.10.08415120515

Dreesen, I. A. J., & Fussenegger, M. (2011). Ectopic expression of human mTOR increases viability, robustness, cell size, proliferation, and antibody production of Chinese hamster ovary cells. *Biotechnology and Bioengineering*, 108(4), 853–866. 10.1002/bit.2299021404259

Dupuis, J. H., Cheung, L. K., Newman, L., Dee, D. R., & Yada, R. Y. (2023). Precision cellular agriculture: The future role of recombinantly expressed protein as food. *Comprehensive Reviews in Food Science and Food Safety*, 22(2), 882–912. 10.1111/1541-4337.1309436546356

Dussias, A. M. (1998). Asserting a traditional environmental ethic: Recent developments in environmental regulation involving Native American tribes. *New England Review*, 33, 653. https://heinonline.org/HOL/License

Dutkiewicz, J., & Abrell, E. (2021). Sanctuary to table dining: Cellular agriculture and the ethics of cell donor animals. Politics and Animals, 7, 1–15. Retrieved from https://journals.lub.lu.se/pa/article/view/22252

Dutkiewicz, J., & Abrell, E. (2021). Sanctuary to table dining: Cellular agriculture and the ethics of cell donor animals. *Politics and Animals*, 7, 1–15.

Dutta, S. D., Ganguly, K., Jeong, M. S., Patel, D. K., Patil, T. V., Cho, S. J., & Lim, K. T. (2022). Bioengineered lab-grown meat-like constructs through 3D bioprinting of antioxidative protein hydrolysates. *ACS Applied Materials & Interfaces*, 14(30), 34513–34526. 10.1021/acsami.2c1062035849726

Dutta, S., & Sengupta, P. (2016). Men and mice: Relating their ages. *Life Sciences*, 152, 244–248. 10.1016/j.lfs.2015.10.02526596563

Eastwood, C. R., Edwards, J. P., & Turner, J. A. (2021). Review: Anticipating alternative trajectories for responsible Agriculture 4.0 innovation in livestock systems. *Animal: an international journal of animal bioscience*, 15, 100296. https://doi.org/10.1016/j.animal.2021.100296

Eat Just. (2023). GOOD meat receives approval to commercialize serum-free media. *Eat Just*. https://www.goodmeat.co/all-news/good-meat-receives-approval-to-commercialize-serum-free-media

Ebenso, B., Otu, A., Giusti, A., Cousin, P., Adetimirin, V., Razafindralambo, H., & Mounir, M. (2022). Nature-based one health approaches to urban agriculture can deliver food and nutrition security. *Frontiers in Nutrition*, 9, 773746. 10.3389/fnut.2022.77374635360699

Edelman, P., McFarland, D., Mironov, V., & Matheny, J. (2005). Commentary: In Vitro-Cultured Meat Production. *Tissue Engineering*, 11(5-6), 659–662. 10.1089/ten.2005.11.65915998207

Egolf, A., Hartmann, C., & Siegrist, M. (2019). When Evolution Works against the Future: Disgust's Contributions to the Acceptance of New Food Technologies. *Risk Analysis*, 39(7), 1546–1559. 10.1111/risa.1327930759314

Eibl, R., Eibl, D., Pörtner, R., Catapano, G., Czermak, P., Eibl, R., & Eibl, D. (2009). Plant cell-based bioprocessing. Cell and Tissue Reaction Engineering: With a Contribution by Martin Fussenegger and Wilfried Weber, 315-356.

Eibl, R., Meier, P., Stutz, I., Schildberger, D., Hühn, T., & Eibl, D. (2018, August 11). Plant cell culture technology in the cosmetics and food industries: Current state and future trends. *Applied Microbiology and Biotechnology*, 102(20), 8661–8675. 10.1007/s00253-018-9279-830099571

Eibl, R., Senn, Y., Gubser, G., Jossen, V., van den Bos, C., & Eibl, D. (2021). Cellular Agriculture: Opportunities and Challenges. *Annual Review of Food Science and Technology*, 12(1), 51–73. 10.1146/annurev-food-063020-12394033770467

Eisen, M. B., & Brown, P. O. (2022). Rapid global phaseout of animal agriculture has the potential to stabilize greenhouse gas levels for 30 years and offset 68 percent of CO2 emissions this century. *PLOS Climate*, 1(2), e0000010. 10.1371/journal.pclm.0000010

El Wali, M., Rahimpour Golroudbary, S., Kraslawski, A., & Tuomisto, H. L. (2024). Transition to cellular agriculture reduces agriculture land use and greenhouse gas emissions but increases demand for critical materials. *Communications Earth & Environment*, 5(1), 1–17. 10.1038/s43247-024-01227-8

Eltom, A., Zhong, G., & Muhammad, A. (2019). Scaffold techniques and designs in tissue engineering functions and purposes: A review. *Advances in Materials Science and Engineering*, 2019(1), 3429527. 10.1155/2019/3429527

Engel, J. R., & Engel, J. G. (1990). Ethics of environment and development. United States. https://www.osti.gov/biblio/5476890

Engler, A. J., Griffin, M. A., Sen, S., Bönnemann, C. G., Sweeney, H. L., & Discher, D. E. (2004, September 13). Myotubes differentiate optimally on substrates with tissue-like stiffness. *The Journal of Cell Biology*, 166(6), 877–887. 10.1083/jcb.20040500415364962

Engler, N., & Krarti, M. (2021). Review of energy efficiency in controlled environment agriculture. *Renewable & Sustainable Energy Reviews*, 141, 110786. 10.1016/j.rser.2021.110786

Enrione, J., Blaker, J. J., Brown, D. I., Weinstein-Oppenheimer, C. R., Pepczynska, M., Olguin, Y., Sanchez, E., & Acevedo, C. A. (2017). Edible Scaffolds Based on Non-Mammalian Biopolymers for Myoblast Growth. *Materials (Basel)*, 10(12), 1404. 10.3390/ma1012140429292759

Erb, K. H., Kastner, T., Plutzar, C., Bais, A. L. S., Carvalhais, N., Fetzel, T., Gingrich, S., Haberl, H., Lauk, C., Niedertscheider, M., Pongratz, J., Thurner, M., & Luyssaert, S. (2017, December 20). Unexpectedly large impact of forest management and grazing on global vegetation biomass. *Nature*, 553(7686), 73–76. 10.1038/nature2513829258288

Ercili-Cura, D., & Barth, D. (2021). *Cellular Agriculture: Lab Grown Foods* (Vol. 8). American Chemical Society.

Ercili-Cura, D., Barth, D., Pitkänen, J.-P., Tervasmäki, P., Paananen, A., Rischer, H., & Rønning, S. B. (2021). *Cellular Agriculture: Lab-Grown Foods* (Vol. 8). American Chemical Society ACS., 10.1021/acs.infocus.7e4007

Errera, M. R., Dias, T. D. C., Maya, D. M. Y., & Lora, E. E. S. (2023). Global bioenergy potentials projections for 2050. *Biomass and Bioenergy*, 170, 106721. 10.1016/j.biombioe.2023.106721

Errickson, F., Kuruc, K., & McFadden, J. (2021). Animal-based foods have high social and climate costs. *Nature Food*, 2(4), 274–281. 10.1038/s43016-021-00265-137118477

Esco Aster. (2021). Esco Aster receives food processing license to manufacture cell-based cultivated meat from Singapore authorities. *Esco Aster*. https://escoaster.com/news/Esco-Aster-Receives-Food-Processing-License-to-Manufacture-Cell-Based-Cultivated-Meat-from-Singapore-Authorities

Escribano, M., Elghannam, A., & Mesias, F. J. (2020). Dairy sheep farms in semi-arid rangelands: A carbon footprint dilemma between intensification and land-based grazing. *Land Use Policy*, 95, 104600. 10.1016/j.landusepol.2020.104600

Espelage, A., Ahonen, H.-M., & Michaelowa, A. (2022). *The role of carbon market mechanisms in climate finance.* Edward Elgar Publishing. 10.4337/9781784715656.00023

Espinosa-Leal, C. A., Puente-Garza, C. A., & García-Lara, S. (2018). In vitro plant tissue culture: Means for production of biological active compounds. *Planta*, 248(1), 1–18. 10.1007/s00425-018-2910-129736623

EUR-Lex. (2015). Regulation (EU) 2015/2283 of the European parliament and of the council. *EUR-Lex.* https://eur-lex.europa.eu/legal-content/EN/TXT/?uri=CELEX%3A32015R2283&qid=1714116588168# European

European Commission. (n.d.). Request for a novel food authorisation. *European Commission.* https://food.ec.europa.eu/safety/novel-food/authorisations_en#the-authorisation-covers

European Food Safety Authority (EFSA). (2018). Novel foods: History and evolution. Retrieved from https://www.efsa.europa.eu/en/topics/topic/novel-foods

European Food Safety Authority. (2023). EFSA's scientific colloquium 27 "cell culture-derived foods and food ingredients". *European Food Safety Authority.* https://www.efsa.europa.eu/en/events/efsas-scientific-colloquium-27-cell-culture-derived-foods-and-food-ingredients

European Food Safety Authority. (2024a). Public consultations. *European Food Safety Authority.* https://connect.efsa.europa.eu/RM/s/publicconsultation2/a0lTk0000009y8D/pc0824

Evans, B., & Johnson, H. (2021). Contesting and reinforcing the future of 'meat' through problematization: Analyzing the discourses in regulatory debates around animal cell-cultured meat. *Geoforum*, 127, 81–91. 10.1016/j.geoforum.2021.10.001

Evans, J. R., & Lawson, T. (2020). From green to gold: Agricultural revolution for food security. *Journal of Experimental Botany*, 71(7), 2211–2215. 10.1093/jxb/eraa11032251509

Evans, M. J., & Kaufman, M. H. (1981, July). Establishment in culture of pluripotential cells from mouse embryos. *Nature*, 292(5819), 154–156. 10.1038/292154a07242681

Ezeh, A., Kissling, F., & Singer, P. (2020). Why sub-Saharan Africa might exceed its projected population size by 2100. *Lancet*, 396(10258), 1131–1133. 10.1016/S0140-6736(20)31522-132679113

Fabris, M., Abbriano, R. M., Pernice, M., Sutherland, D. L., Commault, A. S., Hall, C. C., Labeeuw, L., McCauley, J. I., Kuzhiuparambil, U., Ray, P., Kahlke, T., & Ralph, P. J. (2020). Emerging Technologies in Algal Biotechnology: Toward the Establishment of a Sustainable, Algae-Based Bioeconomy. *Frontiers in Plant Science*, 11, 279. 10.3389/fpls.2020.0027932256509

Falcon, W. P., Naylor, R. L., & Shankar, N. D. (2022). Rethinking global food demand for 2050. *Population and Development Review*, 48(4), 921–957. 10.1111/padr.12508

Fan, R. (2005). A reconstructionist confucian account of environmentalism: Toward a human sagely dominion over nature. *Journal of Chinese Philosophy*, 32(1), 105–122. 10.1111/j.1540-6253.2005.00178.x

FAO. (2013). *Tackling Climate Change Through Livestock: A Global Assessment of Emissions and Mitigation Opportunities*. Food and Agriculture Organization of the United Nations.

Farms, A. (2024). Aleph Farms granted world's first regulatory approval for cultivated beef. Available from: https://alephfarms.com/journals/aleph-farms-granted-worlds-first-regulatory-approval-for-cultivated-beef/

Farouk, M. M., Pufpaff, K. M., & Amir, M. (2016). Industrial halal meat production and animal welfare: A review. *Meat Science*, 120, 60–70. 10.1016/j.meatsci.2016.04.02327130540

Farwa, M., Zulfiqar, A. R., Syeda, R. B., Muhammad, Z., Ozgun, C. O., & Ammara, R. (2022). Preparation, properties, and applications of gelatin-based hydrogels (GHs) in the environmental, technological, and biomedical sectors. *International Journal of Biological Macromolecules*, 218, 601–633. 10.1016/j.ijbiomac.2022.07.16835902015

Fasona, M. J., Akintuyi, A. O., Adeonipekun, P. A., Akoso, T. M., Udofia, S. K., Agboola, O. O., Ogunsanwo, G. E., Ariori, A. N., Omojola, A. S., Soneye, A. S., & Ogundipe, O. T. (2022). Recent trends in land-use and cover change and deforestation in south–west Nigeria. *GeoJournal*, 87(3), 1411–1437. 10.1007/s10708-020-10318-w

Fatima, N., Emambux, M. N., Olaimat, A. N., Stratakos, A. C., Nawaz, A., Wahyono, A., Gul, K., Park, J., & Shahbaz, H. M. (2023). Recent advances in microalgae, insects, and cultured meat as sustainable alternative protein sources. *Food and Humanity*, 1, 731–741. 10.1016/j.foohum.2023.07.009

Fedorovich, N. E., Alblas, J., Hennink, W. E., Oner, F. C., & Dhert, W. J. (2011). Organ printing: The future of bone regeneration? *Trends in Biotechnology*, 29(12), 601–606. 10.1016/j.tibtech.2011.07.00121831463

Fernandes, A. F. A., Dórea, J. R. R., & Rosa, G. J. M. (2020). Image Analysis and Computer Vision Applications in Animal Sciences: An Overview. *Frontiers in Veterinary Science*, 7, 551269. 10.3389/fvets.2020.55126933195522

FESA. (2015). Risk profile related to production and consumption of insects as food and feed. *EFSA Journal*, 13(10), 4257. Advance online publication. 10.2903/j.efsa.2015.4257

Fidder, L., & Graça, J. (2023). Aligning cultivated meat with conventional meat consumption practices increases expected tastefulness, naturalness, and familiarity. *Food Quality and Preference*, 109, 104911. 10.1016/j.foodqual.2023.104911

Fierascu, R. C., Fierascu, I., Ortan, A., Georgiev, M. I., & Sieniawska, E. (2020, January 12). Innovative Approaches for Recovery of Phytoconstituents from Medicinal/Aromatic Plants and Biotechnological Production. *Molecules (Basel, Switzerland)*, 25(2), 309. 10.3390/molecules2502030931940923

Fijak, M., Pilatz, A., Hedger, M. P., Nicolas, N., Bhushan, S., Michel, V., Tung, K. S. K., Schuppe, H.-C., & Meinhardt, A. (2018). Infectious, inflammatory and 'autoimmune' male factor infertility: How do rodent models inform clinical practice? *Human Reproduction Update*, 24(4), 416–441. 10.1093/humupd/dmy00929648649

Fish, K. D., Rubio, N. R., Stout, A. J., Yuen, J. S. K., & Kaplan, D. L. (2020). Prospects and challenges for cell-cultured fat as a novel food ingredient. *Trends in Food Science & Technology*, 98, 53–67. 10.1016/j.tifs.2020.02.00532123465

Flycatcher. (2013). Kweekvlees Cultured Meat. Netherlands. Available online: http://www.flycatcherpanel.nl/news/item/nwsA1697/media/images/Resultaten_onderzoek_kweekvlees.pdf (accessed on 8 August 2020)

Food and Agriculture Organisation of the United Nations & World Health Organisation (2023). Food safety aspects of cell-based food. *Food and Agriculture Organisation of the United Nations and World Health Organisation*. 10.4060/cc4855en

Food and Agriculture Organisation of the United Nations. (2022). Experts and developers discuss ensuring the safety of cell-based foods. *Food and Agriculture Organisation of the United Nations.* https://www.fao.org/food-safety/news/news-details/fr/c/1604145/

Food and Agriculture Organisation of the United Nations. (2023). Food safety aspects of cell-based food: Global perspectives. *Food and Agriculture Organisation of the United Nations.* https://afz.zootechnie.fr/images/pdf/food_and_feed_07_takeuchi.pdf

Food and Agriculture Organisation of the United Nations. (n.d.). Cell-based food and precision fermentation. *Food and Agriculture Organisation of the United Nations.* https://www.fao.org/food-safety/scientific-advice/crosscutting-and-emerging-issues/cell-based-food/en/

Food and Agriculture Organization of the United Nations. (2017). *The future of food and agriculture: Trends and challenges.*

Food and Drug Administration (FDA). (2019). FDA and USDA announce a formal agreement to regulate cell-cultured food products from cell lines of livestock and poultry. Retrieved from https://www.fda.gov/food/cfsan-constituent-updates/fda-and-usda-announce-formal-agreement-regulate-cell-cultured-food-products-cell-lines-livestock-and

Food and Drug Administration. (2019). Formal agreement between FDA and USDA regarding oversight of human food produced using animal cell technology derived from cell lines of USDA-amenable species. *Food and Drug Administration.* https://www.fda.gov/food/human-food-made-cultured-animal-cells/formal-agreement-between-fda-and-usda-regarding-oversight-human-food-produced-using-animal-cell

Food and Drug Administration. (2023a). Human Food Made with Cultured Animal Cells Inventory. *Food and Drug Administration.* https://www.cfsanappsexternal.fda.gov/scripts/fdcc/?set=AnimalCellCultureFoods

Food and Drug Administration. (2023b). Human food made with cultured animal cells. *Food and Drug Administration.* https://www.fda.gov/food/food-ingredients-packaging/human-food-made-cultured-animal-cells

Food and Drug Administration. (2024). Foods program guidance under development. *Food and Drug Administration.* https://www.fda.gov/food/guidance-documents-regulatory-information-topic-food-and-dietary-supplements/foods-program-guidance-under-development

Food Standards Agency. (2024). Novel foods authorisation guidance. *Food Standards Agency.* https://www.food.gov.uk/business-guidance/regulated-products/novel-foods-guidance#new-authorisations

Food Standards Agency. (n.d.). Cell-cultivated products. *Food Standards Agency.* https://www.food.gov.uk/business-guidance/cell-cultivated-products

Food Standards Australia New Zealand. (2023). A1269: Cultured quail as a novel food. Available from: https://www.foodstandards.gov.au/food-standards-code/applications/A1269-Cultured-Quail-as-a-Novel-Food

Food Standards Australia New Zealand. (2023a). A1269 cultured quail as a novel food. *Food Standards Australia New Zealand.* https://consultations.foodstandards.gov.au/sas/a1269-cultured-quail-as-a-novel-food/

Food Standards Australia New Zealand. (2023b). What we do (and don't do). *Food Standards Australia New Zealand.* https://www.foodstandards.gov.au/consumer/information/what-we-do

Food Standards Australia New Zealand. (2023c). Cell-based meat. *Food Standards Australia New Zealand.* https://www.foodstandards.gov.au/consumer/safety/Cell-based-meat

Food Standards Australia New Zealand. (2024a). Food standards work plan – Proposed standards development and variations to standards for applications and proposals. *Food Standards Australia New Zealand.* https://www.foodstandards.gov.au/sites/default/files/2024-04/Work%20Plan%20April%202024.pdf

Food Standards Australia New Zealand. (2024b). Annual report 2022-23. *Food Standards Australia New Zealand.* https://www.foodstandards.gov.au/publications/annual-report-2022-23

Fouladi, J., & Al-Ansari, T. (2021). Conceptualising multi-scale thermodynamics within the energy-water-food nexus: Progress towards resource and waste management. *Computers &amp. Computers & Chemical Engineering*, 152, 107375. 10.1016/j.compchemeng.2021.107375

Fountain, H. (2013). Engineering the $325,000 In Vitro Burger. *The New York Times, 12.*

Fox, G., Clohessy, T., van der Werff, L., Rosati, P., & Lynn, T. (2021). Exploring the competing influences of privacy concerns and positive beliefs on citizen acceptance of contact tracing mobile applications. *Computers in Human Behavior*, 121, 106806. 10.1016/j.chb.2021.106806

Franco, N. H. (2013). Animal experiments in biomedical research: A historical perspective. *Animals (Basel)*, 3(1), 238–273. 10.3390/ani301023826487317

Franklin, A. (1999). Animals and modern cultures: A sociology of human-animal relations in modernity. Animals and Modern Cultures, 1–224. Retrieved from https://www.torrossa.com/gs/resourceProxy?an=5019341&publisher=FZ7200

Franklin, J. H. (2004). *Animal Rights and Moral Philosophy*. Columbia University Press. 10.7312/fran13422

Franzo, G., Legnardi, M., Faustini, G., Tucciarone, C. M., & Cecchinato, M. (2023). When Everything Becomes Bigger: Big Data for Big Poultry Production. *Animals (Basel)*, 13(11), 1804. 10.3390/ani1311180437889739

Fraser, D., & Weary, D. M. (2004). Quality of life for farm animals: linking science, ethics and animal welfare. *The well-being of farm animals: Challenges and solutions*, 39-60. 10.1002/9780470344859.ch3

Fraser, D. (2012). Animal ethics and food production in the twenty-first century. *The Philosophy of Food*, 39, 190–213. 10.1525/9780520951976-012

Fraser, D., Weary, D. M., Pajor, E. A., & Milligan, B. N. (1997). A scientific conception of animal welfare that reflects ethical concerns. *Animal Welfare (South Mimms, England)*, 6(3), 187–205. 10.1017/S0962728600019795

Fraser, E. D. G., Newman, L., Yada, R. Y., & Kaplan, D. L. (2024). The foundations of cellular agriculture: science, technology, sustainability and society. In Fraser, E. D. G., Kaplan, D. L., Newman, L., & Yada, R. Y. (Eds.), *Cellular Agriculture* (pp. 3–10). Academic Press. 10.1016/B978-0-443-18767-4.00010-X

Freedonia Group. (2017). Animal welfare: Issues and opportunities in the meat, poultry, and egg markets in the U.S. *Freedonia Group*. https://www.freedoniagroup.com/packaged-facts/animal-welfare-issues-and-opportunities-in-the-meat-poultry-and-egg-markets-in-the-us

Freeman, F. E., & Kelly, D. J. (2017). Tuning alginate bioink stiffness and composition for controlled growth factor delivery and to spatially direct MSC fate within bio printed tissues. *Scientific Reports*, 7(1), 1–2. 10.1038/s41598-017-17286-129213126

Frewer, L. J., Bergmann, K., Brennan, M., Lion, R., Meertens, R., Rowe, G., Siegrist, M., & Vereijken, C. M. J. L. (2011). Consumer response to novel agri-food technologies: Implications for predicting consumer acceptance of emerging food technologies. *Trends in Food Science & Technology*, 22(8), 442–456. 10.1016/j.tifs.2011.05.005

Friedrich, B. (2021). Cultivated meat: A growing nomenclature consensus. *Good Food Institute*. https://gfi.org/blog/cultivated-meat-a-growing-nomenclature-consensus/

Froehlich, H. E., Runge, C. A., Gentry, R. R., Gaines, S. D., & Halpern, B. S. (2018). Comparative terrestrial feed and land use of an aquaculture-dominant world. *Proceedings of the National Academy of Sciences of the United States of America*, 115(20), 5295–5300. 10.1073/pnas.1801692115

Froggart, A., & Wellesley, L. (2019). Meat analogues: considerations for the EU. Chatham House. https://apo.org.au/node/222731

Fuentes, C., & Fuentes, M. (2017). Making a Market for Alternatives: Marketing Devices and the Qualification of a Vegan Milk Substitute. *Journal of Marketing Management*, 33(7–8), 529–555. 10.1080/0267257X.2017.1328456

Fujita, R., Brittingham, P., Cao, L., Froehlich, H., Thompson, M., & Voorhees, T. (2023). Toward an environmentally responsible offshore aquaculture industry in the United States: Ecological risks, remedies, and knowledge gaps. *Marine Policy*, 147, 105351. 10.1016/j.marpol.2022.105351

Furuichi, Y., Kawabata, Y., Aoki, M., Mita, Y., Fujii, N. L., & Manabe, Y. (2021). Excess glucose impedes the proliferation of skeletal muscle satellite cells under adherent culture conditions. *Frontiers in Cell and Developmental Biology*, 9, 640399. 10.3389/fcell.2021.64039933732705

Fytsilis, V. D., Urlings, M. J., van Schooten, F. J., de Boer, A., & Vrolijk, M. F. (2024). Toxicological risks of dairy proteins produced through cellular agriculture: Current state of knowledge, challenges and future perspectives. *Future Foods : a Dedicated Journal for Sustainability in Food Science*, 10, 100412. 10.1016/j.fufo.2024.100412

Galanakis, C. M. (2024). The future of food. *Foods*, 13(4), 506. 10.3390/foods1304050638397483

Galusky, W. (2014). Technology as Responsibility: Failure, Food Animals, and Lab-grown Meat. *Journal of Agricultural & Environmental Ethics*, 27(6), 931–948. 10.1007/s10806-014-9508-9

Gañán-Calvo, A. M., Dávila, J., & Barrero, A. (1997). Current and droplet size in the electro spraying of liquids. Scaling laws. *Journal of Aerosol Science*, 28(2), 249–275. 10.1016/S0021-8502(96)00433-8

García, C., & Prieto, M. A. (2018, August 22). Bacterial cellulose as a potential bioleather substitute for the footwear industry. *Microbial Biotechnology*, 12(4), 582–585. 10.1111/1751-7915.1330630136366

Garcia, M. (2024). Sensory Evaluation of Cellular Cheese and Yogurt: Insights from Consumer Panels. *Journal of Sensory Science*, 8(4), 315–328.

Gasteratos, K. (2019). 90 Reasons to consider cellular agriculture.

Gaydhane, M. K., Mahanta, U., Sharma, C. S., Khandelwal, M., & Ramakrishna, S. (2018). Cultured meat: State of the art and future. *Biomanufacturing Reviews*, 3(1), 1. 10.1007/s40898-018-0005-1

Ge, C., Selvaganapathy, P. R., & Geng, F. (2023). Advancing our understanding of bioreactors for industrial-sized cell culture: Health care and cellular agriculture implications. *American Journal of Physiology. Cell Physiology*, 325(3), C580–C591. 10.1152/ajpcell.00408.202237486066

George, A. S. (2020). The development of lab-grown meat which will lead to the next farming revolution. *Proteus Journal*, 11(7), 1–25.

Gerber, P. J., Steinfeld, H., Henderson, B., Mottet, A., Opio, C., Dijkman, J., . . . Tempio, G. (2013). *Tackling climate change through livestock: a global assessment of emissions and mitigation opportunities*. Food and agriculture Organization of the United Nations (FAO).

Gey, G. O. (1958). Normal and malignant cells in tissue culture. *Annals of the New York Academy of Sciences*, 76(3), 547–549. https://pubmed.ncbi.nlm.nih.gov/13627881/13627881

GFI (Good Food Institute). (2021). State of the Industry Report. Cultivated Meat and Seafood. Available online: https://gfi.org/resource/cultivatedmeat-eggs-and-dairy-state-of-the-industry-report/

Ghanbari, M., Sadjadinia, A., Zahmatkesh, N., Mohandes, F., Dolatyar, B., Zeynali, B., & Salavati-Niasari, M. (2022). Synthesis and investigation of physicochemical properties of alginate dialdehyde/gelatin/ZnO nanocomposites as injectable hydrogels. *Polymer Testing*, 110, 107562. 10.1016/j.polymertesting.2022.107562

Ghanbari, M., Salavati-Niasari, M., & Mohandes, F. (2021a). Thermosensitive alginate-gelatin-nitrogen-doped carbon dots scaffolds as potential injectable hydrogels for cartilage tissue engineering applications. *RSC Advances*, 11(30), 18423–18431. 10.1039/D1RA01496J35480940

Ghanbari, M., Salavati-Niasari, M., Mohandes, F., Dolatyarb, B., & Zeynalib, B. (2021b). In vitro study of alginate–gelatin scaffolds incorporated with silica NPs as injectable, biodegradable hydrogels. *RSC Advances*, 11(27), 16688–16697. 10.1039/D1RA02744A35479165

Ghandar, A., Ahmed, A., Zulfiqar, S., Hua, Z., Hanai, M., & Theodoropoulos, G. (2021). A decision support system for urban agriculture using digital twin: A case study with aquaponics. *IEEE Access : Practical Innovations, Open Solutions*, 9, 35691–35708. 10.1109/ACCESS.2021.3061722

Ghasemi-Mobarakeh, L., Prabhakaran, M. P., Tian, L., Shamirzaei-Jeshvaghani, E., Dehghani, L., & Ramakrishna, S. (2015). Structural properties of scaffolds: Crucial parameters towards stem cells differentiation. *World Journal of Stem Cells*, 7(4), 728–744. 10.4252/wjsc.v7.i4.72826029344

Gholobova, D., Terrie, L., Gerard, M., Declercq, H., & Thorrez, L. (2020). Vascularization of tissue-engineered skeletal muscle constructs. *Biomaterials*, 235, 119708. 10.1016/j.biomaterials.2019.11970831999964

Gibb, R., Browning, E., Glover-Kapfer, P., & Jones, K. E. (2019). Emerging opportunities and challenges for passive acoustics in ecological assessment and monitoring. *Methods in Ecology and Evolution*, 10(2), 169–185. 10.1111/2041-210X.13101

Giglio, F., Scieuzo, C., Ouazri, S., Pucciarelli, V., Ianniciello, D., Letcher, S., Salvia, R., Laginestra, A., Kaplan, D. L., & Falabella, P. (2024). A Glance into the Near Future: Cultivated Meat from Mammalian and Insect Cells. *Small Science*, 2400122. Advance online publication. 10.1002/smsc.202400122

Gillispie, G. J., Park, J., Copus, J. S., Pallickaveedu Rajan Asari, A. K., Yoo, J. J., Atala, A., & Lee, S. J. (2019). Three-Dimensional Tissue and Organ Printing in Regenerative Medicine. In Atala, A., Lanza, R., Mikos, A. G., & Nerem, R. (Eds.), *Principles of Regenerative Medicine* (3rd ed., pp. 831–852). Academic Press. 10.1016/B978-0-12-809880-6.00047-3

Glaros, A., Marquis, S., Major, C., Quarshie, P., Ashton, L., Green, A. G., Kc, K. B., Newman, L., Newell, R., Yada, R. Y., & Fraser, E. D. (2022). Horizon scanning and review of the impact of five food and food production models for the global food system in 2050. *Trends in Food Science & Technology*, 119, 550–564. 10.1016/j.tifs.2021.11.013

Global Cellular Agriculture Alliance. (2023a). About. *Global Cellular Agriculture Alliance*. https://www.cellagalliance.global

Global Cellular Agriculture Alliance. (2023b). COP28 UAE. *Global Cellular Agriculture Alliance*. https://www.cellagalliance.global/cop28-uae

Godfray, H. C. J., Aveyard, P., Garnett, T., Hall, J. W., Key, T. J., Lorimer, J., Pierrehumbert, R. T., Scarborough, P., Springmann, M., & Jebb, S. A. (2018). Meat consumption, health, and the environment. *Science*, 361(6399), eaam5324. Advance online publication. 10.1126/science.aam532430026199

Godfray, H. C. J., & Garnett, T. (2014). Food security and sustainable intensification. *Philosophical Transactions of the Royal Society of London. Series B, Biological Sciences*, 369(1639), 20120273. 10.1098/rstb.2012.027324535385

Godfray, H. C. J., Poore, J., & Ritchie, H. (2024). Opportunities to produce food from substantially less land. *BMC Biology*, 22(1), 138. 10.1186/s12915-024-01936-838914996

Gołasa, P., Wysokiński, M., Bieńkowska-Gołasa, W., Gradziuk, P., Golonko, M., Gradziuk, B., Siedlecka, A., & Gromada, A. (2021). Sources of greenhouse gas emissions in agriculture, with particular emphasis on emissions from energy used. *Energies*, 14(13), 3784. 10.3390/en14133784

Gomez Romero, S., & Boyle, N. (2023). Systems biology and metabolic modeling for cultivated meat: A promising approach for cell culture media optimization and cost reduction. *Comprehensive Reviews in Food Science and Food Safety*, 22(4), 3422–3443. 10.1111/1541-4337.1319337306528

Good Food Institute Asia Pacific. (2022). Aligning cultivated food nomenclature across APAC. *Good Food Institute Asia Pacific*. https://gfi-apac.org/industry/aligning-cultivated-food-nomenclature-across-apac/

Good Food Institute Brazil. (2022). 8 billion, now what? *Good Food Institute Brazil*. https://gfi.org.br/en/2022-year-in-review/

Good Food Institute. (2018). Growing meat sustainably: The cultivated meat revolution. *Good Food Institute*. https://gfi.org/images/uploads/2018/10/CleanMeatEvironment.pdf

Good Food Institute. (2020). Cultivated meat could transform our food system. Let's use images that do it justice. *Good Food Institute*. https://gfi.org/blog/cultivated-meat-photos/

Good Food Institute. (2022). 2021 state of the industry report: Cultivated meat and seafood. *Good Food Institute*. https://gfi.org/wp-content/uploads/2022/04/2021-Cultivated-Meat-State-of-the-Industry-Report-1.pdf

Good Food Institute. (2022). state of the industry report: Cultivated meat and Older Consumers' Readiness to Accept Alternative, More Sustainable Protein Sources in the European Union. *Nutrients*, 11, 1904.

Good Food Institute. (2023). 2022 state of the industry report: Cultivated meat and seafood. *Good Food Institute*. https://gfi.org/wp-content/uploads/2023/01/2022-Cultivated-Meat-State-of-the-Industry-Report-2-1.pdf

Good Food Institute. (2024). 2023 state of the industry report: Cultivated meat and seafood. *Good Food Institute*. https://gfi.org/wp-content/uploads/2024/04/State-of-the-Industry-report_Cultivated_2023.pdf

Goodland, R., & Anhang, J. (2009). Livestock and climate change. *World Watch*, 22, 10–19.

Goodwin, J. N., & Shoulders, C. W. (2013). The future of meat: A qualitative analysis of cultured meat media coverage. *Meat Science*, 95(3), 445–450. 10.1016/j.meatsci.2013.05.02723793078

Goulart, F. F., Chappell, M. J., Mertens, F., & Soares-Filho, B. (2023). Sparing or expanding? The effects of agricultural yields on farm expansion and deforestation in the tropics. *Biodiversity and Conservation*, 32(3), 1089–1104. 10.1007/s10531-022-02540-4

Government of the Netherlands. (2023). Report code of practice safely conducting tastings cultivated foods prior to EU approval. *Government of the Netherlands*. https://open.overheid.nl/documenten/39127f7e-b18b-4ddf-95a7-0be5ff660aed/file

Graça, J., Godinho, C. A., & Truninger, M. (2019). Reducing meat consumption and following plant-based diets: Current evidence and future directions to inform integrated transitions. *Trends in Food Science & Technology*, 91, 380–390. 10.1016/j.tifs.2019.07.046

Graham, M. L., & Prescott, M. J. (2015). The multifactorial role of the 3Rs in shifting the harm-benefit analysis in animal models of disease. *European Journal of Pharmacology*, 759, 19–29. 10.1016/j.ejphar.2015.03.04025823812

Gray, V. P., Amelung, C. D., Duti, I. J., Laudermilch, E. G., Letteri, R. A., & Lampe, K. J. (2022). Biomaterials via peptide assembly: Design, characterization, and application in tissue engineering. *Acta Biomaterialia*, 140, 43–75. 10.1016/j.actbio.2021.10.03034710626

Green, J. F. (2021). Does carbon pricing reduce emissions? A review of ex-post analyses. *Environmental Research Letters*, 16(4), 43004. 10.1088/1748-9326/abdae9

Gstraunthaler, G. (2003). Alternatives to the use of fetal bovine serum: Serum-free cell culture. *ALTEX*, 20(4), 275–281. 10.14573/altex.2003.4.25714671707

Guan, X., Lei, Q., Yan, Q., Li, X., Zhou, J., Du, G., & Chen, J. (2021). Trends and ideas in technology, regulation and public acceptance of cultured meat. *Future Foods : a Dedicated Journal for Sustainability in Food Science*, 3, 100032. 10.1016/j.fufo.2021.100032

Guan, X., Zhou, J., Du, G., & Chen, J. (2022). Bioprocessing technology of muscle stem cells: Implications for cultured meat. *Trends in Biotechnology*, 40(6), 721–734. 10.1016/j.tibtech.2021.11.00434887105

Guardian. (2013). Would you eat a synthetic beefburger? https://www.theguardian.com/commentisfree/poll/2013/aug/0 5/eat-synthetic-beefburger-poll

Guardian. (2022). All sizzle, no steak: how Singapore became the center of the plant-based meat industry. https://www.theguardian.com/e nvironment/2022/nov/06/all-sizzle-no-steak-how-singapore-became-the-centre-of-the-plant-based-meat-industry

Gu, D., Andreev, K., & Dupre, M. E. (2021). Major trends in population growth around the world. *China CDC Weekly*, 3(28), 604. 10.46234/ccdcw2021.16034594946

Guerrero-Pineda, C., Iacona, G. D., Mair, L., Hawkins, F., Siikamäki, J., Miller, D., & Gerber, L. R. (2022). An investment strategy to address biodiversity loss from agricultural expansion. *Nature Sustainability*, 5(7), 610–618. 10.1038/s41893-022-00871-2

Gulzar, B., Mujib, A., Malik, M. Q., Mamgain, J., Syeed, R., & Zafar, N. (2020). *Plant tissue culture: agriculture and industrial applications*. Elsevier eBooks. 10.1016/B978-0-12-818632-9.00002-2

Guo, X., Wang, D., He, B., Hu, L., & Jiang, G. (2023). 3D Bioprinting of Cultured Meat: A Promising Avenue of Meat Production. *Food and Bioprocess Technology*. Advance online publication. 10.1007/s11947-023-03195-x

Gupta, R. K., Gawad, F. A. E., Ali, E. A. E., Karunanithi, S., Yugiani, P., & Srivastav, P. P. (2024). Nanotechnology: Current applications and future scope in food packaging systems. *Measurement Food*, 13, 100131. 10.1016/j.meafoo.2023.100131

Haberlandt, G. (1902). Kulturversuche mit isolierten Pflanzenzellen. Sitzungser. Mat. Nat. K. *Kais. Akad. Wiss.(Wien)*, 111, 69–92.

Hackett, C., Grim, B. J., Lugo, L., Cooperman, A., Esparza Ochoa, J. C., Gao, C., Connor, P., Shi, A. F., & Kuriakose, N. (2012). The global religious landscape: A report on the size and distribution of the world's major religious groups as of 2010. *Pew Research Center's Forum on Religion & Public Life*. https://assets.pewresearch.org/wp-content/uploads/sites/11/2014/01/global-religion-full.pdf

Hadi, J., & Brightwell, G. (2021). Safety of alternative proteins: Technological, environmental and regulatory aspects of cultured meat, plant-based meat, insect protein and single-cell protein. *Foods*, 10(6), 1226. 10.3390/foods1006122634071292

Hajare, S. T., Chauhan, N. M., & Kassa, G. (2021). Effect of Growth Regulators on *In Vitro* Micropropagation of Potato (*Solanum tuberosum* L.) Gudiene and Belete Varieties from Ethiopia. *The Scientific World Journal*, 5928769, 1–8. Advance online publication. 10.1155/2021/592876933628138

Halder, M., Sarkar, S., & Jha, S. (2019, July 25). Elicitation: A biotechnological tool for enhanced production of secondary metabolites in hairy root cultures. *Engineering in Life Sciences*, 19(12), 880–895. 10.1002/elsc.20190005832624980

Hallman, W. K., Hallman, W. K. II, & Hallman, E. E. (2023). Cell-based, cell-cultured, cell-cultivated, cultured, or cultivated. What is the best name for meat, poultry, and seafood made directly from the cells of animals? *NPJ Science of Food*, 7(1), 62. Advance online publication. 10.1038/s41538-023-00234-x38057390

Hampton, J. O., Hyndman, T. H., Allen, B. L., & Fischer, B. (2021). Animal harms and food production: Informing ethical choices. *Animals (Basel)*, 11(5), 1225. 10.3390/ani11051225 33922738

Hann, E. C., Harland-Dunaway, M., Garcia, A. J., Meuser, J. E., & Jinkerson, R. E. (2023). Alternative carbon sources for the production of plant cellular agriculture: A case study on acetate. *Frontiers in Plant Science*, 14, 1104751. 10.3389/fpls.2023.110475137954996

Hao, Z., Li, H., Wang, Y., Hu, Y., Chen, T., Zhang, S., Guo, X., Cai, L., & Li, J. (2022). Supramolecular peptide nanofiber hydrogels for bone tissue engineering: From multi hierarchical fabrications to comprehensive applications. *Advancement of Science*, 9(11), 2103820. 10.1002/advs.20210382035128831

Haraguchi, Y., & Shimizu, T. (2021). Microalgal culture in animal cell waste medium for sustainable 'cultured food' production. *Archives of Microbiology*, 203(9), 5525–5532. 10.1007/s00203-021-02509-x34426852

Haraguchi, Y., & Shimizu, T. (2021). Three-dimensional tissue fabrication system by co-culture of microalgae and animal cells for production of thicker and healthy cultured food. *Biotechnology Letters*, 43(6), 1117–1129. 10.1007/s10529-021-03106-033689062

Harris, A. F., Lacombe, J., & Zenhausern, F. (2021). The emerging role of decellularized plant-based scafolds as a new biomaterial. *International Journal of Molecular Sciences*, 22(22), 12347. 10.3390/ijms22221234734830229

Harvatine, G. (2023). Lack of nutrition in animals and its affects on animals health. *J Vet Med Allied Sci*, 7(1), 132. 10.35841/2591-7978-7.1.132

Hasnain, A., Naqvi, S. A. H., Ayesha, S. I., Khalid, F., Ellahi, M., Iqbal, S., Hassan, M. Z., Abbas, A., Adamski, R., Markowska, D., Baazeem, A., Mustafa, G., Moustafa, M., Hasan, M. E., & Abdelhamid, M. M. A. (2022). Plants *in vitro* propagation with its applications in food, pharmaceuticals and cosmetic industries; current scenario and future approaches. *Frontiers in Plant Science*, 13, 1009395. 10.3389/fpls.2022.100939536311115

Hassoun, A., Cropotova, J., Trif, M., Rusu, A. V., Bobiş, O., Nayik, G. A., Jagdale, Y. D., Saeed, F., Afzaal, M., Mostashari, P., Khaneghah, A. M., & Regenstein, J. M. (2022). Consumer acceptance of new food trends resulting from the fourth industrial revolution technologies: A narrative review of literature and future perspectives. *Frontiers in Nutrition*, 9, 972154. 10.3389/fnut.2022.97215436034919

Hayek, M. N., Harwatt, H., Ripple, W. J., & Mueller, N. D. (2021). The carbon opportunity cost of animal-sourced food production on land. *Nature Sustainability*, 4(1), 21–24. 10.1038/s41893-020-00603-4

Hedlund, L., & Jensen, P. (2022). Effects of stress during commercial hatching on growth, egg production and feather pecking in laying hens. *PLoS One*, 17(1), e0262307. 10.1371/journal.pone.026230734982788

Helliwell, R., & Burton, R. J. (2021). The promised land? Exploring the future visions and narrative silences of cellular agriculture in news and industry media. *Journal of Rural Studies*, 84, 180–191. 10.1016/j.jrurstud.2021.04.002

Hewson, C. J. (2003). What is animal welfare? Common definitions and their practical consequences. *The Canadian Veterinary Journal. La Revue Veterinaire Canadienne*, 44(6), 496.12839246

Hidayat, R., & Haryanti, I. (2020). The effect of price and taste on the purchase. *Almana/Almana. Jurnal Manajemen Dan Bisnis*, 4(2), 244–252. 10.36555/almana.v4i2.1403

Higuita, N. I. A., LaRocque, R. C., & McGushin, A. (2023). Climate change, industrial animal agriculture, and the role of physicians – Time to act. *The Journal of Climate Change and Health*, 13, 100260. 10.1016/j.joclim.2023.100260

Hinds, S., Tyhovych, N., Sistrunk, C., & Terracio, L. (2013). Improved tissue culture conditions for engineered skeletal muscle sheets. *TheScientificWorldJournal*, 2013(1), 2013. 10.1155/2013/370151

Hirsch, C., & Schildknecht, S. (2019). *In vitro* research reproducibility: Keeping up high standards. *Frontiers in Pharmacology*, 10, 492837. 10.3389/fphar.2019.0148431920667

Hocquette, J. F., Chriki, S., Fournier, D., & Ellies-Oury, M. P. (2024). Will "cultured meat" transform our food system towards more sustainability? Animal, 101145.

Hocquette, A., Lambert, C., Sinquin, C., Peterolff, L., Wagner, Z., Bonny, S., Lebert, A., & Hocquette, J. (2015). Educated consumers don't believe artificial meat is the solution to the problems with the meat industry. *Journal of Integrative Agriculture. Journal of Integrative Agriculture*, 14(2), 273–284. 10.1016/S2095-3119(14)60886-8

Hocquette, É., Liu, J., Ellies-Oury, M.-P., Chriki, S., & Hocquette, J.-F. (2022). Does the Future of Meat in France Depend on Cultured Muscle Cells? Answers from Different Consumer Segments. *Meat Science*, 188, 108776. 10.1016/j.meatsci.2022.10877635245709

Hocquette, J. F. (2016). Is in vitro meat the solution for the future? *Meat Science*, 120, 167–176. 10.1016/j.meatsci.2016.04.03627211873

Hogan, W. W. (2022). Electricity Market Design and Zero-Marginal Cost Generation. *Current Sustainable/Renewable. Energy Reports*, 9(1), 15–26. 10.1007/s40518-021-00200-9

Home, P. D., & Mehta, R. (2021). Insulin therapy development beyond 100 years. *The Lancet Diabetes &amp. Endocrinology*, 9(10), 695–707. 10.1016/S2213-8587(21)00182-034480874

Hong, T. K., Shin, D. M., Choi, J., Do, J. T., & Han, S. G. (2021). Current issues and technical advances in cultured meat production: A review. *Food Science of Animal Resources*, 41(3), 355–372. 10.5851/kosfa.2021.e1434017947

Hopkins, P. D., & Dacey, A. (2008). Vegetarian meat: Could technology save animals and satisfy meat eaters? *Journal of Agricultural & Environmental Ethics*, 21(6), 579–596. 10.1007/s10806-008-9110-0

Hosseini Nezhad, M. (2021). Advances in Microbial Production of Milk and Dairy Products: A Review. *Comprehensive Reviews in Food Science and Food Safety*.

Ho, Y. Y., Lu, H. K., Lim, Z. F. S., Lim, H. W., Ho, Y. S., & Ng, S. K. (2021). Applications and analysis of hydrolysates in animal cell culture. *Bioresources and Bioprocessing*, 8(1), 93. 10.1186/s40643-021-00443-w34603939

Huang, Y.. (2010). Biomaterials and scaffolds in tissue engineering. *Biotechnology Advances*, 28(7), 925–936. 10.1016/j.biotechadv.2010.08.011

Hubalek, S., Post, M. J., & Moutsatsou, P. (2022). Towards resource-efficient and cost-efficient cultured meat. *Current Opinion in Food Science*, 47, 100885. 10.1016/j.cofs.2022.100885

Hubbard, Y. E. (2022). Addressing Public Perceptions About Cell-Based Meat and Cellular Agriculture Through Metaphors (Master's thesis, Old Dominion University).

Hu, D., Jiao, J., Tang, Y., Han, X., & Sun, H. (2021). The effect of global value chain position on green technology innovation efficiency: From the perspective of environmental regulation. *Ecological Indicators*, 121, 107195. 10.1016/j.ecolind.2020.107195

Humbird, D. (2021). Scale-up economics for cultured meat. *Biotechnology and Bioengineering*, 118(8), 3239–3250. 10.1002/bit.2784834101164

Ilea, R. C. (2009). Intensive Livestock Farming: Global Trends, Increased Environmental Concerns, and Ethical Solutions. *Journal of Agricultural & Environmental Ethics*, 22(2), 153–167. 10.1007/s10806-008-9136-3

Imseng, N., Schillberg, S., Schürch, C., Schmid, D., Schütte, K., Gorr, G., ... & Eibl, R. (2014). Suspension culture of plant cells under heterotrophic conditions. *Industrial scale suspension culture of living cells*, 224-258. 10.1002/9783527683321.ch07

Intergovernmental Panel on Climate Change. (2022). *Climate change 2022: Mitigation of climate change. Contribution of working group III to the sixth assessment report of the intergovernmental panel on climate change.* Cambridge University Press. https://www.ipcc.ch/report/ar6/wg3/downloads/report/IPCC_AR6_WGIII_FullReport.pdf

Izhar Ariff Mohd Kashim, M., Abdul Haris, A. A., Abd. Mutalib, S., Anuar, N., & Shahimi, S. (2023). Scientific and Islamic perspectives in relation to the Halal status of cultured meat. *Saudi Journal of Biological Sciences*, 30(1), 103501. 10.1016/j.sjbs.2022.10350136466219

Jagadeesan, P., & Salem, S. (2020). Religious and Regulatory concerns of animal free meat and milk. Science Open *Preprints*, 10.14293/S2199-1006.1.SOR-.PP7B21S.v1

Jahir, N. R., Ramakrishna, S., Abdullah, A. A. A., & Vigneswari, S. (2023). Cultured meat in cellular agriculture: Advantages, applications and challenges. *Food Bioscience*, 53, 102614. 10.1016/j.fbio.2023.102614

Jairath, G., Mal, G., Gopinath, D., & Singh, B. (2021). A holistic approach to access the viability of cultured meat: A review. *Trends in Food Science & Technology*, 110, 700–710. 10.1016/j.tifs.2021.02.024

Jalil, A., Basar, S., Karmaker, S., Ali, A., Choudhury, M. R., & Hoque, S. (2017). Investigation of biogas Generation from the waste of a vegetable and cattle market of Bangladesh. *International Journal of Waste Resources*, 7(1). Advance online publication. 10.4172/2252-5211.1000283

Jamil, U., Bonnington, A., & Pearce, J. M. (2023). The Agrivoltaic Potential of Canada. *Sustainability (Basel)*, 15(4), 3228. 10.3390/su15043228

Jankowski, M., Mozdziak, P., Petitte, J., Kulus, M., & Kempisty, B. (2020). Avian satellite cell plasticity. *Animals (Basel)*, 10(8), 1322. 10.3390/ani1008132232751789

Javaid, M., & Haleem, A. (2019). 4D printing applications in medical field: A brief review. *Clinical Epidemiology and Global Health*, 7(3), 317–321. 10.1016/j.cegh.2018.09.007

Jebari, A., Oyetunde-Usman, Z., McAuliffe, G. A., Chivers, C. A., & Collins, A. L. (2024). Willingness to adopt green house gas mitigation measures: Agricultural land managers in the United Kingdom. *PLoS One*, 19(7), e0306443. 10.1371/journal.pone.030644338976702

Jha, A. (2013, August). First hamburger made from lab-grown meat to be served at press conference. *The Guardian*. https://www.theguardian.com/science/2013/aug/05/first-hamburger-lab-grown-meat-press-conference

Jha, D. N. (2002). The myth of the holy cow. Verso. Retrieved from https://books.google.com/books?hl=en&lr=&id=vTrKWRUOpbcC&oi=fnd&pg=PR9&dq=livestock+like+cows+tied+to+religious+practices+in+india&ots=z-i8Hl4qMe&sig=XTVOzpiXqcq0DsYGuJEQeAzmK3I

Jin, G., & Bao, X. (2024). Tailoring the taste of cultured meat. *eLife*, 13, e98918. Advance online publication. 10.7554/eLife.9891838813866

Jo, B., Nie, M., & Takeuchi, S. (2021). Manufacturing of animal products by the assembly of microfabricated tissues. *Essays in Biochemistry*, 65(3), 611–623. 10.1042/EBC2020009234156065

Johannesen, N. (2022). The global minimum tax. *Journal of Public Economics*, 212, 104709. 10.1016/j.jpubeco.2022.104709

Jones, A. (2024). Animal Welfare Considerations in Cellular Agriculture: A Review. *Journal of Agricultural Ethics*, 18(3), 315–328.

Jones, J. D., Rebello, A. S., & Gaudette, G. R. (2021). Decellularized spinach: An edible scaffold for laboratory-grown meat. Food Bioscience, 41, 1100986 12. 10.1016/j.cofs.2015.08.002

Jones, N. (2023). Lab-grown meat: The science of turning cells into steaks and nuggets. *Nature*, 619(7968), 22–24. 10.1038/d41586-023-02095-6

Jönsson, E. (2020). On breweries and bioreactors: Probing the "present futures" of cellular agriculture. *Transactions of the Institute of British Geographers*, 45(4), 921–936. 10.1111/tran.12392

Jossen, V., van den Bos, C., Eibl, R., & Eibl, D. (2018, March 22). Manufacturing human mesenchymal stem cells at clinical scale: Process and regulatory challenges. *Applied Microbiology and Biotechnology*, 102(9), 3981–3994. 10.1007/s00253-018-8912-x29564526

Juarez, M. (2021). Feasibility of Cellular Milk Production: A Review. *Journal of Cellular Agriculture*, 5(2), 87–102.

Jung, J. W., Lee, J. S., & Cho, D. W. (2016). Computer-aided multiple-head 3D printing system for printing of heterogeneous organ/tissue constructs. *Scientific Reports*, 6(1), 21685. 10.1038/srep21685

Justine, A. K., Kaur, N., Savita, , & Pati, P. K. (2022). Biotechnological interventions in banana: Current knowledge and future prospects. *Heliyon*, 8(11), e11636. 10.1016/j.heliyon.2022.e1163636419664

Kačarević, Ž. P., Rider, P. M., Alkildani, S., Retnasingh, S., Smeets, R., Jung, O., Ivanišević, Z., & Barbeck, M. (2018). An introduction to 3D bioprinting: Possibilities, challenges and future aspects. *Materials (Basel)*, 11(11), 2199. 10.3390/ma1111219930404222

Kadim, I. T., Mahgoub, O., Baqir, S., Faye, B., & Purchas, R. W. (2015). Cultured meat from muscle stem cells: A review of challenges and prospects. *Journal of Integrative Agriculture*, 14(2), 222–234. 10.1016/S2095-3119(14)60881-9

Kaiser, M. (2009). Ethical aspects of livestock genetic engineering. In *Genetic Engineering in Livestock: New Applications and Interdisciplinary Perspectives* (pp. 91-117). Springer Berlin Heidelberg. 10.1007/978-3-540-85843-0_5

Kakaei, H., Nourmoradi, H., Bakhtiyari, S., Jalilian, M., & Mirzaei, A. (2022). Effect of COVID-19 on food security, hunger, and food crisis. In M. H. Dehgani, R. R. Karri, & S. Roy (Eds.), *Covid-19 and the sustainable development goals* (pp. 3–29). Elsevier eBooks. 10.1016/B978-0-323-91307-2.00005-5

Kantono, K., Hamid, N., Malavalli, M. M., Liu, Y., Liu, T., & Seyfoddin, A. (2022). Consumer acceptance and production of in vitro meat: A review. *Sustainability (Basel)*, 14(9), 4910. 10.3390/su14094910

Kantor, B. N., & Kantor, J. (2021). Public attitudes and willingness to pay for cultured meat: A cross-sectional experimental study. *Frontiers in Sustainable Food Systems*, 5, 1–7. 10.3389/fsufs.2021.594650

Kari, G., Rodeck, U., & Dicker, A. P. (2007). Zebrafish: An emerging model system for human disease and drug discovery. *Clinical Pharmacology and Therapeutics*, 82(1), 70–80. 10.1038/sj.clpt.610022317495877

Karthikeyan, L., Chawla, I., & Mishra, A. K. (2020). A review of remote sensing applications in agriculture for food security: Crop growth and yield, irrigation, and crop losses. *Journal of Hydrology (Amsterdam)*, 586, 124905. 10.1016/j.jhydrol.2020.124905

Kaufman, D., McKay, N., Routson, C., Erb, M., Dätwyler, C., Sommer, P. S., Heiri, O., & Davis, B. (2020). Holocene global mean surface temperature, a multi-method reconstruction approach. *Scientific Data*, 7(1), 201. 10.1038/s41597-020-0530-732606396

Keefer, C., Pant, D., Blomberg, L., & Talbot, N. (2007, March). Challenges and prospects for the establishment of embryonic stem cell lines of domesticated ungulates. *Animal Reproduction Science*, 98(1–2), 147–168. 10.1016/j.anireprosci.2006.10.00917097839

Kenigsberg, J. A., & Zivotofsky, A. Z. (2020). A Jewish religious perspective on cellular agriculture. *Frontiers in Sustainable Food Systems*, 128, 128. 10.3389/fsufs.2019.00128

Khan, F., & Tanaka, M. (2018). Designing Smart Biomaterials for Tissue Engineering. *International Journal of Molecular Sciences*, 19(1), 17. 10.3390/ijms1901001729267207

Kiani, A. K., Pheby, D., Henehan, G., Brown, R., Sieving, P., Sykora, P., ... International Bioethics Study Group. (2022). Ethical considerations regarding animal experimentation. *Journal of Preventive Medicine and Hygiene*, 63(2 Suppl 3), E255. 10.15167/2421-4248/jpmh2022.63.2S3.2768

Kilkenny, C., Parsons, N., Kadyszewski, E., Festing, M. F., Cuthill, I. C., Fry, D., Hutton, J., & Altman, D. G. (2009). Survey of the quality of experimental design, statistical analysis and reporting of research using animals. *PLoS One*, 4(11), e7824. 10.1371/journal.pone.000782419956596

Kim, M., Choi, Y. S., Yang, S. H., Hong, H. N., Cho, S. W., Cha, S. M., Pak, J. H., Kim, C. W., Kwon, S. W., & Park, C. J. (2006, November). Erratum to "Muscle regeneration by adipose tissue-derived adult stem cells attached to injectable PLGA spheres" [Biochem. Biophys. Res. Commun. 348 (2006) 386–392]. *Biochemical and Biophysical Research Communications*, 350(2), 499. 10.1016/j.bbrc.2006.09.05115369779

Kim, S. (2024). Bioavailability of Nutrients in Cellular Milk Proteins: Implications for Human Health. *Nutrition Research (New York, N.Y.)*, 28(1), 45–58.

King, J. M., Sayers, A. R., Peacock, C. P., & Kontrohr, E. (1984). Maasai herd and flock structures in relation to livestock wealth, climate and development. *Agricultural Systems*, 13(1), 21–56. 10.1016/0308-521X(84)90054-4

King, L. A. (2003). Behavioral evaluation of the psychological welfare and environmental requirements of agricultural research animals: Theory, measurement, ethics, and practical implications. *ILAR Journal*, 44(3), 211–221. 10.1093/ilar.44.3.21112789022

Kirkwood, J. K., Sainsbury, A. W., & Bennett, P. M. (1994). The welfare of free-living wild animals: Methods of assessment. *Animal Welfare (South Mimms, England)*, 3(4), 257–273. 10.1017/S0962728600017036

Klerkx, L., & Rose, D. (2020). Dealing with the game-changing technologies of Agriculture 4.0: How do we manage diversity and responsibility in food system transition pathways? *Global Food Security*, 24, 100347. 10.1016/j.gfs.2019.100347

Knežić, T., Janjušević, L., Djisalov, M., Yodmuang, S., & Gadjanski, I. (2022). Using vertebrate stem and progenitor cells for cellular agriculture, state-of-the-art, challenges, and future perspectives. *Biomolecules*, 12(5), 699. 10.3390/biom1205069935625626

Knowles, T. G., Kestin, S. C., Haslam, S. M., Brown, S. N., Green, L. E., Butterworth, A., Pope, S. J., Pfeiffer, D., & Nicol, C. J. (2008). Leg disorders in broiler chickens: Prevalence, risk factors and prevention. *PLoS One*, 3(2), e1545. 10.1371/journal.pone.000154518253493

Kodaparthi, A., Kondakindi, V. R., Kehkashaan, L., Belli, M. V., Chowdhury, H. N., Aleti, A., & Chepuri, K. (2024). Environmental Conservation for Sustainable Agriculture. In *Prospects for Soil Regeneration and Its Impact on Environmental Protection* (pp. 15–45). Springer Nature Switzerland. 10.1007/978-3-031-53270-2_2

Kolesky, D. B., Homan, K. A., Skylar-Scott, M. A., & Lewis, J. A. (2016). Three-dimensional bioprinting of thick vascularized tissues. *Proceedings of the National Academy of Sciences of the United States of America*, 113(12), 3179–3184. 10.1073/pnas.1521342113

Kolkmann, A. M., Post, M. J., & Rutjens, M. A. (2020). Cultured meat: The business of biotechnology. *Trends in Biotechnology*, 38(7), 683–689.

Kolkmann, A. M., Post, M. J., Rutjens, M. A., van Essen, A. L. M., & Moutsatsou, P. (2020). Serum-free media for the growth of primary bovine myoblasts. *Cytotechnology*, 72(1), 111–120. 10.1007/s10616-019-00361-y31884572

Kolkmann, A. M., van Essen, A., Post, M. J., & Moutsatsou, P. (2022). Development of a chemically defined medium for in vitro expansion of primary bovine satellite cells. *Frontiers in Bioengineering and Biotechnology*, 10, 895289. 10.3389/fbioe.2022.89528935992337

Kollenda, E., Baldock, D., Hiller, N., & Lorant, A. (2020). Transitioning towards cage-free farming in the EU: Assessment of environmental and socio-economic impacts of increased animal welfare standards. *Policy report by the Institute for European Environmental Policy, Brussels & London*, 1-65. https://ieep.eu/wp-content/uploads/2022/12/Transitioning-towards-cage-free-farming-in-the-EU_Final-report_October_web.pdf

Kolodkin-Gal, I., Dash, O., & Rak, R. (2023). Probiotic cultivated meat: Bacterial-based scaffolds and products to improve cultivated meat. *Trends in Biotechnology*.37805297

Kombolo Ngah, M., Chriki, S., Ellies-Oury, M. P., Liu, J., & Hocquette, J. F. (2023). Consumer perception of "artificial meat" in the educated young and urban population of Africa. *Frontiers in Nutrition*, 10, 1127655. 10.3389/fnut.2023.112765537125051

Kosmatko, J., & Węglarz, A. (2018). Production Technology and Nutritive Value of Iberian Ham (jamón ibérico). Retrieved from https://wz.izoo.krakow.pl/files/WZ_2018_4_art16_en.pdf

Kouřimská, L., & Adámková, A. (2016). Nutritional and sensory quality of edible insects. Elsevier GmbH., 10.1016/j.nfs.2016.07.001

Koyande, A. K., Chew, K. W., Rambabu, K., Tao, Y., Chu, D. T., & Show, P. L. (2019). Microalgae: A potential alternative to health supplementation for humans. In Food Science and Human Wellness (Vol. 8, Issue 1, pp. 16–24). Elsevier B.V. 10.1016/j.fshw.2019.03.001

Kulus, M., Jankowski, M., Kranc, W., Golkar Narenji, A., Farzaneh, M., Dzięgiel, P., Zabel, M., Antosik, P., Bukowska, D., Mozdziak, P., & Kempisty, B. (2023). Bioreactors, scaffolds and microcarriers and in vitro meat production—Current obstacles and potential solutions. *Frontiers in Nutrition*, 10, 1225233. 10.3389/fnut.2023.122523337743926

Kumar, P., Mehta, N., Abubakar, A. A., Verma, A. K., Kaka, U., Sharma, N., Sazili, A. Q., Pateiro, M., Kumar, M., & Lorenzo, J. M. (2022). Potential Alternatives of Animal Proteins for Sustainability in the Food Sector. *Food Reviews International*, 39(8), 5703–5728. 10.1080/87559129.2022.2094403

Kumar, P., Sharma, N., Sharma, S., Mehta, N., Verma, A. K., Chemmalar, S., & Sazili, A. Q. (2021). In-vitro meat: A promising solution for sustainability of meat sector. *Journal of Animal Science and Technology*, 63(4), 693–724. 10.5187/jast.2021.e8534447949

Kupka, R., Siekmans, K., & Beal, T. (2020). The diets of children: Overview of available data for children and adolescents. *Global Food Security*, 27, 100442. 10.1016/j.gfs.2020.100442

Lähteenmäki-Uutela, A., Rahikainen, M., Lonkila, A., & Yang, B. (2021). Alternative proteins and EU food law. *Food Control*, 130, 108336. 10.1016/j.foodcont.2021.108336

Langer, R. (1997). Tissue engineering: A new field and its challenges. *Pharmaceutical Research*, 14(7), 840–841. 10.1023/A:10121313291489244137

Lannutti, J., Reneker, D., Ma, T., Tomasko, D., & Farson, D. (2007). Electrospinning for tissue engineering scaffolds. *Materials Science and Engineering C*, 27(3), 504–509. 10.1016/j.msec.2006.05.019

Laurance, W. F. (2010). Habitat destruction: death by a thousand cuts. Conservation biology for all, 1(9), 73-88.

Laurance, W. F., & Peres, C. A. (Eds.). (2006). *Emerging threats to tropical forests*. University of Chicago Press.

Lazennec, G., & Jorgensen, C. (2008, April 3). Concise Review: Adult Multipotent Stromal Cells and Cancer: Risk or Benefit? *Stem Cells (Dayton, Ohio)*, 26(6), 1387–1394. 10.1634/stemcells.2007-100618388305

Le, B. (2020). Australian Startup Develops Novel Edible Scaffold to Modernize Meat Production, https://www.proteinreport.org/australian-startup-develops-novel-edible-scaffold-modernize-meat-production

Learmonth, M. J. (2020). Human–animal interactions in zoos: What can compassionate conservation, conservation welfare and duty of care tell us about the ethics of interacting, and avoiding unintended consequences? *Animals (Basel)*, 10(11), 2037. 10.3390/ani1011203733158270

Lechanteur, C. (2014). Large-Scale Clinical Expansion of Mesenchymal Stem Cells in the GMP-Compliant, Closed Automated Quantum® Cell Expansion System: Comparison with Expansion in Traditional T-Flasks. *Journal of Stem Cell Research & Therapy*, 04(08). Advance online publication. 10.4172/2157-7633.1000222

Lee, D. K., Kim, M., Jeong, J., Lee, Y. S., Yoon, J. W., An, M. J., & Lee, C. K. (2023). Unlocking the potential of stem cells: Their crucial role in the production of cultivated meat. *Current Research in Food Science*, 7, 100551. 10.1016/j.crfs.2023.10055137575132

Lee, E. K., Jin, Y. W., Park, J. H., Yoo, Y. M., Hong, S. M., Amir, R., Yan, Z., Kwon, E., Elfick, A., Tomlinson, S., Halbritter, F., Waibel, T., Yun, B. W., & Loake, G. J. (2010, October 24). Cultured cambial meristematic cells as a source of plant natural products. *Nature Biotechnology*, 28(11), 1213–1217. 10.1038/nbt.169320972422

Lee, J. D., Shin, D., & Roh, J.-L. (2018). Development of an in vitro cell-sheet cancer model for chemotherapeutic screening. *Theranostics*, 8(14), 3964–3973. 10.7150/thno.2643930083273

Lee, M., Park, S., Choi, B., Kim, J., Choi, W., Jeong, I., Han, D., Koh, W.-G., & Hong, J. (2022). Tailoring a gelatin/agar matrix for the synergistic effect with cells to produce high-quality cultured meat. *ACS Applied Materials & Interfaces*, 14(33), 38235–38245. 10.1021/acsami.2c1098835968689

Lee, S. (2022). Production of Cellular Butter and Cream: Process Optimization and Quality Evaluation. *Journal of Food Engineering*, 28(1), 45–58.

Lee, S. (2023). Transforming the Global Food System: The Potential of Cellular Agriculture. *Food Policy*, 28(2), 123–136.

Lee, S. Y., Jeong, J. W., Kim, J. H., Yun, S. H., Mariano, E.Jr, Lee, J., & Hur, S. J. (2023). Current technologies, regulation, and future perspective of animal product analogs—A review. *Animal Bioscience*, 36(10), 1465–1487. 10.5713/ab.23.002937170512

Lee, S. Y., Lee, D. Y., Mariano, E. J., Yun, S. H., Lee, J., Park, J., Choi, Y., Han, D., Kim, J. S., Joo, S.-T., & Hur, S. J. (2023). Study on the current research trends and future agenda in animal products: An Asian perspective. *Journal of Animal Science and Technology*, 65(6), 1124–1150. 10.5187/jast.2023.e12138616880

Lee, S. Y., Yun, S. H., Lee, J., Mariano, E.Jr, Park, J., Choi, Y., & Hur, S. J. (2024). Current technology and industrialization status of cell-cultivated meat. *Journal of Animal Science and Technology*, 66(1), 1–30. 10.5187/jast.2023.e10738618028

Lee, S.-H., & Yoon, K.-H. (2021). A Century of Progress in Diabetes Care with Insulin: A History of Innovations and Foundation for the Future. *Diabetes & Metabolism Journal*, 45(5), 629–640. 10.4093/dmj.2021.016334610718

Lee, S., Trinh, T. H., Yoo, M., Shin, J., Lee, H., Kim, J., Hwang, E., Lim, Y., & Ryou, C. (2019). Self-assembling peptides and their application in the treatment of diseases. *International Journal of Molecular Sciences*, 20(23), 5850. 10.3390/ijms2023585031766475

Lei, Z. L., Li MingSheng, L. M., Ma ZhongRen, M. Z., & Feng YuPing, F. Y. (2017). Preparation of DEAE-soybean starch microspheres for anchorage-dependent mammal cell culture.

Leitzmann, C. (2003). Nutrition ecology: The contribution of vegetarian diets. *The American Journal of Clinical Nutrition*, 78(3, Suppl), 657s–659s. 10.1093/ajcn/78.3.657S

Leprivier, G., & Rotblat, B. (2020). How does mTOR sense glucose starvation? AMPK is the usual suspect. *Cell Death Discovery*, 6(1), 27. 10.1038/s41420-020-0260-932351714

Lerman, M. J., Lembong, J., Muramoto, S., Gillen, G., & Fisher, J. P. (2018). The evolution of polystyrene as a cell culture material. *Tissue Engineering. Part B, Reviews*, 24(5), 359–372. 10.1089/ten.teb.2018.005629631491

Levi, S., Yen, F. C., Baruch, L., & Machluf, M. (2022). Scaffolding technologies for the engineering of cultured meat: Towards a safe, sustainable, and scalable production. *Trends in Food Science & Technology*, 126, 13–25. 10.1016/j.tifs.2022.05.011

Leytem, A. B., Dungan, R. S., Bjorneberg, D. L., & Koehn, A. C. (2011). Emissions of ammonia, methane, carbon dioxide, and nitrous oxide from dairy cattle housing and manure management systems. *Journal of Environmental Quality*, 40(5), 1383–1394. 10.2134/jeq2009.0515

Li, G., Hu, L., Liu, J., Huang, J., Yuan, C., Takaki, K., & Hu, Y. (2022a). A Review on 3D Printable Food Materials: Types and Development Trends. *International Journal of Food Science & Technology*, 2022(57), 164–172. 10.1111/ijfs.15391

Li, L., Chen, L., Chen, X., Chen, Y., Ding, S., Fan, X., Liu, Y., Xu, X., Zhou, G., Zhu, B., Ullah, N., & Feng, X. (2022Chitosan-sodium alginate-collagen/gelatin three-dimensional edible scaffolds for building a structured model for cell cultured meat. *International Journal of Biological Macromolecules*, 209, 668–679. 10.1016/j.ijbiomac.2022.04.05235413327

Li, L., Sun, N., Zhang, L., Xu, G., Liu, J., Hu, J., & Han, L. (2020). Fast food consumption among young adolescents aged 12–15 years in 54 low-and middle-income countries. *Global Health Action*, 13(1), 1795438. 10.1080/16549716.2020.17954 3832762333

Lim, C. H., Choi, Y., Kim, M., Jeon, S. W., & Lee, W. K. (2017). Impact of deforestation on agro-environmental variables in cropland, North Korea. *Sustainability (Basel)*, 9(8), 1354. 10.3390/su9081354

Lioutas, E. D., Charatsari, C., & De Rosa, M. (2021). Digitalization of agriculture: A way to solve the food problem or a trolley dilemma? *Technology in Society*, 67, 101744. 10.1016/j.techsoc.2021.101744

Li, Q., Lin, H., Du, Q., Liu, K., Wang, O., Evans, C., Christian, H., Zhang, C., & Lei, Y. (2018). Scalable and physiologically relevant microenvironments for human pluripotent stem cell expansion and differentiation. *Biofabrication*, 10(2), 025006. 10.1088/1758-5090/aaa6b529319535

Liu, C., Xia, Z., & Czernuszka, J. T. (2007). Design and development of three-dimensional scaffolds for tissue engineering. *Chemical Engineering Research & Design*, 85(7), 1051–1064. 10.1205/cherd06196

Liu, F., Zhang, S. Y., Chen, K. X., & Zhang, Y. (2023). Fabrication, in-vitro digestion and pH-responsive release behavior of soy protein isolate glycation conjugates-based hydrogels. *Food Research International*, 169, 112884. 10.1016/j.foodres.2023.11288437254332

Liu, J., Chriki, S., Kombolo, M., Santinello, M., Pflanzer, S. B., Hocquette, É., Ellies-Oury, M.-P., & Hocquette, J.-F. (2023). Consumer perception of the challenges facing livestock production and meat consumption. *Meat Science*, 200, 109144. 10.1016/j.meatsci.2023.10914436863253

Liu, J., Xiao, D., Liu, Y., & Huang, Y. (2023). A Pig Mass Estimation Model Based on Deep Learning without Constraint. *Animals (Basel)*, 13(8), 1376. 10.3390/ani13081376 37106939

Liu, W. (2019). A Review on the Genetic Regulation of Myogenesis and Muscle Development. *American Journal of Biochemistry and Biotechnology*, 15(1), 1–12. Advance online publication. 10.3844/ajbbsp.2019.1.12

Liu, X., & Ma, P. X. (2009). Phase separation, pore structure, and properties of nanofibrous gelatin scaffolds. *Biomaterials*, 30(25), 4094–4103. 10.1016/j.biomaterials.2009.04.02419481080

Liu, Y., Dong, X., Wang, B., Tian, R., Li, J., Liu, L., Du, G., & Chen, J. (2021). Food synthetic biology-driven protein supply transition: From animal-derived production to microbial fermentation. *Chinese Journal of Chemical Engineering*, 30, 29–36. 10.1016/j.cjche.2020.11.014

Liu, Y., Meyer, C., Xu, C., Weng, H., Hellerbrand, C., ten Dijke, P., & Dooley, S. (2013). Animal models of chronic liver diseases. *American Journal of Physiology. Gastrointestinal and Liver Physiology*, 304(5), G449–G468. 10.1152/ajpgi.00199.201223275613

Liu, Y., Wang, R., Ding, S., Deng, L., Zhang, Y., Li, J., Shi, Z., Wu, Z., Liang, K., Yan, X., Liu, W., & Du, Y. (2022). Engineered meatballs via scalable skeletal muscle cell expansion and modular micro-tissue assembly using porous gelatin microcarriers. *Biomaterials*, 287, 121615. 10.1016/j.biomaterials.2022.12161535679644

Liu, Z., Zhang, M., Bhandari, B., & Wang, Y. (2017). 3D printing: Printing precision and application in food sector. *Trends in Food Science & Technology*, 69, 83–94. 10.1016/j.tifs.2017.08.018

Li, W., Han, Y., Yang, H., Wang, G., Lan, R., & Wang, J. Y. (2016). Preparation of microcarriers based on zein and their application in cell culture. *Materials Science and Engineering C*, 58, 863–869. 10.1016/j.msec.2015.09.04526478381

Li, X. (2023). Optimization of Bioreactor Systems for Cellular Milk Protein Production. *Journal of Biotechnology*, 45(2), 210–225.

Li, X., Liu, B., Pei, B., Chen, J., Zhou, D., Peng, J., Zhang, X., Jia, W., & Xu, T. (2020). Inkjet Bioprinting of Biomaterials. *Chemical Reviews*, 120(19), 10793–10833. 10.1021/acs.chemrev.0c0000832902959

Li, Y., Liu, W., Li, S., Zhang, M., Yang, F., & Wang, S. (2021). Porcine skeletal muscle tissue fabrication for cultured meat production using three-dimensional bioprinting technology. *J Future Foods.*, 1(1), 88–97. 10.1016/j.jfutfo.2021.09.005

Logan, T. J. (2020). Sustainable Agriculture and Water Quality. In *Sustainable Agricultural Systems* (pp. 582–613). CRC Press. 10.1201/9781003070474-40

Lovett, M., Lee, K., Edwards, A., & Kaplan, D. L. (2009). Vascularization strategies for tissue engineering. *Tissue Engineering. Part B, Reviews*, 15(3), 353–370. 10.1089/ten.teb.2009.0085

Luiz Morais-da-Silva, R., Glufke Reis, G., Sanctorum, H., & Forte Maiolino Molento, C. (2022). The social impacts of a transition from conventional to cultivated and plant- **based** meats: Evidence from Brazil. *Food Policy*, 111, 102337. 10.1016/j.foodpol.2022.102337

Lu, T. Y., Li, Y., & Chen, T. (2013). Techniques for fabrication and construction of threedimensional scaffolds for tissue engineering. *International Journal of Nanomedicine*, 337, 337. Advance online publication. 10.2147/IJN.S3863523345979

Macias-Corral, M., Samani, Z., Hanson, A., Smith, G., Funk, P., Yu, H., & Longworth, J. (2008). Anaerobic digestion of municipal solid waste and agricultural waste and the effect of co-digestion with dairy cow manure. *Bioresource Technology*, 99(17), 8288–8293. 10.1016/j.biortech.2008.03.05718482835

MacLeod, M. J., Hasan, M. R., Robb, D. H., & Mamun-Ur-Rashid, M. (2020). Quantifying greenhouse gas emissions from global aquaculture. *Scientific Reports*, 10(1), 11679. 10.1038/s41598-020-68231-832669630

MacQueen, L. A., Alver, C. G., Chantre, C. O., Ahn, S., Cera, L., Gonzalez, G. M., ... Parker, K. K. (2019). Muscle tissue engineering in fibrous gelatin: implications for meat analogs. NPJ Science of Food, 3(1), 20.

MacQueen, L. A., Alver, C. G., Chantre, C. O., Ahn, S., Cera, L., Gonzalez, G. M., O'Connor, B. B., Drennan, D. J., Peters, M. M., Motta, S. E., Zimmerman, J. F., & Parker, K. K. (2019). Muscle tissue engineering in fibrous gelatin: Implications for meat analogs. *NPJ Science of Food*, 3(1), 20. 10.1038/s41538-019-0054-831646181

Madzingira, O. (2018). Animal welfare considerations in food-producing animals. *Animal Welfare (South Mimms, England)*, 99, 171–179. 10.5772/intechopen.78223

Majlis Ugama Islam Singapura. (2024). Fatwa on cultivated meat. *Majlis Ugama Islam Singapura*. https://www.muis. gov.sg/Media/Media-Releases/2024/2/3-Feb-24-Fatwa-on-Cultivated-Meat

Mäkinen, O. E., Wanhalinna, V., Zannini, E., & Arendt, E. K. (2016). Foods for Special Dietary Needs: Non-dairy Plant-based Milk Substitutes and Fermented Dairy-type Products. *Critical Reviews in Food Science and Nutrition*, 56(3), 339–349. 10.1080/10408398.2012.76195025575046

Ma, L., Seager, M., Wittmann, M., Jacobson, M., Bickel, D., Burno, M., Jones, K., Graufelds, V. K., Xu, G., Pearson, M., McCampbell, A., Gaspar, R., Shughrue, P., Danziger, A., Regan, C., Flick, R., Pascarella, D., Garson, S., Doran, S., & Ray, W. J. (2009). Selective activation of the M1 muscarinic acetylcholine receptor achieved by allosteric potentiation. *Proceedings of the National Academy of Sciences of the United States of America*, 106(37), 15950–15955. 10.1073/pnas.090090310619717450

Malav, O. P., Talukder, S., Gokulakrishnan, P., & Chand, S. (2015). Meat Analog: A Review. *Critical Reviews in Food Science and Nutrition*, 55(9), 1241–1245. 10.1080/10408398.2012.68938124915320

Malda, J., Woodfield, T. B. F., van der Vloodt, F., Wilson, C., Martens, D. E., Tramper, J., van Blitterswijk, C. A., & Riesle, J. (2004). The effect of PEGT/PBT scaffold architecture on oxygen gradients in tissue engineered cartilaginous constructs. *Biomaterials*, 25(26), 5773–5780. 10.1016/j.biomaterials.2004.01.02815147823

Malekpour, A., & Chen, X. (2022). Printability and Cell Viability in Extrusion-Based Bioprinting from Experimental, Computational, and Machine Learning Views. *Journal of Functional Biomaterials*, 13(2), 40. 10.3390/jfb1302004035466222

Malerich, M., & Bryant, C. (2022). Nomenclature of cell-cultivated meat & seafood products. *NPJ Science of Food*, 6(1), 56. Advance online publication. 10.1038/s41538-022-00172-036496502

Malik, S., Bhushan, S., Sharma, M., & Ahuja, P. S. (2014, October 16). Biotechnological approaches to the production of shikonins: A critical review with recent updates. *Critical Reviews in Biotechnology*, 36(2), 327–340. 10.3109/07388551.2014.96100325319455

Mancini, M. C., & Antonioli, F. (2019). Exploring consumers' attitude towards cultured meat in Italy. Meat Science, 150, 101–110. https://doi.org/.Meatscience ,12.01410.1016/J

Mancini, C. (2017). Towards an animal-centred ethics for Animal–Computer Interaction. *International Journal of Human-Computer Studies*, 98, 221–233. 10.1016/j.ijhcs.2016.04.008

Mancini, M. C., & Antonioli, F. (2022). The future of cultured meat between sustainability expectations and socio-economic challenges. In *Future Foods* (pp. 331–350). Elsevier. 10.1016/B978-0-323-91001-9.00024-4

Mancuso, J. (2021). What are food grade solvents. Accessed: https:// ecolink.com/info/what-are-food-grade-solvents/

Mannully, C. T., Bruck-Haimson, R., Zacharia, A., Orih, P., Shehadeh, A., Saidemberg, D., Kogan, N. M., Alfandary, S., Serruya, R., Dagan, A., Petit, I., & Moussaieff, A. (2022). Lipid desaturation regulates the balance between self-renewal and differentiation in mouse blastocyst-derived stem cells. *Cell Death & Disease*, 13(12), 1027. 10.1038/s41419-022-05263-036477438

Mano, A. (2024, March). Brazil slaughtering records make it global meat king, IBGE data shows. *Reuters*. https://www.reuters.com/markets/commodities/brazil-slaughtering-records-make-it-global-meat-king-ibge-data-shows-2024-03-14/

Manuel Jesús, H.-O., & Garzón-Moreno, J. (2022). Risk management methodology in the supply chain: A case study applied. *Annals of Operations Research*, 313(2), 1051–1075. Advance online publication. 10.1007/s10479-021-04220-y

Manyi-Loh, C., Mamphweli, S., Meyer, E., & Okoh, A. (2018). Antibiotic use in agriculture and its consequential resistance in environmental sources: Potential public health implications. *Molecules (Basel, Switzerland)*, 23(4), 795. 10.3390/molecules2304079529601469

Marchant-Forde, J. N., & Boyle, L. A. (2020). COVID-19 Effects on Livestock Production: A One Welfare Issue. *Frontiers in Veterinary Science*, 7, 585787. 10.3389/fvets.2020.585787

Marga, F.. (2007). Development of tissue engineered meat: Experiments on myoblast cells on collagen mesh. *Meat Science*, 75(1), 18–24. 10.1016/j.meatsci.2006.06.004

Marlow, H. J., Hayes, W. K., Soret, S., Carter, R. L., Schwab, E. R., & Sabaté, J. (2009). Diet and the environment: Does what you eat matter? *The American Journal of Clinical Nutrition*, 89(5), 1699s–1703s. 10.3945/ajcn.2009.26736Z

Martinez, A., Donoso, E., Hernández, R. O., Sanchez, J. A., & Romero, M. H. (2024). Assessment of animal welfare in fattening pig farms certified in good livestock practices. *Journal of applied animal welfare science. Journal of Applied Animal Welfare Science*, 27(1), 33–45. 10.1080/10888705.2021.202153238314792

Martin, I., Wendt, D., & Heberer, M. (2004). The role of bioreactors in tissue engineering. *Trends in Biotechnology*, 22(2), 80–86. 10.1016/j.tibtech.2003.12.00114757042

Martin-Ortega, J., Rothwell, S. A., Anderson, A., Okumah, M., Lyon, C., Sherry, E., Johnston, C., Withers, P. J. A., & Doody, D. G. (2022). Are stakeholders ready to transform phosphorus use in food systems? A transdisciplinary study in a livestock intensive system. *Environmental Science & Policy*, 131, 177–187. 10.1016/j.envsci.2022.01.01135505912

Martins, B., Bister, A., Dohmen, R. G., Gouveia, M. A., Hueber, R., Melzener, L., Messmer, T., Papadopoulos, J., Pimenta, J., Raina, D., Schaeken, L., Shirley, S., Bouchet, B. P., & Flack, J. E. (2024). Advances and challenges in cell biology for cultured meat. *Annual Review of Animal Biosciences*, 12(1), 345–368. 10.1146/annurev-animal-021022-05513237963400

Matheny, G., & Leahy, C. (2007). Farm-animal welfare, legislation, and trade. *Law & Contemp. Probs.*, 70, 325. https://scholarship.law.duke.edu/lcp/vol70/iss1/10

Mathieu, C., Martens, P.-J., & Vangoitsenhoven, R. (2021). One hundred years of insulin therapy. *Nature Reviews. Endocrinology*, 17(12), 715–725. 10.1038/s41574-021-00542-w34404937

Matsuishi, M., Fujimori, M., & Okitani, A. (2001). Wagyu beef aroma in Wagyu (Japanese Black cattle) beef preferred by the Japanese over imported beef. Nihon Chikusan Gakkaiho, 72(6), 498–504. Retrieved from https://www.jstage.jst.go.jp/article/chikusan1924/72/6/72_6_498/_article/-char/ja/

Mattick, C. S. (2018, January 2). Cellular agriculture: The coming revolution in food production. *Bulletin of the Atomic Scientists*, 74(1), 32–35. 10.1080/00963402.2017.1413059

Mattick, C. S., Landis, A. E., Allenby, B. R., & Genovese, N. J. (2015). Anticipatory life cycle analysis of in vitro biomass cultivation for cultured meat production in the United States. *Environmental Science & Technology*, 49(19), 11941–11949. 10.1021/acs.est.5b0161426383898

Ma, X., Qu, J., Mei, H., Zhao, Y., & Jin, W. (2021). Effects of oxygen concentration on cell growth and differentiation in tissue engineering. *Journal of Biomedical Materials Research*, 109(5), 714–726.

May, J., & Kumar, V. (2023). Harnessing moral psychology to reduce meat consumption. *Journal of the American Philosophical Association*, 9(2), 367–387. https://www.cambridge.org/core/journals/journal-of-the-american-philosophical-association/article/harnessing-moral-psychology-to-reduce-meat-consumption/C3DD02CA582919B2BC7F25B126CD9DD5. 10.1017/apa.2022.2

Mazzucato, M. (2023). Financing the Sustainable Development Goals through mission-oriented development banks. UN DESA Policy Brief Special issue. New York: UN Department of Economic and Social Affairs; UN High-level Advisory Board on Economic and Social Affairs; University College London Institute for Innovation and Public Purpose.

Mbow, C., Rosenzweig, C., Barioni, L. G., Benton, T. G., Herrero, M., Krishnapillai, M., Liwenga, E., Pradhan, P., Rivera-Ferre, M. G., Sapkota, T., Tubiello, F. N., & Xu, Y. (2019). Food security. In *Climate change and land: An IPCC special report on climate change, desertification, land degradation, sustainable land management, food security, and greenhouse gas fluxes in terrestrial ecosystems* (pp. 437–550). https://www.ipcc.ch/site/assets/uploads/sites/4/2022/11/SRCCL_Chapter_5.pdf

Mbow, C., Rosenzweig, C. E., Barioni, L. G., Benton, T. G., Herrero, M., Krishnapillai, M., & Diouf, A. A. (2020). *Food security (No. GSFC-E-DAA-TN78913)*. IPCC.

MBRAPA. (2023). Brasil está na vanguarda no desenvolvimento de carne cultivada - Portal Embrapa. https://www.embrapa.br/busca-de-noticias/-/noticia/77704192/brasil-esta-na-vanguarda-no-desenvolvimento-de-carne-cultivada

McCulloch, S. P. (2013). A critique of FAWC's five freedoms as a framework for the analysis of animal welfare. *Journal of Agricultural & Environmental Ethics*, 26(5), 959–975. 10.1007/s10806-012-9434-7

McMillan, F. D. (2002). Development of a mental wellness program for animals. *Journal of the American Veterinary Medical Association*, 220(7), 965–972. 10.2460/javma.2002.220.96512420769

McNamara, E. (n.d.). Consumer education on the food safety of cultivated meat. *Good Food Institute*. https://gfi.org/solutions/consumer-education-food-safety-cultivated-meat/

Meatable (2024). Press release: First EU tasting. *Meatable*. https://meatable.com/news-room/

Mee, J. F. (2013). Why do so many calves die on modern dairy farms and what can we do about calf welfare in the future? *Animals (Basel)*, 3(4), 1036–1057. 10.3390/ani304103626479751

Meissner, H., Blignaut, J., Smith, H. & Du Toit, L. (2023). *The broad-based eco-economic impact of beef and dairy production: A global review.* https://doi.org/10.4314/sajas.v53i2.11

Mekonnen, M. M., & Hoekstra, A. Y. (2012). A global assessment of the water footprint of farm animal products. *Ecosystems (New York, N.Y.)*, 15(3), 401–415. 10.1007/s10021-011-9517-8

Melzener, L., Verzijden, K. E., Buijs, A. J., Post, M. J., & Flack, J. E. (2021). Cultured beef: From small biopsy to substantial quantity. *Journal of the Science of Food and Agriculture*, 101(1), 7–14. 10.1002/jsfa.1066332662148

Mendes, A. C., Stephansen, K., & Chronakis, I. S. (2017). Electrospinning of food proteins and polysaccharides. *Food Hydrocolloids*, 68, 53–68. 10.1016/j.foodhyd.2016.10.022

Mendly-Zambo, Z., Powell, L. J., & Newman, L. L. (2021). Dairy 3.0: Cellular agriculture and the future of milk. *Food, Culture, & Society*, 24(5), 675–693. 10.1080/15528014.2021.1888411

Mengistie, D. (2020). Lab-growing meat production from stem cell. *Journal of Nutrition & Food Sciences*, 3(1), 100015.

Mia, N., Rahman, M. M., & Hashem, M. A. (2023). Effect of heat stress on meat quality: A review. *Meat Research*, 3(6). Advance online publication. 10.55002/mr.3.6.73

Milford, A. B., Le Mouël, C., Bodirsky, B. L., & Rolinski, S. (2019). Drivers of meat consumption. *Appetite*, 141, 104313. 10.1016/j.appet.2019.06.00531195058

Mimuma, S., Kimura, N., Hirata, M., Tateyama, D., Hayashida, M., Umezawa, A., & Furue, M. K. (2011). Growth factor-defined culture medium for human mesenchymal stem cells. *The International Journal of Developmental Biology*, 55(2), 181–187. 10.1387/ijdb.103232sm21305471

Ministry of Food and Drug Safety. (2023). [Press Release] Republic of Korea elected as inaugural chair of APFRAS. *Ministry of Food and Drug Safety*. https://www.mfds.go.kr/eng/brd/m_61/view.do?seq=149

Ministry of Health. (2024a). First in the world: The Ministry of Health has approved cattle-based cultivated meat. *Ministry of Health*. https://www.gov.il/en/pages/17012024-02

Ministry of Health. (2024b). New food in Israel. *Ministry of Health*. https://www.gov.il/he/pages/novel-food

Misawa, M. (1977). Production of natural substances by plant cell cultures described in Japanese patents. In *Plant Tissue Culture and Its Bio-technological Application: Proceedings of the First International Congress on Medicinal Plant Research, Section B, held at the University of Munich, Germany September 6–10, 1976* (pp. 17-26). Springer Berlin Heidelberg.

Modarresi Chahardehi, A., Arsad, H., & Lim, V. (2020). Zebrafish as a successful animal model for screening toxicity of medicinal plants. *Plants*, 9(10), 1345. 10.3390/plants910134533053800

Modulevsky, D. J., Cuerrier, C. M., & Pelling, A. E. (2016). Biocompatibility of subcutaneously implanted plant-derived cellulose biomaterials. *PLoS One*, 11(6), e0157894. 10.1371/journal.pone.015789427328066

Mohamed, M. (2023). Agricultural Sustainability in the Age of Deep Learning: Current Trends, Challenges, and Future Trajectories. *Sustainable Machine Intelligence Journal*, 4, 2–1. 10.61185/SMIJ.2023.44102

Mohanty, S., Larsen, L. B., Trifol, J., Szabo, P., Burri, H. V. R., Canali, C., Dufva, M., Emnéus, J., & Wolff, A. (2015). Fabrication of scalable and structured tissue engineering scaffolds using water dissolvable sacrificial 3D printed moulds. *Materials Science and Engineering C*, 55, 569–578. 10.1016/j.msec.2015.06.002

Mohorčich, J., & Reese, J. (2019). Cell-cultured meat: Lessons from GMO adoption and resistance. *Appetite*, 143, 104408. 10.1016/j.appet.2019.10440831449883

Möhring, N., Ingold, K., Kudsk, P., Martin-Laurent, F., Niggli, U., Siegrist, M., & Finger, R. (2020). Pathways for advancing pesticide policies. *Nature Food*, 1(9), 535–540. 10.1038/s43016-020-00141-437128006

Molossi, L., Hoshide, A. K., Pedrosa, L. M., Oliveira, A. S. D., & Abreu, D. C. D. (2020). Improve pasture or feed grain? Greenhouse gas emissions, profitability, and resource use for nelore beef cattle in Brazil's Cerrado and Amazon Biomes. *Animals (Basel)*, 10(8), 1386. 10.3390/ani1008138632785150

Monamy, V. (2017). *Animal experimentation: A guide to the issues*. Cambridge University Press. 10.1017/9781316678329

Montanari, M. (2006). *Food is culture*. Columbia University Press. Retrieved from https://books.google.com/books?hl=en&lr=&id=SRloQL52eysC&oi=fnd&pg=PR5&dq=Food+is+a+fundamental+component+of+cultural+identity,+and+conventional+meals+often+incorporate+specialized+ways+of+animal+rearing,+slaughtering,+and+processing.+&ots=8qPu4AEpw7&sig=zOjfzqFsFX9r2LmMLmNCfw1PzK0

Moraes, K. C., & Montagne, J. (2021). *Drosophila melanogaster*: A powerful tiny animal model for the study of metabolic hepatic diseases. *Frontiers in Physiology*, 12, 728407. 10.3389/fphys.2021.72840734603083

Morais, A. Í., Vieira, E. G., Afewerki, S., Sousa, R. B., Honorio, L. M., Cambrussi, A. N., Santos, J. A., Bezerra, R. D. S., Furtini, J. A. O., Silva-Filho, E. C., Webster, T. J., & Lobo, A. O. (2020). Fabrication of polymeric microparticles by electrospray: The impact of experimental parameters. *Journal of Functional Biomaterials*, 11(1), 4. 10.3390/jfb1101000431952157

Morais-da-Silva, R. L., Villar, E. G., Reis, G. G., Sanctorum, H., & Molento, C. F. M. (2022). The expected impact of cultivated and plant-based meats on jobs: The views of experts from Brazil, the United States and europe. Humanities and Social Sciences Communications, 9(1), 1–14. 10.1057/s41599-022-01316-z

Morales-Rubio, M. E., Espinosa-Leal, C., & Garza-Padrón, R. A. (2016). Cultivation of plant tissues and their application in natural products. In Rivas-Morales, C., Oranday-Cardenas, M. A., & Verde-Star, M. J. (Eds.), *Investigación en plantas de importancia médica* (1st ed., pp. 351–410)., 10.3926/oms.315

Mora, O., Le Mouël, C., de Lattre-Gasquet, M., Donnars, C., Dumas, P., Réchauchère, O., & Marty, P. (2020). Exploring the future of land use and food security: A new set of global scenarios. *PLoS One*, 15(7), e0235597. 10.1371/journal.pone.023559732639991

Morice, C. P., Kennedy, J. J., Rayner, N. A., Winn, J. P., Hogan, E., Killick, R. E., Dunn, R. J. H., Osborn, T. J., Jones, P. D., & Simpson, I. R. (2021). An Updated Assessment of Near-Surface Temperature Change From 1850: The HadCRUT5 Data Set. *Journal of Geophysical Research. Atmospheres*, 126(3), e2019JD032361. Advance online publication. 10.1029/2019JD032361

Moritz, J., Tuomisto, H. L., & Ryynänen, T. (2022). The transformative innovation potential of cellular agriculture: Political and policy stakeholders' perceptions of cultured meat in Germany. *Journal of Rural Studies*, 89, 54–65. 10.1016/j.jrurstud.2021.11.018

Morrone, S., Dimauro, C., Gambella, F., & Cappai, M. G. (2022). Industry 4.0 and Precision Livestock Farming (PLF): An up to Date Overview across Animal Productions. *Sensors (Basel)*, 22(12), 4319. 10.3390/s2212431935746102

Moses, A., & Tomaselli, P. (2017). Industrial animal agriculture in the United States: Concentrated animal feeding operations (CAFOs). *International farm animal, wildlife and food safety law*, 185-214. 10.1007/978-3-319-18002-1_6

Motoyama, M., Sasaki, K., & Watanabe, A. (2016). Wagyu and the factors contributing to its beef quality: A Japanese industry overview. *Meat Science*, 120, 10–18. 10.1016/j.meatsci.2016.04.02627298198

Moyano-Fernández, C. (2023). The moral pitfalls of cultivated meat: Complementing utilitarian perspective with eco-republican justice approach. *Journal of Agricultural & Environmental Ethics*, 36(1), 23. 10.1007/s10806-022-09896-136467858

Mrabet, R. (2023). Sustainable agriculture for food and nutritional security. In *Sustainable agriculture and the environment* (pp. 25–90). Academic Press. 10.1016/B978-0-323-90500-8.00013-0

Mridul, A. (2021). Cultured meat to hit UK menus by 2023, says cell-based startup Ivy Farm. The Vegan Review. https://theveganreview.com/cultured-meat-ukmenus-2023-cell-based-startup-ivy-farm/

Mridul, A. (2024, January). Aleph Farms: Israel awards the world's first regulatory approval for cultivated beef. *Green Queen*.https://www.greenqueen.com.hk/aleph-farms-israel-cultured-meat-regulatory-approval-beef/

Muehleder, S., Ovsianikov, A., Zipperle, J., Redl, H., & Holnthoner, W. (2014). Connections matter: Channeled hydrogels to improve vascularization. *Frontiers in Bioengineering and Biotechnology*, 2, 52. 10.3389/fbioe.2014.00052

Munteanu, C., Mireşan, V., Răducu, C., Ihuţ, A., Uiuiu, P., Pop, D., Neacşu, A., Cenariu, M., & Groza, I. (2021). Can cultured meat be an alternative to farm animal production for a sustainable and healthier lifestyle? *Frontiers in Nutrition*, 8, 749298. 10.3389/fnut.2021.74929834671633

Murthy, H. N., Georgiev, M. I., Kim, Y. S., Jeong, C. S., Kim, S. J., Park, S. Y., & Paek, K. Y. (2014, May 25). Ginsenosides: Prospective for sustainable biotechnological production. *Applied Microbiology and Biotechnology*, 98(14), 6243–6254. 10.1007/s00253-014-5801-924859520

Murthy, H. N., Georgiev, M. I., Park, S. Y., Dandin, V. S., & Paek, K. Y. (2015, June). The safety assessment of food ingredients derived from plant cell, tissue and organ cultures: A review. *Food Chemistry*, 176, 426–432. 10.1016/j.foodchem.2014.12.07525624252

Mysko, L., Minviel, J.-J., Veysset, P., & Veissier, I. (2024). How to concurrently achieve economic, environmental, and animal welfare performances in French suckler cattle farms. *Agricultural Systems*, 218, 103956. 10.1016/j.agsy.2024.103956

Nacef, M., Lelièvre-Desmas, M., Symoneaux, R., Jombart, L., Flahaut, C., & Chollet, S. (2019). Consumers' expectation and liking for cheese: Can familiarity effects resulting from regional differences be highlighted within a country? *Food Quality and Preference*, 72, 188–197. 10.1016/j.foodqual.2018.10.004

Nadathur, S., Wanasundara, J. P., Marinangeli, C. P. F., & Scanlin, L. (2024). Proteins in Our Diet: Challenges in Feeding the Global Population. In Sustainable Protein Sources (pp. 1-29). Academic Press.

Naing, A. H., Kim, S. H., Chung, M. Y., Park, S. K., & Kim, C. K. (2019). In vitro propagation method for production of morphologically and genetically stable plants of different strawberry cultivars. *Plant Methods*, 15(1), 36. 10.1186/s13007-019-0421-031011361

Nakyinsige, K., Man, Y. B. C., & Sazili, A. Q. (2012). Halal authenticity issues in meat and meat products. *Meat Science*, 91(3), 207–214. 10.1016/j.meatsci.2012.02.01522405913

Nannoni, E., & Mancini, C. (2024). Toward an integrated ethical review process: An animal-centered research framework for the refinement of research procedures. *Frontiers in Veterinary Science*, 11, 1343735. 10.3389/fvets.2024.134373538694478

Napolitano, A. P., Chai, P., Dean, D. M., & Morgan, J. R. (2007). Dynamics of the self-assembly of complex cellular aggregates on micromolded nonadhesive hydrogels. *Tissue Engineering*, 13(8), 2087–2094. 10.1089/ten.2006.019017518713

Naser, M. M., & Pearce, P. (2022). Evolution of the International Climate Change Policy and Processes: UNFCCC to Paris Agreement. In *Oxford Research Encyclopedia of Environmental Science*. Oxford University Press. 10.1093/acrefore/9780199389414.013.422

National Research Council. (1999). *The Use of Drugs in Food Animals: Benefits and Risks*. The National Academies Press. 10.17226/5137

National Research Council. (2006). *Guidelines for the Humane Transportation of Research Animals*. The National Academies Press. 10.17226/11557

National Research Council. (2011). *Guide for the Care and Use of Laboratory Animals* (8th ed.). The National Academies Press. 10.17226/12910

Navarro, M., Soto, D. A., Pinzon, C. A., Wu, J., & Ross, P. J. (2019). Livestock pluripotency is finally captured in vitro. *Reproduction, Fertility, and Development*, 32(2), 11–39. 10.1071/RD19272321885555

Neethirajan, S. (2023). Artificial Intelligence and Sensor Technologies in Dairy Livestock Export: Charting a Digital Transformation. *Sensors (Basel)*, 23(16), 7045. 10.3390/s2316704537631580

Neethirajan, S. (2024). Artificial Intelligence and Sensor Innovations: Enhancing Livestock Welfare with a Human-Centric Approach. *Hum-Cent Intell Syst*, 4(1), 77–92. 10.1007/s44230-023-00050-2

Nepstad, D., McGrath, D., Stickler, C., Alencar, A., Azevedo, A., Swette, B., & Hess, L. (2014). Slowing Amazon deforestation through public policy and interventions in beef and soy supply chains. *Science*, 344(6188), 1118-1123.

Newell, R., Newman, L., & Mendly-Zambo, Z. (2021). The role of incubators and accelearators in the fourth agricultural revolution: A case study of Canada. *Agriculture*, 11(11), 1066. 10.3390/agriculture11111066

Newman, L. (2020). *The Promise and Peril of "Cultured Meat"; McGill-Queen's Uuniversity Press*. QU, Canada.

Newman, L., Fraser, E., Newell, R., Bowness, E., Newman, K., & Glaros, A. (2023). Cellular agriculture and the sustainable development goals. In *Genomics and the global bioeconomy* (pp. 3–23). Academic Press. 10.1016/B978-0-323-91601-1.00010-9

Newman, L., Newell, R., Dring, C., Glaros, A., Fraser, E., Mendly-Zambo, Z., & Kc, K. B. (2023). Agriculture for the Anthropocene: Novel applications of technology and the future of food. *Food Security*, 15(3), 613–627. 10.1007/s12571-023-01356-6

Newman, L., Newell, R., Mendly-Zambo, Z., & Powell, L. (2021). Bioengineering, telecoupling, and alternative dairy: Agricultural land use futures in the Anthropocene. *The Geographical Journal*, 188(3), 342–357. 10.1111/geoj.12392

Newton, P., & Blaustein-Rejto, D. (2021). Social and economic opportunities and challenges of plant-based and cultured meat for rural producers in the US. *Frontiers in Sustainable Food Systems*, 5, 624270. 10.3389/fsufs.2021.624270

Nickerson, C. A., Ott, C. M., Mister, S. J., Morrow, B. J., Burns-Keliher, L., & Pierson, D. L. (2004). Microgravity as a novel environmental signal affecting Salmonella enterica serovar Typhimurium virulence. *Infection and Immunity*, 72(4), 2247–2256.10816456

Nielsen, B. L., Dybkjær, L., & Herskin, M. S. (2011). Road transport of farm animals: Effects of journey duration on animal welfare. *Animal*, 5(3), 415–427. 10.1017/S175173111000198922445408

Nielsen, H. B., Sonne, A., Grunert, K. G., Bánáti, D., Pollák-Tóth, A., Lakner, Z., Olsen, N. V., Žontar, T. P., & Peterman, M. (2009). Consumer perception of the use of high-pressure processing and pulsed electric field technologies in food production. *Appetite*, 52(1), 115–126. 10.1016/j.appet.2008.09.01018845196

Nie, W., Peng, C., Zhou, X., Chen, L., Wang, W., Zhang, Y., Ma, P. X., & He, C. (2017). Three dimensional porous scaffold by self-assembly of reduced graphene oxide and nano-hydroxyapatite composites for bone tissue engineering. *Carbon*, 116, 325–337. 10.1016/j.carbon.2017.02.013

Nijdam, D., Rood, T., & Westhoek, H. (2012). The price of protein: Review of land use and carbon footprints from life cycle assessments of animal food products and their substitutes. *Food Policy*, 37(6), 760–770. 10.1016/j.foodpol.2012.08.002

Nogueira, D. E. S., Rodrigues, C. A. V., Carvalho, M. S., Miranda, C. C., Hashimura, Y., Jung, S., Lee, B., & Cabral, J. M. S. (2019, September 14). Strategies for the expansion of human induced pluripotent stem cells as aggregates in single-use Vertical-Wheel™ bioreactors. *Journal of Biological Engineering*, 13(1), 74. 10.1186/s13036-019-0204-131534477

Nordlund, E., Lille, M., Silventoinen, P., Nygren, H., Seppänen-Laakso, T., Mikkelson, A., Aura, A. M., Heiniö, R. L., Nohynek, L., Puupponen-Pimiä, R., & Rischer, H. (2018, May). Plant cells as food – A concept taking shape. *Food Research International*, 107, 297–305. 10.1016/j.foodres.2018.02.04529580489

Novellino, A., Brown, T. J., Bide, T., Th c Anh, N. T., Petavratzi, E., & Kresse, C. (2021). Using Satellite Data to Analyse Raw Material Consumption in Hanoi, Vietnam. *Remote Sensing (Basel)*, 13(3), 334. 10.3390/rs13030334

Nowak, V., Persijn, D., Rittenschober, D., & Charrondiere, U. R. (2016). Review of food composition data for edible insects. *Food Chemistry*, 193, 39–46. 10.1016/j.foodchem.2014.10.11426433285

Nyika, J., Mackolil, J., Workie, E., Adhav, C., & Ramadas, S. (2021). Cellular agriculture research progress and prospects: Insights from bibliometric analysis. *Current Research in Biotechnology*, 3, 215–224. 10.1016/j.crbiot.2021.07.001

Nyyssölä, A., Suhonen, A., Ritala, A., & Oksman-Caldentey, K. M. (2022). The role of single cell protein in cellular agriculture. *Current Opinion in Biotechnology*, 75, 102686. 10.1016/j.copbio.2022.10268635093677

O'Neill, E. N., Ansel, J. C., Kwong, G. A., Plastino, M. E., Nelson, J., Baar, K., & Block, D. E. (2022). Spent media analysis suggests cultivated meat media will require species and cell type optimization. *NPJ Science of Food*, 6(1), 46. 10.1038/s41538-022-00157-z36175443

O'Neill, E. N., Cosenza, Z. A., Baar, K., & Block, D. E. (2021). Considerations for the development of cost-effective cell culture media for cultivated meat production. *Comprehensive Reviews in Food Science and Food Safety*, 20(1), 686–709. 10.1111/1541-4337.1267833325139

Odeleye, A. O. O., Baudequin, T., Chui, C. Y., Cui, Z., & Ye, H. (2020). An additive manufacturing approach to bioreactor design for mesenchymal stem cell culture. *Biochemical Engineering Journal*, 156, 107515. 10.1016/j.bej.2020.107515

Oecd, F. A. O. (2022). OECD-FAO agricultural outlook 2022-2031. Retrieved from https://policycommons.net/artifacts/2652558/oecd-fao-agricultural-outlook-2022-2031/3675435/

Ogrodnik, M. (2021). Cellular aging beyond cellular senescence: Markers of senescence prior to cell cycle arrest in vitro and in vivo. *Aging Cell*, 20(4), e13338. 10.1111/acel.1333833711211

Okamoto, Y., Haraguchi, Y., Yoshida, A., Takahashi, H., Yamanaka, K., Sawamura, N., Asahi, T., & Shimizu, T. (2022). Proliferation and differentiation of primary bovine myoblasts using Chlorella vulgaris extract for sustainable production of cultured meat. *Biotechnology Progress*, 38(3), e3239. 10.1002/btpr.323935073462

Olowe, V. (2021). Africa 2100: How to feed Nigeria in 2100 with 800 million inhabitants. *Organic Agriculture*, 11(2), 199–208. 10.1007/s13165-020-00307-1

Olsson, I. A. S., Nielsen, B. L., Camerlink, I., Pongrácz, P., Golledge, H. D., Chou, J. Y., Ceballos, M. C., & Whittaker, A. L. (2022). An international perspective on ethics approval in animal behaviour and welfare research. *Applied Animal Behaviour Science*, 253, 105658. 10.1016/j.applanim.2022.105658

Ong, K. J., Johnston, J., Datar, I., Sewalt, V., Holmes, D., & Shatkin, J. A. (2021). Food safety considerations and research priorities for the cultured meat and seafood industry. *Comprehensive Reviews in Food Science and Food Safety*, 20(6), 5421–5448. 10.1111/1541-4337.1285334633147

Ong, S., Choudhury, D., & Naing, M. W. (2020). Cell-based meat: Current ambiguities with nomenclature. *Trends in Food Science & Technology*, 102, 223–231. 10.1016/j.tifs.2020.02.010

Orellana, N., Sánchez, E., Benavente, D., Prieto, P., Enrione, J., & Acevedo, C. A. (2020). A New Edible Film to Produce In Vitro Meat. *Foods*, 9(2), 185. 10.3390/foods902018532069986

Ozbolat, I. T., & Hospodiuk, M. (2016). Current Advances and Future Perspectives in Extrusion-Based Bioprinting. *Biomaterials*, 76, 321–343. 10.1016/j.biomaterials.2015.10.07626561931

Padilha, L., Malek, L., & Umberger, W. (2021). Food choice drivers of potential lab-grown meat consumers in australia. *British Food Journal*, 123(9), 3014–3031. 10.1108/BFJ-03-2021-0214

Paek, K. Y., Chakrabarty, D., & Hahn, E. J. (2005, June). Application of bioreactor systems for large scale production of horticultural and medicinal plants. *Plant Cell, Tissue and Organ Culture*, 81(3), 287–300. 10.1007/s11240-004-6648-z

Pajčin, I., Knežić, T., Savic Azoulay, I., Vlajkov, V., Djisalov, M., Janjušević, L., Grahovac, J., & Gadjanski, I. (2022). Bioengineering outlook on cultivated meat production. *Micromachines*, 13(3), 402. 10.3390/mi1303040235334693

Palmer, A., Greenhough, B., Hobson-West, P., Message, R., Aegerter, J. N., Belshaw, Z., Dennison, N., Dickey, R., Lane, J., Lorimer, J., Millar, K., Newman, C., Pullen, K., Reynolds, S. J., Wells, D. J., Witt, M. J., & Wolfensohn, S. (2020). Animal research beyond the laboratory: Report from a workshop on places other than licensed establishments (POLEs) in the UK. *Animals (Basel)*, 10(10), 1868. 10.3390/ani1010186833066272

Panchasara, H., Samrat, N. H., & Islam, N. (2021). Greenhouse gas emissions trends and mitigation measures in Australian agriculture sector—. *Revista de Agricultura (Piracicaba)*, 11(2), 85.

Pandurangan, M., & Kim, D. H. (2015). A novel approach for in vitro meat production. *Applied Microbiology and Biotechnology*, 99(13), 5391–5395. 10.1007/s00253-015-6671-5

Papastavrou, V., & Ryan, C. (2023). Ethical standards for research on marine mammals. *Research Ethics*, 19(4), 390–408. 10.1177/17470161231182066

Parihar, S. S., Saini, K. P. S., Lakhani, G. P., Jain, A., Roy, B., Ghosh, S., & Aharwal, B. (2019). Livestock waste management: A review. *Journal of Entomology and Zoology Studies*, 7(3), 384–393.

Park, J. A., Yoon, S., Kwon, J., Kim, Y. K., Kim, W. J., Yoo, J. Y., & Jung, S. (2017). Freeform micropatterning of living cells into cell culture medium using direct inkjet printing. *Scientific Reports*, 7(1), 14610. 10.1038/s41598-017-14726-w29097768

Park, S., Jung, S., Choi, M., Lee, M., Choi, B., Koh, W. G., Lee, S., & Hong, J. (2021). Gelatin MAGIC powder as nutrient-delivering 3D spacer for growing cell sheets into cost-effective cultured meat. *Biomaterials*, 278, 121155. 10.1016/j.biomaterials.2021.12115534607049

Park, Y. H., Gong, S. P., Kim, H. Y., Kim, G. A., Choi, J. H., Ahn, J. Y., & Lim, J. M. (2013). Development of a serum-free defined system employing growth factors for preantral follicle culture. *Molecular Reproduction and Development*, 80(9), 725–733. 10.1002/mrd.2220423813589

Parsons, E. C. M., Baulch, S., Bechshoft, T., Bellazzi, G., Bouchet, P., Cosentino, A. M., Godard-Codding, C. A. J., Gulland, F., Hoffmann-Kuhnt, M., Hoyt, E., Livermore, S., MacLeod, C. D., Matrai, E., Munger, L., Ochiai, M., Peyman, A., Recalde-Salas, A., Regnery, R., Rojas-Bracho, L., & Sutherland, W. J. (2015). Key research questions of global importance for cetacean conservation. *Endangered Species Research*, 27(2), 113–118. 10.3354/esr00655

Pathirana, R., & Carimi, F. (2024). Plant Biotechnology-An Indispensable Tool for Crop Improvement. *Plants*, 13(8), 1133. 10.3390/plants13081133338674542

Pawlak, K., & Kołodziejczak, M. (2020). The role of agriculture in ensuring food security in developing countries: Considerations in the context of the problem of sustainable food production. *Sustainability (Basel)*, 12(13), 5488. 10.3390/su12135488

Penuelas, J., Coello, F., & Sardans, J. (2023). A better use of fertilizers is needed for global food security and environmental sustainability. *Agriculture &Agriculture & Food Security*, 12(1), 5. Advance online publication. 10.1186/s40066-023-00409-5

Percie du Sert, N., Alfieri, A., Allan, S. M., Carswell, H. V., Deuchar, G. A., Farr, T. D., Flecknell, P., Gallagher, L., Gibson, C. L., Haley, M. J., Macleod, M. R., McColl, B. W., McCabe, C., Morancho, A., Moon, L. D. F., O'Neill, M. J., Pérez de Puig, I., Planas, A., Ragan, C. I., & Macrae, I. M. (2017). The IMPROVE guidelines (ischaemia models: Procedural refinements of in vivo experiments). *Journal of Cerebral Blood Flow and Metabolism*, 37(11), 3488–3517. 10.1177/0271678X1770918528797196

Percival, B., & Percival, F. (n.d.). Reinventing the Wheel. In *Milk, Microbes, and the Fight for Real Cheese*. University of California Press. https://doi.org/doi:10.1525/9780520964464

Pereira, L. M., Drimie, S., Maciejewski, K., Tonissen, P. B., & Biggs, R. (2020). Food system transformation: Integrating a political–economy and social–ecological approach to regime shifts. *International Journal of Environmental Research and Public Health*, 17(4), 1313. 10.3390/ijerph1704131332085576

Perino, G., & Schwickert, H. (2023). Animal welfare is a stronger determinant of public support for meat taxation than climate change mitigation in Germany. *Nature Food*, 4(2), 160–169. 10.1038/s43016-023-00696-y37117860

Perreault, L. R., Thyden, R., Kloster, J., Jones, J. D., Nunes, J., Patmanidis, A. A., Reddig, D., Dominko, T., & Gaudette, G. R. (2023). Repurposing agricultural waste as low-cost cultured meat scaffolds. *Frontiers in Food Science and Technology*, 3, 1208298. 10.3389/frfst.2023.1208298

Phelan, M. A., Kruczek, K., Wilson, J. H., Brooks, M. J., Drinnan, C. T., Regent, F., Gerstenhaber, J. A., Swaroop, A., Lelkes, P. I., & Li, T. (2020). Soy protein nanofiber scaffolds for uniform maturation of human induced pluripotent stem cell-derived retinal pigment epithelium. *Tissue Engineering. Part C, Methods*, 26(8), 433–446. 10.1089/ten.tec.2020.007232635833

Picouet, P. A., Fernandez, A., Realini, C. E., & Lloret, E. (2014). Influence of PA6 nanocomposite films on the stability of vacuum-aged beef loins during storage in modified atmospheres. *Meat Science*, 96(1), 574–580. 10.1016/j.meatsci.2013.07.020

Pimentel, D., & Pimentel, M. (2003). Sustainability of meat-based and plant-based diets and the environment1. *The American Journal of Clinical Nutrition*, 78(3), 660S–663S. 10.1093/ajcn/78.3.660S

Piwowar, A. (2020). Farming practices for reducing ammonia emissions in Polish agriculture. *Atmosphere (Basel)*, 11(12), 1353. 10.3390/atmos11121353

Plec, E. (2024). The Animal People. In *Animal Activism On and Off Screen* (pp. 65–88). Sydney University Press. 10.2307/jj.16110774.7

Poirier, N. (2022). On the Intertwining of Cellular Agriculture and Animal Agriculture: History, Materiality, Ideology, and Collaboration. *Frontiers in Sustainable Food Systems*, 6, 907621. 10.3389/fsufs.2022.907621

Pond, W. G., Bazer, F. W., & Rollin, B. E. (2011). Animal welfare in animal agriculture. CRC Press. Retrieved from https://api.taylorfrancis.com/content/books/mono/download?identifierName=doi&identifierValue=10.1201/b11679&type=googlepdf

Porto, L., & Berti, F. (2022). Carne Cultivada: Perspectivas e Oportunidades Para o Brasil. Available online: https://gfi.org.br/wp-content/uploads/2022/06/WP-Carne-Cultivada-no-Brasil-GFI-Brasil-05_2022_.pdf

Post, M. J. (2012). Cultured meat from stem cells: Challenges and prospects. *Meat Science*, 92(3), 297–301. 10.1016/j.meatsci.2012.04.00822543115

Post, M. J. (2014). Cultured beef: Medical technology to produce food. *Journal of the Science of Food and Agriculture*, 94(6), 1039–1041. 10.1002/jsfa.647424214798

Post, M., Levenberg, S., Kaplan, D., Genovese, N., Fu, J., Bryant, C., Negowetti, N., Verzijden, K., & Moutsatsou, P. (2020). Scientific, sustainability and regulatory challenges of cultured meat. *Nature Food*, 1(7), 403–415. 10.1038/s43016-020-0112-z

Potopová, V., Musiolková, M., Gaviria, J. A., Trnka, M., Havlík, P., Boere, E., Trifan, T., Muntean, N., & Chawdhery, M. R. A. (2023). Water Consumption by Livestock Systems from 2002–2020 and Predictions for 2030–2050 under Climate Changes in the Czech Republic. *Agriculture*, 13(7), 1291. 10.3390/agriculture13071291

Prasertsan, S., & Sajjakulnukit, B. (2006). Biomass and biogas energy in Thailand: Potential, opportunity and barriers. *Renewable Energy*, 31(5), 599–610. 10.1016/j.renene.2005.08.005

Premalatha, M., Abbasi, T., Abbasi, T., & Abbasi, S. A. (2011). Energy-efficient food production to reduce global warming and ecodegradation: The use of edible insects. In Renewable and Sustainable Energy Reviews (Vol. 15, Issue 9, pp. 4357–4360). 10.1016/j.rser.2011.07.115

Profeta, A., Siddiqui, S. A., Smetana, S., Hossaini, S. M., Heinz, V., & Kircher, C. (2021). The Impact of Corona Pandemic on Consumer's Food Consumption. *Journal für Verbraucherschutz und Lebensmittelsicherheit*, 16(4), 305–314. 10.1007/s00003-021-01341-134421498

Purdy, C. (2019). The first cell-cultured meat will cost about $50. https://qz.com/1598076/the-first-cell-cultured-meat-will-cost-about-50

Quesada-Salas, M. C., Delfau-Bonnet, G., Willig, G., Préat, N., Allais, F., & Ioannou, I. (2021). Article optimization and comparison of three cell disruption processes on lipid extraction from microalgae. *Processes (Basel, Switzerland)*, 9(2), 1–20. 10.3390/pr9020369

Quinlan, R. J., Rumas, I., Naisikye, G., Quinlan, M. B., & Yoder, J. (2016). Searching for symbolic value of cattle: Tropical livestock units, market price, and cultural value of Maasai livestock. *Ethnobiology Letters*, 7(1), 76–86. https://www.jstor.org/stable/26423652. 10.14237/ebl.7.1.2016.621

Qu, J., Zhang, W., Yu, X., & Jin, M. (2005, April). Instability of anthocyanin accumulation in Vitis vinifera L. var. Gamay Fréaux suspension cultures. *Biotechnology and Bioprocess Engineering; BBE*, 10(2), 155–161. 10.1007/BF02932586

Radaei, P., Mashayekhan, S., & Vakilian, S. (2017). Modeling and optimization of gelatin- chitosan micro-carriers preparation for soft tissue engineering: Using Response Surface Methodology. 10.1016/j.msec.2017.02.108

Rafiq, Q. A., Ruck, S., Hanga, M. P., Heathman, T. R., Coopman, K., Nienow, A. W., Williams, D. J., & Hewitt, C. J. (2018, July). Qualitative and quantitative demonstration of bead-to-bead transfer with bone marrow-derived human mesenchymal stem cells on microcarriers: Utilising the phenomenon to improve culture performance. *Biochemical Engineering Journal*, 135, 11–21. 10.1016/j.bej.2017.11.005

Raihan, A., & Tuspekova, A. (2022). Dynamic impacts of economic growth, energy use, urbanization, tourism, agricultural value-added, and forested area on carbon dioxide emissions in Brazil. *Journal of Environmental Studies and Sciences*, 12(4), 794–814. 10.1007/s13412-022-00782-w

Ramachandraiah, K., Han, S. G., & Chin, K. B. (2015). Nanotechnology in meat processing and packaging: Potential applications - a review. *Asian-Australasian Journal of Animal Sciences*, 28(2), 290–302. 10.5713/ajas.14.0607

Ramani, S., Ko, D., Kim, B., Cho, C., Kim, W., Jo, C., & Park, S. (2021). Technical requirements for cultured meat production: A review. *Journal of Animal Science and Technology*, 63(4), 681–692. 10.5187/jast.2021.e4534447948

Rana Khalid, I., Darakhshanda, I., & Rafi, R. (2019, July 5). 3D Bioprinting: An attractive alternative to traditional organ transplantation. *Archive of Biomedical Science and Engineering*, 5(1), 7–18.

Ranga, A.. (2014). Muscle tissue engineering: From cell biology to cell assembly. *Advanced Drug Delivery Reviews*, 84, 107–124. 10.1016/j.addr.2014.03.004

Rasche, S., Herwartz, D., Schuster, F., Jablonka, N., Weber, A., Fischer, R., & Schillberg, S. (2016, March 18). More for less: Improving the biomass yield of a pear cell suspension culture by design of experiments. *Scientific Reports*, 6(1), 23371. 10.1038/srep2337126988402

Räty, N., Tuomisto, H. L., & Ryynänen, T. (2023). On what basis is it agriculture?: A qualitative study of farmers' perceptions of cellular agriculture. *Technological Forecasting and Social Change*, 196, 122797. 10.1016/j.techfore.2023.122797

Reddy, M. S. B., Ponnamma, D., Choudhary, R., & Sadasivuni, K. K. (2021). A comparative review of natural and synthetic biopolymer composite scaffolds. *Polymers*, 13(7), 1105. 10.3390/polym1307110533808492

Reeds, P., Schaafsma, G., Tomé, D., & Young, V. (2000). Criteria and Significance of Dietary Protein Sources in Humans Summary of the Workshop with Recommendations 1.

Rehman, N., Edkins, V., & Ogrinc, N. (2024). Is sustainable consumption a sufficient motivator for consumers to adopt meat alternatives? a consumer perspective on plant-based, cell-culture-derived, and insect-based alternatives. *Foods*, 13(11), 1627. 10.3390/foods1311162738890856

Reis, G. G., Heidemann, M. S., Borini, F. M., & Molento, C. F. M. (2020). Livestock value chain in transition: Cultivated (cell-based) meat and the need for breakthrough capabilities. *Technology in Society*, 62, 101286. 10.1016/j.techsoc.2020.101286

Reiss, J., Robertson, S., & Suzuki, M. (2021). Cell sources for cultivated meat: Applications and considerations throughout the production workflow. *International Journal of Molecular Sciences*, 22(14), 7513. 10.3390/ijms2214751334299132

Ren, Z., Dong, Y., Lin, D., Zhang, L., Fan, Y., & Xia, X. (2022). Managing energy-water-carbon-food nexus for cleaner agricultural greenhouse production: A control system approach. *The Science of the Total Environment*, 848, 157756. 10.1016/j.scitotenv.2022.15775635926594

Resare Sahlin, K., Carolus, J., von Greyerz, K., Ekqvist, I., & Röös, E. (2022). Delivering "less but better" meat in practice—A case study of a farm in agroecological transition. *Agronomy for Sustainable Development*, 42(2), 24. 10.1007/s13593-021-00737-5

Ressurreição, A., Gibbons, J., Kaiser, M., Dentinho, T. P., Zarzycki, T., Bentley, C., Austen, M., Burdon, D., Atkins, J., Santos, R. S., & Edwards-Jones, G. (2012). Different cultures, different values: The role of cultural variation in public's WTP for marine species conservation. *Biological Conservation*, 145(1), 148–159. 10.1016/j.biocon.2011.10.026

Rilov, G., Mazaris, A. D., Stelzenmüller, V., Helmuth, B., Wahl, M., Guy-Haim, T., Mieszkowska, N., Ledoux, J.-B., & Katsanevakis, S. (2019). Adaptive marine conservation planning in the face of climate change: What can we learn from physiological, ecological and genetic studies? *Global Ecology and Conservation*, 17, e00566. 10.1016/j.gecco.2019.e00566

Rischer, H., Szilvay, G. R., & Oksman-Caldentey, K. M. (2020, February). Cellular agriculture — Industrial biotechnology for food and materials. *Current Opinion in Biotechnology*, 61, 128–134. 10.1016/j.copbio.2019.12.00331926477

Risner, D., Li, F., Fell, J. S., Pace, S. A., Siegel, J. B., Tagkopoulos, I., & Spang, E. S. (2020). Preliminary Techno-Economic Assessment of Animal Cell-Based Meat. *Foods*, 10(1), 3. 10.3390/foods1001000333374916

Ritala, A., Häkkinen, S. T., Toivari, M., & Wiebe, M. G. (2017, October 13). Single Cell Protein—State-of-the-Art, Industrial Landscape and Patents 2001–2016. *Frontiers in Microbiology, 8*.

Rodríguez Escobar, M. I., Cadena, E., Nhu, T. T., Cooreman-Algoed, M., De Smet, S., & Dewulf, J. (2021). Analysis of the Cultured Meat Production System in Function of Its Environmental Footprint: Current Status, Gaps and Recommendations. *Foods*, 10(12), 2941. 10.3390/foods1012294134945492

Roe, S., Streck, C., Beach, R., Busch, J., Chapman, M., Daioglou, V., Deppermann, A., Doelman, J., Emmet-Booth, J., Engelmann, J., Fricko, O., Frischmann, C., Funk, J., Grassi, G., Griscom, B., Havlik, P., Hanssen, S., Humpenöder, F., Landholm, D., & Lawrence, D. (2021). Land-based measures to mitigate climate change: Potential and feasibility by country. *Global Change Biology*, 27(23), 6025–6058. 10.1111/gcb.1587334636101

Rogers, A. D., Baco, A., Escobar-Briones, E., Currie, D., Gjerde, K., Gobin, J., Jaspars, M., Levin, L., Linse, K., Rabone, M., Ramirez-Llodra, E., Sellanes, J., Shank, T. M., Sink, K., Snelgrove, P. V. R., Taylor, M. L., Wagner, D., & Harden-Davies, H. (2021). Marine genetic resources in areas beyond national jurisdiction: Promoting marine scientific research and enabling equitable benefit sharing. *Frontiers in Marine Science*, 8, 667274. 10.3389/fmars.2021.667274

Rojas-Tavara, A. (2023). Microalgae in lab-grown meat production. *Czech Journal of Food Sciences*, 41(6), 406–418. 10.17221/69/2023-CJFS

Rolland, N. C. M., Markus, C. R., & Post, M. J. (2020). The effect of information content on acceptance of cultured meat in a tasting context. *PLoS One*, 15(4), e0231176. 10.1371/journal.pone.023117632298291

Rollin, B. E. (2006). The regulation of animal research and the emergence of animal ethics: A conceptual history. *Theoretical Medicine and Bioethics*, 27(4), 285–304. 10.1007/s11017-006-9007-816937023

Rombach, M., Dean, D., Vriesekoop, F., de Koning, W., Aguiar, L. K., Anderson, M., Mongondry, P., Oppong-Gyamfi, M., Urbano, B., Gómez Luciano, C. A., Hao, W., Eastwick, E., Jiang, Z. V., & Boereboom, A. (2022). Is cultured meat a promising consumer alternative? Exploring key factors determining consumer's willingness to try, buy and pay a premium for cultured meat. *Appetite*, 179, 106307. 10.1016/j.appet.2022.10630736089124

Röös, E., Bajželj, B., Smith, P., Patel, M., Little, D., & Garnett, T. (2017, November). Greedy or needy? Land use and climate impacts of food in 2050 under different livestock futures. *Global Environmental Change*, 47, 1–12. 10.1016/j.gloenvcha.2017.09.001

Rosenfeld, D. L., & Tomiyama, A. J. (2023). Toward consumer acceptance of cultured meat. *Trends in Cognitive Sciences*, 27(8), 689–691. 10.1016/j.tics.2023.05.00237246026

Rothwell, S. A., Doody, D. G., Johnston, C., Forber, K. J., Cencic, O., Rechberger, H., & Withers, P. J. A. (2020). Phosphorus stocks and flows in an intensive livestock dominated food system. *Resources, Conservation and Recycling*, 163, 105065. 10.1016/j.resconrec.2020.10506533273754

Rowe, B. D. (2011). Understanding animals-becoming-meat: Embracing a disturbing education. Critical Education, 2(7). Retrieved from https://ices.library.ubc.ca/index.php/criticaled/article/view/182311

Roy, B., Hagappa, A., Ramalingam, Y. D., & Mahalingam, N. (2021). A review on lab-grown meat: Advantages and disadvantages. *Quest International Journal of Medical and Health Sciences*, 4(1), 19–24.

Rubio, N. R., Fish, K. D., Trimmer, B. A., & Kaplan, D. L. (2019). In Vitro Insect Muscle for Tissue Engineering Applications. *ACS Biomaterials Science & Engineering*, 5(2), 1071–1082. 10.1021/acsbiomaterials.8b0126133405797

Rubio, N. R., Xiang, N., & Kaplan, D. L. (2020). Plant-based and cell-based approaches to meat production. *Nature Communications*, 11(1), 1–11. 10.1038/s41467-020-20061-y33293564

Rubio, N., Datar, I., Stachura, D., Kaplan, D., & Krueger, K. (2019, June 11). Cell-Based Fish: A Novel Approach to Seafood Production and an Opportunity for Cellular Agriculture. *Frontiers in Sustainable Food Systems*, 3, 3. 10.3389/fsufs.2019.00043

Rumpold, B. A., & Schlüter, O. K. (2013). Potential and challenges of insects as an innovative source for food and feed production. In *Innovative Food Science and Emerging Technologies* (Vol. 17, pp. 1–11). Elsevier Ltd., 10.1016/j.ifset.2012.11.005

Sadigov, R. (2022). Rapid growth of the world population and its socioeconomic results. *TheScientificWorldJournal*, 2022(1), 8110229. 10.1155/2022/811022935370481

Safdar, B., Zhou, H., Li, H., Cao, J., Zhang, T., Ying, Z., & Liu, X. (2022). Prospects for plant-based meat: Current standing, consumer perceptions, and shifting trends. *Foods*, 11(23), 3770. 10.3390/foods1123377036496577

Sagaradze, G., Grigorieva, O., Nimiritsky, P., Basalova, N., Kalinina, N., Akopyan, Z., & Efimenko, A. (2019, April 3). Conditioned Medium from Human Mesenchymal Stromal Cells: Towards the Clinical Translation. *International Journal of Molecular Sciences*, 20(7), 1656. 10.3390/ijms2007165630987106

Sahirman, S., & Ardiansyah. (2014). Assessment Of Tofu Carbon Footprint In Banyumas, Indonesia - Towards 'Greener' Tofu. Proceeding of International Conference on Research, Implementation And Education Of Mathematics And Sciences, 18–20.

Sahoo, D. R., & Biswal, T. (2021). Alginate and its application to tissue engineering. *SN Applied Sciences*, 3(1), 30–90. 10.1007/s42452-020-04096-w

Samal, P., Babu, S. C., Mondal, B., & Mishra, S. N. (2022). The global rice agriculture towards 2050: An inter-continental perspective. *Outlook on Agriculture*, 51(2), 164–172. 10.1177/00307270221088338

Samanta, I., Joardar, S. N., & Das, P. K. (2018). Biosecurity strategies for backyard poultry: a controlled way for safe food production. In *Food control and biosecurity* (pp. 481–517). Academic Press. 10.1016/B978-0-12-811445-2.00014-3

Sandvig, I., Karstensen, K., Rokstad, A. M., Aachmann, F. L., Formo, K., Sandvig, A., Skjåk-Bræk, G., & Strand, B. L. (2015). RGD-peptide modified alginate by a chemoenzymatic strategy for tissue engineering applications. *Journal of Biomedical Materials Research. Part A*, 103(3), 896–906. 10.1002/jbm.a.3523024826938

Santeramo, F. G., Carlucci, D., De Devitiis, B., Seccia, A., Stasi, A., Viscecchia, R., & Nardone, G. (2018). Emerging trends in European food, diets and food industry. *Food Research International*, 104, 39–47. 10.1016/j.foodres.2017.10.03929433781

Santo, R. E., Kim, B. F., Goldman, S. E., Dutkiewicz, J., Biehl, E. M., Bloem, M. W., Neff, R. A., & Nachman, K. E. (2020). Considering plant-based meat substitutes and cell-based meats: A public health and food systems perspective. *Frontiers in Sustainable Food Systems*, 4, 134. 10.3389/fsufs.2020.00134

Santos, A. C. A., Camarena, D. E. M., Roncoli Reigado, G., Chambergo, F. S., Nunes, V. A., Trindade, M. A., & Stuchi Maria-Engler, S. (2023). Tissue engineering challenges for cultivated meat to meet the real demand of a global market. International Journal of Molecular Sciences, 24(7), 6033. https://gfi.org/wp-content/uploads/2023/01/2022-Cultivated-MeatState-of-the-Industry-Report.pdf(2023)

Santovito, S., Campo, R., Rosato, P., & Khuc, L. D. (2023). Impact of faith on food marketing and consumer behaviour: A review. *British Food Journal*, 125(13), 462–481. 10.1108/BFJ-02-2023-0112

Sapkota, A. R., Curriero, F. C., Gibson, K. E., & Schwab, K. J. (2007). Antibiotic-resistant enterococci and fecal indicators in surface water and groundwater impacted by a concentrated Swine feeding operation. *Environmental Health Perspectives*, 115(7), 1040–1045. 10.1289/ehp.9770

Sapontzis, S. (2004). Food for thought: The debate over eating meat. *Environmental Values*, 15(2), 264–267.

Saunders, R. E., & Derby, B. (2014). Inkjet printing biomaterials for tissue engineering: Bioprinting. *International Materials Reviews*, 59(8), 430–448. 10.1179/1743280414Y.0000000040

Schaefer, G. O., & Savulescu, J. (2014). The ethics of producing in vitro meat. *Journal of Applied Philosophy*, 31(2), 188–202. 10.1111/japp.1205625954058

Schirmaier, C., Jossen, V., Kaiser, S. C., Jüngerkes, F., Brill, S., Safavi-Nab, A., Siehoff, A., van den Bos, C., Eibl, D., & Eibl, R. (2014, March 20). Scale-up of adipose tissue-derived mesenchymal stem cell production in stirred single-use bioreactors under low-serum conditions. *Engineering in Life Sciences*, 14(3), 292–303. 10.1002/elsc.201300134

Schumacher, H. M., Westphal, M., & Heine-Dobbernack, E. (2015). Cryopreservation of plant cell lines. *Cryopreservation and Freeze-Drying Protocols*, 423-429. https://doi.org/10.1007/978-1-4939-2193-5_21

Schuppli, C. A. (2011). Decisions about the use of animals in research: Ethical reflection by Animal Ethics Committee members. *Anthrozoos*, 24(4), 409–425. 10.2752/175303711X13159027359980

Searchinger, T., Waite, R., Hanson, C., Ranganathan, J., Dumas, P., & Matthews, E. (2018). Creating a sustainable food future. https://research.wri.org/sites/default/files/2019-07/ WRR_Food_Full_Report_0.pdf

Segovia, M., Yu, N., & Loo, E. (2022). The effect of information nudges on online purchases of meat alternatives. *Applied Economic Perspectives and Policy*, 45(1), 106–127. 10.1002/aepp.13305

Sekaran, U., Lai, L., Ussiri, D. A., Kumar, S., & Clay, S. (2021). Role of integrated crop-livestock systems in improving agriculture production and addressing food security–A review. *Journal of Agriculture and Food Research*, 5, 100190. 10.1016/j.jafr.2021.100190

Sexton, A. E. (2016). Alternative proteins and the (non)stuff of "meat.". *Gastronomica*, 16(3), 66–78. 10.1525/gfc.2016.16.3.66

Sexton, A. E., Garnett, T., & Lorimer, J. (2019). Framing the future of food: The contested promises of alternative proteins. *Environment and Planning. E, Nature and Space*, 2(1), 47–72. 10.1177/2514848619827009320393432

Shahin-Shamsabadi, A., & Selvaganapathy, P. R. (2022). Engineering Murine Adipocytes and Skeletal Muscle Cells in Meat-like Constructs Using Self-Assembled Layer-by-Layer Biofabrication: A Platform for Development of Cultivated Meat. *Cells, Tissues, Organs*, 2022(211), 304–312. 10.1159/00051176433440375

Sharma, C., Dhiman, R., Rokana, N., & Panwar, H. (2017). Nanotechnology: An Untapped Resource for Food Packaging. *Frontiers in Microbiology*, 8, 1735. 10.3389/fmicb.2017.01735

Sharma, S., Thind, S. S., & Kaur, A. (2015). In vitro meat production system: Why and how? *Journal of Food Science and Technology*, 52(12), 7599–7607. 10.1007/s13197-015-1972-326604337

Shekaran, A., Lam, A., Sim, E., Jialing, L., Jian, L., Wen, J. T. P., Chan, J. K. Y., Choolani, M., Reuveny, S., Birch, W., & Oh, S. (2016, October). Biodegradable ECM-coated PCL microcarriers support scalable human early MSC expansion and in vivo bone formation. *Cytotherapy*, 18(10), 1332–1344. 10.1016/j.jcyt.2016.06.01627503763

Shen, C. R., Chen, Y. U. S., Yang, C. J., Chen, J. K., & Liu, C. L. (2010). Colloid chitin azure is a dispersible, low-cost substrate for chitinase measurements in a sensitive, fast, reproducible assay. *Journal of Biomolecular Screening*, 15(2), 213–217. 10.1177/108705710935505720042532

Shiomi, M. (2009). Rabbit as a model for the study of human diseases. *Rabbit biotechnology*, 49-63. 10.1007/978-90-481-2227-1_7

Shi, X., & Garry, D. J. (2006). Muscle stem cells in development, regeneration, and disease. *Genes & Development*, 20(13), 1692–1708. 10.1101/gad.1419406

Shi, Z., Zhang, Y., Phillips, G. O., & Yang, G. (2014). Utilization of bacterial cellulose in food. *Food Hydrocolloids*, 35, 539–545. 10.1016/j.foodhyd.2013.07.012

Shyh-Chang, N., & Ng, H. H. (2017). The metabolic programming of stem cells. *Genes & Development*, 31(4), 336–346. 10.1101/gad.293167.11628314766

Siddiqui, S. A., Bahmid, N. A., Karim, I., Mehany, T., Gvozdenko, A. A., Blinov, A. V., Nagdalian, A. A., Arsyad, M., & Lorenzo, J. M. (2022). Cultured meat: Processing, packaging, shelf life, and consumer acceptance. *Lebensmittel-Wissenschaft + Technologie*, 172, 114192. https://doi.org/https://doi.org/10.1016/j.lwt.2022.114192. 10.1016/j.lwt.2022.114192

Siddiqui, S. A., Zannou, O., Karim, I., Kasmiati, , Awad, N. M. H., Gołaszewski, J., Heinz, V., & Smetana, S. (2022). Avoiding Food Neophobia and Increasing Consumer Acceptance of New Food Trends—A Decade of Research. *Sustainability (Basel)*, 14(16), 10391. 10.3390/su141610391

Siegrist, M., Sütterlin, B., & Hartmann, C. (2018). Perceived naturalness and evoked disgust influence acceptance of cultured meat. *Meat Science*, 139, 213–219. 10.1016/j.meatsci.2018.02.00729459297

Silvy, N. J., Lopez, R. R., & Peterson, M. J. (2005). Wildlife marking techniques. *Techniques for wildlife investigations and management, 6*, 339-376.

Simsa, R., Yuen, J., Stout, A., Rubio, N., Fogelstrand, P., & Kaplan, D. L. (2019). Extracellular Heme Proteins Influence Bovine Myosatellite Cell Proliferation and the Color of Cell-Based Meat. *Foods*, 8(10), 521. Advance online publication. 10.3390/foods8100521

Singapore Food Agency (SFA). (2020). How are alternative proteins regulated in Singapore? https://www.sfa.gov.sg/food-information/risk-at-a-glance/safety-of-alternativeprotein(2020)

Singapore Food Agency. (2020). SFA grants first regulatory approval for cultured meat. Retrieved from https://www.sfa.gov.sg/docs/default-source/default-document-library/sfa-press-release---sfa-grants-first-regulatory-approval-for-cultured-meat_011220.pdf

Singapore Food Agency. (2021). A growing culture of safe, sustainable meat. *Singapore Food Agency*. https://www.sfa.gov.sg/food-for-thought/article/detail/a-growing-culture-of-safe-sustainable-meat

Singapore Food Agency. (2023a). The codex alimentarius commission and Singapore's involvement. *Singapore Food Agency*. https://www.sfa.gov.sg/food-information/singapore's-national-codex-office/the-codex-alimentarius-commission-and-singapore-s-involvement

Singapore Food Agency. (2023b). Requirements for the safety assessment of novel foods and novel food ingredients. *Singapore Food Agency*. https://www.sfa.gov.sg/docs/default-source/food-information/requirements-for-the-safety-assessment-of-novel-foods-and-novel-food-ingredients.pdf

Singapore Food Agency. (2023c). Our Singapore food story. *Singapore Food Agency*. https://www.sfa.gov.sg/food-farming/sgfoodstory/our-singapore-food-story

Singapore Food Agency. (2024a). Novel food. *Singapore Food Agency*. https://www.sfa.gov.sg/food-information/novel-food

Singapore Food Agency. (2024b). Roundtable on novel food regulation 2023. *Singapore Food Agency*. https://www.sfa.gov.sg/docs/default-source/default-document-library/roundtable-on-novel-food-regulations-2023_post-event-summary-and-full-report.pdf

Singh, A. (2023). Environmental Impact Assessment of Cellular Dairy Production. *Journal of Environmental Management, 45*(2), 210–225.

Singh, A., Kumar, V., Singh, S. K., Gupta, J., Kumar, M., Sarma, D. K., & Verma, V. (2023). Recent advances in bioengineered scaffold for in vitro meat production. *Cell and Tissue Research*, 391(2), 235–247. 10.1007/s00441-022-03718-636526810

Singh, A., Singh, S. K., Kumar, V., Gupta, J., Kumar, M., Sarma, D. K., Singh, S., Kumawat, M., & Verma, V. (2023). Derivation and characterization of novel cytocompatible decellularized tissue scaffold for myoblast growth and differentiation. *Cells*, 13(1), 41. 10.3390/cells1301004138201245

Sinke, P., Swartz, E., Sanctorum, H., van der Giesen, C., & Odegard, I. (2023). Ex-ante life cycle assessment of commercial-scale cultivated meat production in 2030. *The International Journal of Life Cycle Assessment*, 28(3), 234–254. 10.1007/s11367-022-02128-8

Slade, P. (2018). If you build it, will they eat it? Consumer preferences for plant-based and cultured meat burgers. *Appetite*, 125, 428–437. 10.1016/j.appet.2018.02.03029501683

Smaoui, S., Chérif, I., Ben Hlima, H., Khan, M. U., Rebezov, M., Thiruvengadam, M., Sarkar, T., Shariati, M. A., & Lorenzo, J. M. (2023). Zinc oxide nanoparticles in meat packaging: A systematic review of recent literature. *Food Packaging and Shelf Life*, 36, 101045. 10.1016/j.fpsl.2023.101045

Smetana, S., Mathys, A., Knoch, A., & Heinz, V. (2015). Meat alternatives: Life cycle assessment of most known meat substitutes. *The International Journal of Life Cycle Assessment*, 20(9), 1254–1267. 10.1007/s11367-015-0931-6

Smith, D. J., Helmy, M., Lindley, N. D., & Selvarajoo, K. (2022). The transformation of our food system using cellular agriculture: What lies ahead and who will lead it? *Trends in Food Science & Technology*, 127, 368–376. 10.1016/j.tifs.2022.04.015

Smith, L. W. (2005). Helping industry ensure animal well-being. *Agricultural Research*, 53(3), 2–3.

Smith, R. (2019). Probiotic Properties of Cellular Yogurt: A Comparative Study. *Journal of Food Microbiology*, 8(4), 315–328. 10.1016/0740-0020(86)90015-8

Sneddon, L. U., Halsey, L. G., & Bury, N. R. (2017). Considering aspects of the 3Rs principles within experimental animal biology. *The Journal of Experimental Biology*, 220(17), 3007–3016. 10.1242/jeb.14705828855318

Snyman, C., Goetsch, K. P., Myburgh, K. H., & Niesler, C. U. (2013). Simple silicone chamber system for 3D skeletal muscle tissue formation. *Frontiers in Physiology*, 4, 65276. 10.3389/fphys.2013.00349

Soice, E., & Johnston, J. (2021). How cellular agriculture systems can promote food security. *Frontiers in Sustainable Food Systems*, 5, 753996. 10.3389/fsufs.2021.753996

Sokołowski, Ł. M. (2020). *The placing of novel foods on the EU market in the light of new EU regulations*. Adam Mickiewicz University Press. 10.14746/amup.9788323236153

Soulsbury, C., Gray, H., Smith, L., Braithwaite, V., Cotter, S., Elwood, R. W., Wilkinson, A., & Collins, L. M. (2020). The welfare and ethics of research involving wild animals: A primer. *Methods in Ecology and Evolution*, 11(10), 1164–1181. 10.1111/2041-210X.13435

Soyer, G., & Yilmaz, E. (2020). Waste management in dairy cattle farms in Aydın region. Potential of energy application. *Sustainability (Basel)*, 12(4), 1614. 10.3390/su12041614

Specht, L., Welch, D., Rees Clayton, E., & Lagally, C. D. (2018). Opportunities for applying biomedical production and manufacturing methods to the development of the clean meat industry. *Biochemical Engineering Journal*, 132, 161–168. 10.1016/j.bej.2018.01.015

Spier, M. R., Vandenberghe, L. P. D. S., Medeiros, A. B. P., & Soccol, C. R. (2011). *Application of different types of bioreactors in bioprocesses. Bioreactors: Design, properties and applications*. Nova Science Publishers, Inc.

Spier, M., Vandenberghe, L., Medeiros, A., & Soccol, C. (2011). Application of different types of bioreactors in bioprocesses. In *Bioreactors: Design, Properties, and Applications* (Vol. 1, pp. 53–87). Nova Science Publishers, Inc.

Stanton, M. M., Tzatzalos, E., Donne, M., Kolundzic, N., Helgason, I., & Ilic, D. (2018, September 24). Prospects for the Use of Induced Pluripotent Stem Cells in Animal Conservation and Environmental Protection. *Stem Cells Translational Medicine*, 8(1), 7–13. 10.1002/sctm.18-004730251393

State of the Industry Report—Cultivated Meat. (2022). Available online: https://gfi.org/wp-content/uploads/2021/04/COR-SOTIRCultivated-Meat-2021-0429.pdf

Stavi, I., Paschalidou, A., Kyriazopoulos, A. P., Halbac-Cotoara-Zamfir, R., Siad, S. M., Suska-Malawska, M., Savic, D., Roque de Pinho, J., Thalheimer, L., Williams, D. S., Hashimshony-Yaffe, N., van der Geest, K., Cordovil, C. M. S., & Ficko, A. (2021). Multidimensional Food Security Nexus in Drylands under the Slow Onset Effects of Climate Change. *Land (Basel)*, 10(12), 1350. 10.3390/land10121350

Steidl, R. J., Hayes, J. P., & Schauber, E. (1997). Statistical power analysis in wildlife research. *The Journal of Wildlife Management*, 61(2), 270–279. 10.2307/3802582

Steinfeld, H., Gerber, P., Wassenaar, T. D., Castel, V., & De Haan, C. (2006). *Livestock's long shadow: environmental issues and options*. Food & Agriculture Org.

Stephens, N. (2022). Join Our Team, Change the World: Edibility, Producibility and Food Futures in Cultured Meat Company Recruitment Videos. *Food, Culture, & Society*, 25(1), 32–48. 10.1080/15528014.2021.188478735177960

Stephens, N., Di Silvio, L., Dunsford, I., Ellis, M., Glencross, A., & Sexton, A. (2018, August). Bringing cultured meat to market: Technical, socio-political, and regulatory challenges in cellular agriculture. *Trends in Food Science & Technology*, 78, 155–166. 10.1016/j.tifs.2018.04.01030100674

Stephens, N., & Ellis, M. (2020, October 12). Cellular agriculture in the UK: A review. *Wellcome Open Research*, 5, 12. 10.12688/wellcomeopenres.15685.132090174

Stephens, N., Sexton, A. E., & Driessen, C. (2019). Making sense of making meat: Key moments in the first 20 years of tissue engineering muscle to make food. *Frontiers in Sustainable Food Systems*, 45, 45. Advance online publication. 10.3389/fsufs.2019.0004534250447

Stout, A. J., Mirliani, A. B., Rittenberg, M. L., Shub, M., White, E. C., Yuen, J. S.Jr., & Kaplan, D. L. (2022). Simple and effective serum-free medium for sustained expansion of bovine satellite cells for cell cultured meat. *Communications Biology*, 5(1), 466. 10.1038/s42003-022-03423-835654948

Su, L., Jing, L., Zeng, X., Chen, T., Liu, H., Kong, Y., Wang, X., Yang, X., Fu, C., Sun, J., & Huang, D. (2023). 3D-Printed prolamin scaffolds for cell-based meat culture. *Advanced Materials*, 35(2), 2207397. 10.1002/adma.20220739736271729

Sun, C., Zhan, Y., & Du, G. (2020). Can value-added tax incentives of new energy industry increase firm's profitability? Evidence from financial data of China's listed companies. *Energy Economics*, 86, 104654. 10.1016/j.eneco.2019.104654

Sun, C., Zhan, Y., & Gao, X. (2023). Does environmental regulation increase domestic value-added in exports? An empirical study of cleaner production standards in China. *World Development*, 163, 106154. 10.1016/j.worlddev.2022.106154

Sun, J., Zhou, W., Yan, L., Huang, D., & Lin, L. Y. (2018). Extrusion-based food printing for digitalized food design and nutrition control. *Journal of Food Engineering*, 220, 1–1. 10.1016/j.jfoodeng.2017.02.028

Suntornnond, R., Yap, W. S., Lim, P. Y., & Choudhury, D. (2024). Redefining the Plate: Biofabrication in Cultivated Meat for Sustainability, Cost-Efficiency, Nutrient Enrichment, and Enhanced Organoleptic Experiences. *Current Opinion in Food Science*, 101164, 101164. Advance online publication. 10.1016/j.cofs.2024.101164

Surya, S., Smarak, B., & Bhat, R. (2023). Sustainable polysaccharide and protein hydrogel-based packaging materials for food products: A review. *International Journal of Biological Macromolecules*, 248, 125845. 10.1016/j.ijbiomac.2023.12584537473880

Su, X., Xian, C., Gao, M., Liu, G., & Wu, J. (2021). Edible materials in tissue regeneration. *Macromolecular Bioscience*, 21(8), 2100114. 10.1002/mabi.20210011434117831

Swallow, J., Anderson, D., Buckwell, A. C., Harris, T., Hawkins, P., Kirkwood, J., Lomas, M., Meacham, S., Peters, A., Prescott, M., Owen, S., Quest, R., Sutcliffe, R., & Thompson, K. (2005). Guidance on the transport of laboratory animals. *Laboratory Animals*, 39(1), 1–39. 10.1258/002367705288649315703122

Szyndler-Nedza, M., Nowicki, J., & Malopolska, M. (2019). The production system of high-quality pork products–an example. Annals of Warsaw University of Life Sciences-SGGW. *Animal Science (Penicuik, Scotland)*, 58. https://agro.icm.edu.pl/agro/element/bwmeta1.element.agro-813b019a-1c97-4a01-b46a-1be10f806300

Tabak, M. A., Norouzzadeh, M. S., Wolfson, D. W., Sweeney, S. J., VerCauteren, K. C., Snow, N. P., Halseth, J. M., Di Salvo, P. A., Lewis, J. S., White, M. D., Teton, B., Beasley, J. C., Schlichting, P. E., Boughton, R. K., Wight, B., Newkirk, E. S., Ivan, J. S., Odell, E. A., Brook, R. K., & Miller, R. S. (2019). Machine learning to classify animal species in camera trap images: Applications in ecology. *Methods in Ecology and Evolution*, 10(4), 585–590. 10.1111/2041-210X.13120

Tabassum-Abbasi & Abbasi, S. A. (2016). Reducing the global environmental impact of livestock production: The minilivestock option. In Journal of Cleaner Production (Vol. 112, pp. 1754–1766). Elsevier Ltd. 10.1016/j.jclepro.2015.02.094

Tacon, A. G. J., & Metian, M. (2008). Global overview on the use of fish meal and fish oil in industrially compounded aquafeeds: Trends and future prospects. *Aquaculture (Amsterdam, Netherlands)*, 285(1–4), 146–158. 10.1016/j.aquaculture.2008.08.015

Takahashi, K., & Yamanaka, S. (2006). Induction of pluripotent stem cells from mouse embryonic and adult fibroblast cultures by defined factors. *Cell*, 126(4), 663–676. 10.1016/j.cell.2006.07.02416904174

Talwar, R., Freymond, M., Beesabathuni, K., & Lingala, S. (2024). Current and Future Market Opportunities for Alternative Proteins in Low- and Middle-Income Countries. *Current Developments in Nutrition*, 8, 102035. 10.1016/j.cdnut.2023.10203538476721

Tamimah, T., Herianingrum, S., Ratih, I. S., Rofi'ah, K., & Kulsum, U. (2018). Halalan thayyiban: The key of successful halal food industry development. *Ulumuna: Jurnal Studi Keislaman/Ulumuna, 4*(2), 171–186. 10.36420/ju.v4i2.3501

Tan, C. (2024, April). Singapore approves lab-grown quail for consumption. *The Straits Times*. https://www.straitstimes.com/singapore/s-pore-approves-lab-grown-quail-for-consumption-here

Tan, J., & Joyner, H. S. (2020). Characterizing wear behaviors of edible hydrogels by kernel-based statistical modeling. J. Food Eng. 275, 109850. doi:.jfoodeng.2019.10985010.1016/j

Tassoni, A., Durante, L., & Ferri, M. (2012, May). Combined elicitation of methyljasmonate and red light on stilbene and anthocyanin biosynthesis. *Journal of Plant Physiology*, 169(8), 775–781. 10.1016/j.jplph.2012.01.01722424571

Tauscher, M., Wolf, F., Lode, A., & Gelinsky, M. (2023). Bioreactor design for tissue engineering: A review. *Bioreactor Design and Operation*, 54(1), 27–45.

Temple, D., & Manteca, X. (2020). Animal welfare in extensive production systems is still an area of concern. *Frontiers in Sustainable Food Systems*, 4, 545902. 10.3389/fsufs.2020.545902

Tessmar, J. K. A. M., & Göpferich, A. M. (2007). Customized PEG-derived copolymers for tissue- € engineering applications. *Macromolecular Bioscience*, 7(1), 23–39. 10.1002/mabi.20060009617195277

The Good Food Institute. (2023). GOOD Meat and UPSIDE Foods approved to sell cultivated chicken following landmark USDA action - The Good Food Institute. Retrieved May 22, 2024, from https://gfi.org/press/good-meat-and-upside-foods-approved-to-sell-cultivated-chicken-following-landmark-usda action/#:~:text=WASHINGTON%20(June%2021%2C%202023),history%20of%20food%20and%20agriculture

Thisted, E. V., & Thisted, R. V. (2019). The diffusion of carbon taxes and emission trading schemes: The emerging norm of carbon pricing. *Environmental Politics*, 29(5), 804–824. 10.1080/09644016.2019.1661155

Thompson, P. B., & Nardone, A. (1999). Sustainable livestock production: Methodological and ethical challenges. *Livestock Production Science*, 61(2–3), 111–119. 10.1016/S0301-6226(99)00061-5

Thyden, R., Perreault, L. R., Jones, J. D., Notman, H., Varieur, B. M., Patmanidis, A. A., Dominko, T., & Gaudette, G. R. (2022). An edible, decellularized plant derived cell carrier for lab grown meat. *Applied Sciences (Basel, Switzerland)*, 12(10), 5155. 10.3390/app12105155

Tian, X., Engel, B. A., Qian, H., Hua, E., Sun, S., & Wang, Y. (2021). Will reaching the maximum achievable yield potential meet future global food demand? *Journal of Cleaner Production*, 294, 126285. 10.1016/j.jclepro.2021.126285

TingWei, JingWen, XinRui, GuoQiang, XueLiang, GuoCheng, Jian, & XiuLan. (2019). Research progress on lab-grown meat risk prevention and safety management norms. *Shipin Yu Fajiao Gongye*, 45, 254–258.

Tobita, K., Liu, X., & Lo, C. W. (2010). Imaging modalities to assess structural birth defects in mutant mouse models. *Birth Defects Research. Part C, Embryo Today*, 90(3), 176–184. 10.1002/bdrc.2018720860057

To, K. V., Comer, C. C., O'Keefe, S. F., & Lahne, J. (2024). A taste of cell-cultured meat: A scoping review. *Frontiers in Nutrition*, 11, 1332765. Advance online publication. 10.3389/fnut.2024.133276538321991

Tomiyama, A. J., Kawecki, N. S., Rosenfeld, D. L., Jay, J. A., Rajagopal, D., & Rowat, A. C. (2020). Bridging the gap between the science of cultured meat and public perceptions. *Trends in Food Science & Technology*, 104, 144–152. 10.1016/j.tifs.2020.07.019

Touzdjian Pinheiro Kohlrausch Távora, F., de Assis Dos Santos Diniz, F., de Moraes Rêgo-Machado, C., Chagas Freitas, N., Barbosa Monteiro Arraes, F., Chumbinho de Andrade, E., Furtado, L. L., Osiro, K. O., Lima de Sousa, N., Cardoso, T. B., Márcia Mertz Henning, L., Abrão de Oliveira Molinari, P., Feingold, S. E., Hunter, W. B., Fátima Grossi de Sá, M., Kobayashi, A. K., Lima Nepomuceno, A., Santiago, T. R., & Correa Molinari, H. B. (2022). CRISPR/Cas- and Topical RNAi-Based Technologies for Crop Management and Improvement: Reviewing the Risk Assessment and Challenges Towards a More Sustainable Agriculture. *Frontiers in Bioengineering and Biotechnology*, 10, 913728. 10.3389/fbioe.2022.91372835837551

Treich, N. (2021). Cultured Meat: Promises and Challenges. *Environmental and Resource Economics*, 79(1), 33–61. 10.1007/s10640-021-00551-333758465

Tsuruwaka, Y., & Shimada, E. (2022). Reprocessing seafood waste: challenge to develop aquatic clean meat from fish cells. NPJ Science of Food, 6(1), 7.

Tsvakirai, C. Z., & Nalley, L. L. (2023). The coexistence of psychological drivers and deterrents of consumers' willingness to try cultured meat hamburger patties: Evidence from South Africa. *Agricultural and Food Economics*, 11(1), 1–17. 10.1186/s40100-023-00293-4

Tsvakirai, C., Nalley, L., Rider, S., Van Loo, E., & Tshehla, M. (2023). The Alternative Livestock Revolution: Prospects for Consumer Acceptance of Plant-based and Cultured Meat in South Africa. *Journal of Agricultural and Applied Economics*, 55(4), 710–729. 10.1017/aae.2023.36

Tubiello, F. N., Salvatore, M., Ferrara, A. F., House, J., Federici, S., Rossi, S., & Smith, P. (2015). The contribution of agriculture, forestry and other land use activities to global warming, 1990–2012. *Global Change Biology*, 21(7), 2655–2660. 10.1111/gcb.1286525580828

Tuomisto, H. L. (2019). The eco-friendly burger: Could cultured meat improve the environmental sustainability of meat products? *EMBO Reports*, 20(1), e47395. 10.15252/embr.20184739530552146

Tuomisto, H. L., Allan, S. J., & Ellis, M. J. (2022). Prospective life cycle assessment of a bioprocess design for cultured meat production in hollow fiber bioreactors. *The Science of the Total Environment*, 851, 158051. 10.1016/j.scitotenv.2022.15805135985596

Tuomisto, H. L., Ellis, M. J., & Haastrup, P. (2015). *Environmental impacts of cultured meat: alternative production scenarios*. EU Sci. Hub-Eur. Comm.

Tuomisto, H. L., & Teixeira de Mattos, M. J. (2011). Environmental impacts of cultured meat production. *Environmental Science & Technology*, 45(14), 6117–6123. 10.1021/es200130u21682287

Turck, D., Castenmiller, J., De Henauw, S., Hirsch-Ernst, K. I., Kearney, J., Maciuk, A., Mangelsdorf, I., McArdle, H. J., Naska, A., Pelaez, C., Pentieva, K., Siani, A., Thies, F., Tsabouri, S., Vinceti, M., Cubadda, F., Frenzel, T., Heinonen, M., Marchelli, R., & Knutsen, H. K.EFSA Panel on Nutrition. (2020). Safety of the extension of use of plant sterol esters as a novel food pursuant to Regulation (EU) 2015/2283. *EFSA Journal.EFSA Journal*, 18(6), e06135–e06135. 10.2903/j.efsa.2020.6135

Tutundjian, S., Clarke, M., Egal, F., Dixson-Decleve, S., Candotti, S. W., Schmitter, P., & Lovins, L. H. (2020). Future food systems: challenges and consequences of the current food system. The Palgrave Handbook of Climate Resilient Societies, 1-29.

Twine, R. (2021). Emissions from animal agriculture—16.5% is the new minimum figure. *Sustainability (Basel)*, 13(11), 6276. 10.3390/su13116276

Tyagi, J., Ahmad, S., & Malik, M. (2022). Nitrogenous fertilizers: Impact on environment sustainability, mitigation strategies, and challenges. *International Journal of Environmental Science and Technology*, 19(11), 11649–11672. 10.1007/s13762-022-04027-9

U.S. Food And Drug Administration. (2023, March 21). F. F. S. A. A. Human Food Made with Cultured Animal Cells. https://www.fda.gov/food/food-ingredients-packaging/human-food-made-cultured-animal-cells

United Nations Environment Programme. (2023). What's cooking? An assessment of the potential impacts of selected novel alternatives to conventional animal products. *Frontiers*. Advance online publication. 10.59117/20.500.11822/44236

United Nations. (n.d.). Your guide to climate action: Food. *United Nations*. https://www.un.org/en/actnow/food

Unsworth, B. R., & Lelkes, P. I. (1998). Growing tissues in microgravity. *Nature Medicine*, 4(8), 901–907. 10.1038/nm0898-9019701241

USDA. (2022). ERS - Meat Price Spreads.

Ushiyama, K. (1991). *Large scale culture of ginseng*. Plant Cell Culture in Japan. CMC.

Uwizeye, A., de Boer, I. J. M., Opio, C. I., Schulte, R. P. O., Falcucci, A., Tempio, G., Teillard, F., Casu, F., Rulli, M., Galloway, J. N., Leip, A., Erisman, J. W., Robinson, T. P., Steinfeld, H., & Gerber, P. J. (2020). Nitrogen emissions along global livestock supply chains. *Nature Food*, 1(7), 437–446. 10.1038/s43016-020-0113-y

Valpey, K. R. (2020). Cow care in Hindu animal ethics. Springer Nature. Retrieved from https://library.oapen.org/handle/20.500.12657/22832

van den Heuvel, E., Newbury, A., & Appleton, K. M. (2019). The psychology of nutrition with advancing age: Focus on food Neophobia. *Nutrients*, 11(1), 151. 10.3390/nu1101015130642027

van der Valk, J., Brunner, D., De Smet, K., Fex Svenningsen, Å., Honegger, P., Knudsen, L., & Gstraunthaler, G. (2010). Optimization of chemically defined cell culture media—Replacing fetal bovine serum in mammalian in vitro methods. *Toxicology In Vitro*, 24(4), 1053–1063. 10.1016/j.tiv.2010.03.01620362047

van der Weele, C., & Tramper, J. (2014). Cultured meat: Every village its own factory? *Trends in Biotechnology*, 32(6), 294–296. 10.1016/j.tibtech.2014.04.00924856100

Van Dijk, M., Morley, T., Rau, M. L., & Saghai, Y. (2021). A meta-analysis of projected global food demand and population at risk of hunger for the period 2010–2050. *Nature Food*, 2(7), 494–501. 10.1038/s43016-021-00322-937117684

Van Eenennaam, A. L., & Werth, S. J. (2021). Animal board invited review: Animal agriculture and alternative meats–learning from past science communication failures. *Animal*, 15(10), 100360. 10.1016/j.animal.2021.10036034563799

Van Loo, E. J., Caputo, V., & Lusk, J. L. (2020). Consumer Preferences for Farm-Raised Meat, Lab-Grown Meat, and Plant-Based Meat alternatives: Does Information or Brand Matter? *Food Policy*, 95, 101931. 10.1016/j.foodpol.2020.101931

Vandeweyer, D., Lievens, B., & Van Campenhout, L. (2020). Identification of bacterial endospores and targeted detection of foodborne viruses in industrially reared insects for food. *Nature Food*, 1(8), 511–516. 10.1038/s43016-020-0120-z37128070

Varley, M. C., Markaki, A. E., & Brooks, R. A. (2017, June). Effect of Rotation on Scaffold Motion and Cell Growth in Rotating Bioreactors. *Tissue Engineering. Part A*, 23(11–12), 522–534. 10.1089/ten.tea.2016.035728125920

Vegaconomist. (2022). Cultured Meat in Europe: Which Country Is Leading the Race? 2022. Available online: https://vegconomist. com/cultivated-cell-cultured-biotechnology/cultured-meat-in-europe-which-country-is-leading-the-race/

Venmans, F., Ellis, J., & Nachtigall, D. (2020). Carbon pricing and competitiveness: Are they at odds? *Climate Policy*, 20(9), 1070–1091. 10.1080/14693062.2020.1805291

Verbeke, W., Marcu, A., Rutsaert, P., Gaspar, R., Seibt, B., Fletcher, D., & Barnett, J. (2015). 'Would you eat cultured meat?': Consumers' reactions and attitude formation in Belgium, Portugal and the United Kingdom. *Meat Science*, 102, 49–58. 10.1016/j.meatsci.2014.11.01325541372

Verbeke, W., Sans, P., & Van Loo, E. J. (2015). Challenges and prospects for consumer acceptance of cultured meat. *Journal of Integrative Agriculture*, 14(2), 285–294. 10.1016/S2095-3119(14)60884-4

Verbruggen, S., Luining, D., van Essen, A., & Post, M. J. (2017, May 3). Bovine myoblast cell production in a microcarriers-based system. *Cytotechnology*, 70(2), 503–512. 10.1007/s10616-017-0101-828470539

Verbruggen, S., Luining, D., van Essen, A., & Post, M. J. (2018). Bovine myoblast cell production in a microcarriers-based cultivation system for large-scale cultured meat production. *Frontiers in Sustainable Food Systems*, 2, 79.

Vergeer, R. (2021). TEA of cultivated meat. Future projections of different scenarios.

Vermeulen, S. J., Park, T., Khoury, C. K., & Béné, C. (2020). Changing diets and the transformation of the global food system. *Annals of the New York Academy of Sciences*, 1478(1), 3–17. 10.1111/nyas.1444632713024

Verni, M., Demarinis, C., Rizzello, C. G., & Pontonio, E. (2023). Bioprocessing to Preserve and Improve Microalgae Nutritional and Functional Potential: Novel Insight and Perspectives. In Foods (Vol. 12, Issue 5). MDPI. 10.3390/foods12050983

Verzijden, K. (2019). *Regulatory pathways for clean meat in the EU and the US- differences & analogies*. Food Health Legal.

vGuttenberg, H. (1943, July). Kulturversuche mit isolierten Pflanzenzellen. *Planta*, 33(4), 576–588. 10.1007/BF01916543

Viana, C. M., Freire, D., Abrantes, P., Rocha, J., & Pereira, P. (2022). Agricultural land systems importance for supporting food security and sustainable development goals: A systematic review. *The Science of the Total Environment*, 806, 150718. 10.1016/j.scitotenv.2021.15071834606855

Viceconti, M., Henney, A., & Morley-Fletcher, E. (2016). *In silico* clinical trials: How computer simulation will transform the biomedical industry. *International Journal of Clinical Trials*, 3(2), 37–46. 10.18203/2349-3259.ijct20161408

Voigt, C. A. (2020). Synthetic biology 2020–2030: Six commercially-available products that are changing our world. *Nature Communications*, 11(1), 1–6. 10.1038/s41467-020-20122-233311504

Volden, J., & Wethal, U. (2021). What happens when cultured meat meets meat culture. Changing Meat Cultures: Food Practices, Global Capitalism, and the Consumption of Animals, 185–206. Retrieved from https://books.google.com/books?hl=en&lr=&id=M3ZPEAAAQBAJ&oi=fnd&pg=PA185&dq=Transitioning+to+lab-grown+meat+may+result+into+loss+of+cultural+history+and+identity.+&ots=lN0v0cRUm2&sig=at7ihJOgFAMCmaCgeS_5qG8ACvk

Vollset, S. E., Goren, E., Yuan, C. W., Cao, J., Smith, A. E., Hsiao, T., & Murray, C. J. (2020). Fertility, mortality, migration, and population scenarios for 195 countries and territories from 2017 to 2100: A forecasting analysis for the Global Burden of Disease Study. *Lancet*, 396(10258), 1285–1306. 10.1016/S0140-6736(20)30677-232679112

Von Keyserlingk, M. A., & Hötzel, M. J. (2015). The ticking clock: Addressing farm animal welfare in emerging countries. *Journal of Agricultural & Environmental Ethics*, 28(1), 179–195. 10.1007/s10806-014-9518-7

Vonthron, S., Perrin, C., & Soulard, C. T. (2020). Foodscape: A scoping review and a research agenda for food security-related studies. *PLoS One*, 15(5), e0233218. 10.1371/journal.pone.023321832433690

Voss, M., Valle, C., Calcio Gaudino, E., Tabasso, S., Forte, C., & Cravotto, G. (2024). Unlocking the Potential of Agrifood Waste for Sustainable Innovation in Agriculture. *Recycling*, 9(2), 25. 10.3390/recycling9020025

Walsh, J. E., Ballinger, T. J., Euskirchen, E. S., Hanna, E., Mård, J., Overland, J. E., Tangen, H., & Vihma, T. (2020). Extreme weather and climate events in northern areas: A review. *Earth-Science Reviews*, 209, 103324. 10.1016/j.earscirev.2020.103324

Waltz, E. (2021). Club-goers take first bites of lab-made chicken. *Nature Biotechnology*, 39(3), 257–258. 10.1038/s41587-021-00855-133692516

Wang, Y., Tibbetts, S. M., & McGinn, P. J. (2021). Microalgae as sources of high-quality protein for human food and protein supplements. In Foods (Vol. 10, Issue 12). MDPI. 10.3390/foods10123002

Wang, D., Liu, W., Han, B., & Xu, R. (2005). The bioreactor: A powerful tool for large-scale culture of animal cells. *Current Pharmaceutical Biotechnology*, 6(5), 397–403. 10.2174/1389201057743705801624881

Wang, J., & Azam, W. (2024). Natural resource scarcity, fossil fuel energy consumption, and total greenhouse gas emissions in top emitting countries. *Geoscience Frontiers*, 15(2), 101757. 10.1016/j.gsf.2023.101757

Wang, S., Hausfather, Z., Davis, S., Lloyd, J., Olson, E. B., Liebermann, L., Núñez-Mujica, G. D., & McBride, J. (2023). Future demand for electricity generation materials under different climate mitigation scenarios. *Joule*, 7(2), 309–332. 10.1016/j.joule.2023.01.001

Wang, S., Li, K., Gao, H., Liu, Z., Shi, S., Tan, Q., & Wang, Z. (2021). Ubiquitin-specific peptidase 8 regulates proliferation and early differentiation of sheep skeletal muscle satellite cells. *Czech Journal of Animal Science*, 66(3), 87–96. Advance online publication. 10.17221/105/2020-CJAS

Wang, Y. (2024). Development of Enhanced Cell Culture Media for Cellular Dairy Products. *Food Science Research*, 18(3), 315–328.

Wang, Y., Xiao, X., & Wang, L. (2020). In vitro characterization of goat skeletal muscle satellite cells. *Animal Biotechnology*, 31(2), 115–121. 10.1080/10495398.2018.155123030602329

Wanibuchi, H., Salim, E. I., Kinoshita, A., Shen, J., Wei, M., Morimura, K., Yoshida, K., Kuroda, K., Endo, G., & Fukushima, S. (2004). Understanding arsenic carcinogenicity by the use of animal models. *Toxicology and Applied Pharmacology*, 198(3), 366–376. 10.1016/j.taap.2003.10.03215276416

Webb, L., Fleming, A., Ma, L., & Lu, X. (2021). Uses of cellular agriculture in plant-based meat analogues for improved palatability. *ACS Food Science & Technology*, 1(10), 1740–1747. 10.1021/acsfoodscitech.1c00248

Weishaupt, A., Ekardt, F., Garske, B., Stubenrauch, J., & Wieding, J. (2020). Land Use, Livestock, Quantity Governance, and Economic Instruments—Sustainability Beyond Big Livestock Herds and Fossil Fuels. *Sustainability (Basel)*, 12(5), 2053. 10.3390/su12052053

Welin, S. (2013). Introducing the new meat. Problems and prospects. *Etikk i Praksis*, 7(1), 24–37. 10.5324/eip.v7i1.1788

Wells, M. L., Potin, P., Craigie, J. S., Raven, J. A., Merchant, S. S., Helliwell, K. E., Smith, A. G., Camire, M. E., & Brawley, S. H. (2017). Algae as nutritional and functional food sources: revisiting our understanding. In Journal of Applied Phycology (Vol. 29, Issue 2, pp. 949–982). Springer Netherlands. 10.1007/s10811-016-0974-5

Welty, J. (2007). Humane slaughter laws. *Law and Contemporary Problems*, 70(1), 175–206.

White, M., & Barquera, S. (2020). Mexico adopts food warning labels, why now? *Health Systems and Reform*, 6(1), e1752063. 10.1080/23288604.2020.175206332 486930

Widdowson, R. W. (2013). *Towards holistic agriculture: A scientific approach*. Elsevier.

Wikandari, R., Manikharda, , Baldermann, S., Ningrum, A., & Taherzadeh, M. J. (2021). Application of cell culture technology and genetic engineering for production of future foods and crop improvement to strengthen food security. *Bioengineered*, 12(2), 11305–11330. 10.1080/21655979.2021.200366534779353

Wilks, M., & Phillips, C. J. C. (2017). Attitudes to in Vitro Meat: A Survey of Potential Consumers in the United States. *PLoS One*, 2017(12), e0171904. 10.1371/journal.pone.017190428207878

Willett, W., Rockström, J., Loken, B., Springmann, M., Lang, T., Vermeulen, S., Garnett, T., Tilman, D., DeClerck, F., Wood, A., Jonell, M., Clark, M., Gordon, L. J., Fanzo, J., Hawkes, C., Zurayk, R., Rivera, J. A., De Vries, W., Majele Sibanda, L., & Murray, C. J. L. (2019, February). Food in the Anthropocene: The EAT–Lancet Commission on healthy diets from sustainable food systems. *Lancet*, 393(10170), 447–492. 10.1016/S0140-6736(18)31788-430660336

Williams, L. A., Davis-Dusenbery, B. N., & Eggan, K. C. (2012). Snapshot: Directed differentiation of pluripotent stem cells. *Cell*, 149(5), 1174–1174. 10.1016/j.cell.2012.05.01522632979

Wilschut, K. J., Jaksani, S., Van Den Dolder, J., Haagsman, H. P., & Roelen, B. A. (2008, September 26). Isolation and characterization of porcine adult muscle-derived progenitor cells. *Journal of Cellular Biochemistry*, 105(5), 1228–1239. 10.1002/jcb.2192118821573

Włodarczyk-Biegun, M. K., Farbod, K., Werten, M. W., Slingerland, C. J., De Wolf, F. A., Van Den Beucken, J. J., Leeuwenburgh, S. C. G., Cohen Stuart, M. A., & Kamperman, M. (2016). Fibrous hydrogels for cell encapsulation: A modular and supramolecular approach. *PLoS One*, 11(5), e0155625. 10.1371/journal.pone.015562527223105

Wolfensohn, S. (2010). Euthanasia and other fates for laboratory animals. *The UFAW handbook on the care and management of laboratory and other research animals*, 219-226. 10.1002/9781444318777.ch17

Wolk, A. (2017). Potential health hazards of eating red meat. *Journal of Internal Medicine*, 281(2), 106–122. 10.1111/joim.1254327597529

World Food Situation—FAO Food Price Index. (2022). https://www.fao.org/worldfoodsituation/foodpricesindex/en/

Worm, B., Barbier, E. B., Beaumont, N., Duffy, J. E., Folke, C., Halpern, B. S., Jackson, J. B., Lotze, H. K., Micheli, F., Palumbi, S. R., Sala, E., Selkoe, K. A., Stachowicz, J. J., & Watson, R. (2006). Impacts of biodiversity loss on ocean ecosystem services. *Science*, 314(5800), 787–790. 10.1126/science.1132294

Worobey, M., Han, G. Z., & Rambaut, A. (2014). A synchronized global sweep of the internal genes of modern avian influenza virus. *Nature*, 508(7495), 254–257. 10.1038/nature13016

Wurm, F. M. (2004). Production of recombinant protein therapeutics in cultivated mammalian cells. *Nature Biotechnology*, 22(11), 1393–1398. 10.1038/nbt102615529164

Xiang, N., Yuen, J. S.Jr, Stout, A. J., Rubio, N. R., Chen, Y., & Kaplan, D. L. (2022). 3D porous scaffolds from wheat glutenin for cultured meat applications. *Biomaterials*, 285, 121543. Advance online publication. 10.1016/j.biomaterials.2022.12154335533444

Xu, E., Niu, R., Lao, J., Zhang, S., Li, J., Zhu, Y., ... Liu, D. (2023). Tissue-like cultured fish fillets through a synthetic food pipeline. NPJ Science of Food, 7(1), 17.

Xu, B., Wang, W., Guo, L., Chen, G., Wang, Y., Zhang, W., & Li, Y. (2021). Evaluation of Deep Learning for Automatic Multi-View Face Detection in Cattle. *Agriculture*, 11(11), 1062. 10.3390/agriculture11111062

Xu, X., Sharma, P., Shu, S., Lin, T.-S., Ciais, P., Tubiello, F. N., Smith, P., Campbell, N., & Jain, A. K. (2021). Global greenhouse gas emissions from animal-based foods are twice those of plant-based foods. *Nature Food*, 2(9), 724–732. 10.1038/s43016-021-00358-x37117472

Yang, X., & Khan, I. (2021). Dynamics among economic growth, urbanization, and environmental sustainability in IEA countries: The role of industry value-added. *Environmental Science and Pollution Research International*, 29(3), 4116–4127. 10.1007/s11356-021-16000-z34402019

Yang, Z., Yuan, S., Liang, B., Liu, Y., Choong, C., & Pehkonen, S. O. (2014). Chitosan Microsphere Scaffold Tethered with RGD-Conjugated Poly (methacrylic acid) Brushes as Effective Carriers for the Endothelial Cells. *Macromolecular Bioscience*, 14(9), 1299–1311. 10.1002/mabi.20140013624895289

Ye, L., Yao, F., & Li, J. (2023). Chapter 6 - peptide and protein-based hydrogels. In Sustainable hydrogels. Elsevier.

Yegappan, R., Selvaprithiviraj, V., Amirthalingam, S., & Jayakumar, R. (2018). Carrageenan based hydrogels for drug delivery, tissue engineering and wound healing. *Carbohydrate Polymers*, 198, 385–400. 10.1016/j.carbpol.2018.06.08630093014

Ye, Y., Zhou, J., Guan, X., & Sun, X. (2022). Commercialization of cultured meat products: Current status, challenges, and strategic prospects. *Future Foods : a Dedicated Journal for Sustainability in Food Science*, 6, 100177. 10.1016/j.fufo.2022.100177

Young, S. (2022). Celling Meat—Is Cultivated Meat Really Here to Stay? Available online: https://thefarmermagazine.com.au/cellingmeat-is-cultivated-meat-here-to-stay/

Yuen, K. K.-C. (2017). New sustainable models of open innovation to accelerate technology development in cellular agriculture (PhD Thesis). Massachusetts Institute of Technology. Retrieved from https://dspace.mit.edu/handle/1721.1/113537

Yue, W., Ming, Q. L., Lin, B., Rahman, K., Zheng, C. J., Han, T., & Qin, L. P. (2014, June 25). Medicinal plant cell suspension cultures: Pharmaceutical applications and high-yielding strategies for the desired secondary metabolites. *Critical Reviews in Biotechnology*, 36(2), 215–232. 10.3109/07388551.2014.92398624963701

Zappelli, C., Barbulova, A., Apone, F., & Colucci, G. (2016, November 18). Effective Active Ingredients Obtained through Biotechnology. *Cosmetics*, 3(4), 39. 10.3390/cosmetics3040039

Zarbà, C., Chinnici, G., & D'Amico, M. (2020). Novel Food: The Impact of Innovation on the Paths of the Traditional Food Chain. *Sustainability (Basel)*, 12(2), 555. 10.3390/su12020555

Zeltinger, J., Sherwood, J. K., Graham, D. A., Müeller, R., & Griffith, L. G. (2001). Effect of pore size and void fraction on cellular adhesion, proliferation, and matrix deposition. *Tissue Engineering*, 7(5), 557–572. 10.1089/10763270175321318311 694190

Zeraatpisheh, M., Bakhshandeh, E., Hosseini, M., & Alavi, S. M. (2020). Assessing the effects of deforestation and intensive agriculture on the soil quality through digital soil mapping. *Geoderma*, 363, 114139. 10.1016/j.geoderma.2019.114139

Zerbe, S. (2022). Global Land-Use Development Trends: Traditional Cultural Landscapes Under Threat. In *Landscape Series* (pp. 129–199). Springer International Publishing. 10.1007/978-3-030-95572-4_4

Zhang, G. (2020). Development of a serum-free medium for in vitro cultivation of muscle stem cells. *Frontiers in Bioengineering and Biotechnology*, 8, 654. 10.3389/fbioe.2020.00654

Zhang, G., Zhao, X., Li, X., Du, G., Zhou, J., & Chen, J. (2020). Challenges and possibilities for bio-manufacturing cultured meat. *Trends in Food Science & Technology*, 97, 443–450. 10.1016/j.tifs.2020.01.026

Zhang, H. (2023). Nutritional Analysis of Cellular Cheese and Yogurt: A Comparative Study. *Journal of Food Chemistry*, 12(4), 415–428.

Zhang, H., Li, Z., Dai, C., Wang, P., Fan, S., Yu, B., & Qu, Y. (2021). Antibacterial properties and mechanism of selenium nanoparticles synthesized by Providencia sp. DCX. *Environmental Research*, 194, 110630. 10.1016/j.envres.2020.11063033345899

Zhang, J., Liu, L., Liu, H., Shi, A., Hu, H., & Wang, Q. (2017). Research advances on food extrusion equipment, technology and its mechanism. *Nongye Gongcheng Xuebao (Beijing)*, 2017(33), 275–283.

Zhang, L., Hu, Y., Badar, I. H., Xia, X., Kong, B., & Chen, Q. (2021). Prospects of artificial meat: Opportunities and challenges around consumer acceptance. *Trends in Food Science & Technology*, 116, 434–444. 10.1016/j.tifs.2021.07.010

Zhang, L., Tian, H., Shi, H., Pan, S., Chang, J., Dangal, S. R., & Jackson, R. B. (2022). A 130-year global inventory of methane emissions from livestock: Trends, patterns, and drivers. *Global Change Biology*, 28(17), 5142–5158. 10.1111/gcb.1628035642457

Zhang, S., Xing, M., & Li, B. (2018). Biomimetic Layer-by-Layer Self-Assembly of Nanofilms, Nanocoatings, and 3D Scaffolds for Tissue Engineering. *International Journal of Molecular Sciences*, 2018(19), 1641. 10.3390/ijms1906164129865178

Zhao, J. R., Zheng, R., Tang, J., Sun, H. J., & Wang, J. (2022). A mini-review on building insulation materials from perspective of plastic pollution: Current issues and natural fibres as a possible solution. *Journal of Hazardous Materials*, 438, 129449. 10.1016/j.jhazmat.2022.12944935792430

Zhao, Y., & Wang, S. (2022). Experimental and biophysical modeling of transcription and translation dynamics in bacterial-and mammalian-based cell-free expression systems. *SLAS Technology*.35231628

Zheng, Y.-Y., Zhu, H.-Z., Wu, Z.-Y., Song, W.-J., Tang, C.-B., Li, C.-B., Ding, S.-J., & Zhou, G.-H. (2021). Evaluation of the effect of smooth muscle cells on the quality of cultured meat in a model for cultured meat. *Food Research International*, 150, 110786. 10.1016/j.foodres.2021.11078634865801

Zhichkin, K., Nosov, V., Zhichkina, L., Allen, G., Kotar, O., & Fasenko, T. (2021, September). Efficiency of regional agriculture state support. *IOP Conference Series. Earth and Environmental Science*, 839(2), 022042. 10.1088/1755-1315/839/2/022042

Zhou, X. H., Yin, L., Yang, B. S., Chen, C. Y., Chen, W. H., Xie, Y., Yang, X., Pham, J. T., Liu, S., & Xue, L. J. (2021). Programmable local orientation of micropores by mold-assisted ice templating. *Small Methods*, 5(2), 2000963. Advance online publication. 10.1002/smtd.20200096334927890

Żuk-Gołaszewska, K., Gałęcki, R., Obremski, K., Smetana, S., Figiel, S., & Gołaszewski, J. (2022). Edible Insect Farming in the Context of the EU Regulations and Marketing-An Overview. *Insects*, 13(5), 446. 10.3390/insects1305044635621781

About the Contributors

Ahmed M. Hamad is a renowned food science and nutrition expert with years of experience in teaching, research, and industry. Dr. Ahmed has published numerous papers and books on food hygiene, nutrition, and bioscience and has worked with various government and non-government organizations on food safety and nutrition projects.

Tanima Bhattacharya is a highly accomplished academician and researcher in the field of Applied Science, currently serving as a faculty member at Lincoln University College in Petaling Jaya, Malaysia. Dr. Tanima Bhattacharya has always been driven by a passion for learning and a desire to make a positive impact on society. Dr. Tanima Bhattacharya academic journey began with a Bachelor of Science in Biotechnology from the University of Dhaka, Bangladesh. She then went on to pursue a Master of Science in Biotechnology from the same institution, followed by a Ph.D. in Biotechnology from the University of Malaya, Malaysia. Her doctoral research focused on the development of a novel biosensor for the detection of cancer biomarkers, which was successfully completed in 2016. Dr. Tanima Bhattacharya's professional career began as a lecturer at the University of Dhaka, Bangladesh, where she taught various courses in biotechnology and conducted research in the field of cancer biomarkers. In 2017, she joined Lincoln University College as a faculty member in the Department of Applied Science, where she currently serves as a Senior Lecturer. Dr. Tanima Bhattacharya's research interests include biotechnology, cancer biomarkers, biosensors, and nanotechnology. She has published numerous research articles in esteemed journals and presented her work at various international conferences. She is also a reviewer for several reputed scientific journals and has received several awards for her outstanding contributions to the field of biotechnology. Dr. Tanima Bhattacharya's dedication to her work has earned her several accolades, including the Best Researcher Award at Lincoln University College in 2018 and 2020. She has also received the Excellent Service Award from the University of Dhaka, Bangladesh, and the Outstanding Young Scientist Award from the Bangladesh Academy of Sciences. Dr. Tanima Bhattacharya is a member of several professional organizations, including the Malaysian Society for Biochemistry and Molecular Biology, the Asian Pacific Association of Biotechnology, and the International Association of Advanced Materials. She is also a certified trainer for the International Association of Advanced Materials and has conducted several workshops and training programs in biotechnology and nanotechnology. Dr. Tanima Bhatt Acharya is a highly accomplished academician, researcher, and community servant who has made significant contributions to the field of Applied Science, particularly in the areas of biotechnology, cancer biomarkers, and biosensors. Her dedication to her work, passion for learning, and commitment to community service make her an inspiration to her students and colleagues alike.

* * *

Idris A. Ahmed is a life scientist with over ten years of teaching and research experience in the field of natural products R & D and biological sciences including natural products characterization, toxicology, pharmacology evaluation, and product commercialization. Dr. Idris is currently the Deputy Dean for Postgraduate Studies at the Faculty of Applied Science, Lincoln University College. He has published more than 48 research articles in high-impact peer-reviewed journals. Dr. Idris has also actively participated in many international and local conferences as an oral and poster presenter and has been awarded at least three times gold/best awards. Dr. Idris has written over seventeen book chapters, co-authored a book, and contributes regularly to issues of national discourse through opinion letters in national dailies such as the New Strait Times and The Star. Dr. Idris previously worked as a Visiting Research Fellow at the prestigious Universiti Malaya. Dr. Idris is passionately interested in Communicable and non-communicable Diseases; Molecular Dynamics, Terrestrial Ecology and Biodiversity; Molecular docking & Molecular dynamics; Nanotechnology & Drug delivery; Halal Products and Audit; Natural Products and Medicinal Plants. He is the Founder and Chairman of Pure Heart Relief Organization (https://phrfnigeria.org)) and a Co-Founder of Mimia Sdn. Bhd (https://mimiaskincare.com/). Dr. Idris is also a Certified Halal Executive and a Certified Halal Lead Auditor.

Rathimalar Ayakannu is a PhD holder in Medical Sciences, has successfully completed her PhD degree at the University of Malaya, Faculty of Medicine. She is a registered Graduate Technologist (GT) awarded by the Malaysian Board of Technologist (MBOT). She is a highly analytical person and with extensive experience in biological research, documentation and clinical laboratory environments. Her PhD project consists of clinical research in human subjects with molecular research work in genetic profiling and tissue culture. She is also familiar with compiling patient information data and implementing Good Clinical Practice (GCP). She has graduated with a Master in Biotechnology at University Malaya in 2011, and have completed her Degree in Bio health science from University of Malaya as well. Currently, she is attached with Lincoln University College as Lecturer at Department of Biotechnology, Faculty of Applied Sciences. She has acquired and developed learning and teaching skills and have a strong knowledge in her subjects (Molecular Biology, Genetics, Cell Biology). She also has significant working experience as Laboratory Scientist at Research Lab, International Medical University (IMU). She has published articles in indexed and nonindexed journals. Her areas of expertise are immunology, genetics and molecular biology. These past years' experiences, have strengthened a combination of skills that gives her a solid foundation upon which to make an immediate and meaningful contribution at education and research field.

Asita Elengoe is a head of Biotechnology Department at Faculty of Applied Science, Lincoln University College Malaysia. She is also a professional technologist (Ts.). She obtained her PhD in Bioscience (2015) and Master of Science in Biotechnology (2010) from Universiti Teknologi Malaysia. She has received my Bachelor of Science Biotechnology (Hons) from Universiti Tunku Abdul Rahman. Her research interest includes cancer, gene therapy, computational biology, natural products and medicinal plants. She is a member of Royal Society of Biology (MRSB) United Kingdom. She is editor for Current Issues on Molecular Biology (Q2, Web of Science), SCIREA Journal of Biology and assistant executive editor for International Journal of Advancement in Life Sciences Research. She is reviewers for indexed and non-indexed journals. She is actively involved in research and publications. She has published articles in indexed, non-indexed journals and book chapters. She has published six books. She is editor for 'Obesity and its impact on health' book (Springer Nature). She has received award of the 'Young Researcher in Biotechnology' at International Healthcare Awards 2018, Chennai, India. Currently, she is carrying out research on drug designing using computational biology approach. She is supervisor for undergraduate and postgraduate (Master and PhD) students.

Jayasree S. Kanathasan is Program Coordinator in Biomedical Science Department at the Faculty of Applied Sciences, Lincoln University. Dr Jayasree completed her undergraduate degree in Biomedical Science from University Tunku Abdul Rahman. She pursued her postgraduate studies in Master of Engineering Science (Research) and Doctor of Philosophy in Biomedical Engineering at Monash University Malaysia sponsored by Ministry of Higher Education (MOHE) and Monash University scholarship. She designed her PhD project on peptide functionalized fluorescent porous silicon and carbon dot nanoparticles for breast cancer imaging which was awarded with FRGS grant and known as one of the outstanding projects in the round of application for the year 2016. Her work was presented in International Conference on Advances in Functional Materials, University of California, Los Angeles.

Calisa Lim is the Project Manager at the APAC Society for Cellular Agriculture. Her area of focus are monitoring and managing of regional regulations and policies, alongside key trends that could impact the industry across Asia. Through relationship building between key stakeholders and the development of science-based reports, advocacy tools and communication materials, she aims to enhance the industry's role as a trusted partner in the development of robust regulatory frameworks and standards to advance the cellular agriculture space. Calisa has experience working in a food industry association and managed the issues of nutrition, fiscal measures and ultra processed foods etc. She holds a BSocSci in Sociology from Nanyang Technological University, Singapore.

Dipali Saxena is an accomplished Assistant Professor with over 12years of academic experience in Food Sciences and Technology. She holds a Ph.D. in Food Technology from Jiwaji University Gwalior M. P. with a specialization of Nutraceuticals and functional foods. She is a university topper and MSc. Gold Medalist in Food Technology from Jiwaji University Gwalior Madhya Pradesh. Currently she is working as an Assistant professor of Food Sciences in Shri Vaishnav Institute of Home Science Shri Vaishnav Vidyapeeth Vishwavidyalaya, Indore. Dr Saxena is an MP SLET qualified and was also selected in a panel of technical officers from FSSAI New Delhi. Previously, she worked as a lecturer at Centre for Food Technology Jiwaji University Gwalior for almost 8 years. In addition to her academic expertise, she has worked with various research organizations and food enterprises such as Madya Pradesh Vigyan Sabha Bhopal, ICAR-Central Institute of Agricultural Engineering Bhopal, Shreeji foods, and PepsiCo Mathura. Dr. Saxena has a long list of publications to her name, including 25 research papers published in journals and conferences both nationally and internationally. She also serves on the advisory boards of several journals, including Asian Journal of Dairy and Food Research ARCC periodicals, International Journal of Food, Nutrition and Dietetics, Food Therapy and Health Care, and others. In the last three years, she has organized numerous conferences, faculty development programs, webinars, and workshops under the auspices of shri Vaishnav Vidyapeeth Vishwavidyalaya.

Aya Tayel is a Teaching Assistant at Food Hygiene and Control Department, Faculty of Veterinary Medicine, Benha University.

Peter Yu is the Program Director at the APAC Society for Cellular Agriculture – an association and coalition of 10+ cultivated meat and seafood companies across the APAC region. Through the society, Peter works as a drive-force to both promote the industry as well as to encourage the establishment efficient and transparent regulations and policies. Peter is actively engaging towards creating a harmonised cultivated food industry and have experience in driving various of industry moving projects across the globe. Peter is the Co-chair of the Enterprise Singapore Pro-term committee for setting Novel food standards and has been a convener for the first MoU regarding regional Nomenclature alignment in Asia. Moreover, Peter was appointed as a technical expert to FAO/WHO on safety guidelines for the industry. Peter holds triple Masters' degrees (Business Administration, Molecular Biosciences and Bioengineering, and Laws).

Index

A

animal welfare 25, 26, 44, 45, 67, 68, 69, 70, 72, 73, 74, 76, 77, 78, 79, 80, 81, 82, 84, 85, 86, 87, 88, 89, 91, 94, 99, 110, 115, 117, 120, 124, 128, 129, 130, 131, 132, 133, 139, 140, 141, 142, 145, 146, 147, 149, 150, 178, 185, 208, 209, 211, 212, 213, 216, 220, 221, 222, 225, 228, 263, 281, 305, 308, 309, 314

Asia-Pacific Society for Cellular Agriculture 299, 300, 302, 303, 310, 311, 322

B

Biodiversity Conservation 51, 191

Bioreactor 3, 8, 9, 10, 13, 14, 15, 16, 20, 26, 27, 36, 37, 46, 49, 62, 93, 97, 99, 101, 103, 109, 115, 116, 117, 126, 136, 138, 139, 141, 143, 144, 145, 148, 168, 169, 174, 210, 236, 259, 270, 297

Bioreactors 1, 5, 7, 8, 9, 12, 14, 15, 16, 20, 21, 22, 29, 32, 33, 37, 44, 48, 53, 57, 59, 60, 62, 73, 97, 101, 116, 117, 121, 122, 123, 124, 130, 134, 136, 137, 142, 143, 145, 162, 166, 168, 198, 203, 239, 246, 248, 264, 265, 278, 285

C

Cell Line 3, 156, 176, 210, 236, 300

Cellular Agriculture 1, 2, 3, 4, 5, 11, 16, 18, 20, 21, 22, 24, 25, 26, 29, 30, 31, 32, 33, 34, 35, 36, 38, 41, 43, 44, 50, 51, 52, 53, 54, 55, 56, 57, 58, 59, 60, 61, 62, 63, 64, 65, 66, 67, 69, 73, 74, 77, 84, 85, 89, 92, 95, 97, 98, 99, 101, 105, 106, 108, 109, 118, 119, 120, 126, 128, 129, 130, 131, 132, 133, 134, 135, 138, 139, 140, 141, 144, 146, 147, 148, 181, 194, 203, 208, 209, 210, 211, 212, 213, 214, 216, 217, 218, 219, 220, 221, 222, 223, 224, 225, 226, 227, 228, 230, 231, 232, 233, 234, 235, 236, 237, 238, 239, 240, 241, 242, 243, 244, 245, 246, 247, 248, 249, 250, 251, 252, 255, 256, 257, 258, 259, 260, 262, 264, 265, 266, 267, 268, 269, 270, 271, 272, 273, 274, 276, 277, 278, 279, 280, 281, 282, 283, 284, 288, 299, 300, 302, 303, 306, 308, 310, 311, 314, 315, 319, 320, 322

Cellular Milk 128, 129, 134, 135, 136, 137, 138, 139, 140, 147, 148

Climate Change 2, 25, 33, 50, 52, 54, 63, 80, 89, 95, 104, 123, 129, 141, 172, 185, 209, 231, 232, 234, 255, 258, 259, 260, 262, 264, 265, 270, 274, 275, 276, 282, 283, 284, 288, 289, 303, 305, 316, 317

Consumer Perception 37, 195, 197, 288, 300, 301, 302, 304, 306, 307, 318

Consumer Preference 94, 307

Cultivated Meat 26, 28, 29, 30, 31, 35, 36, 37, 38, 39, 40, 42, 45, 46, 47, 48, 65, 73, 83, 87, 104, 111, 114, 130, 152, 158, 164, 168, 176, 182, 188, 189, 192, 193, 195, 201, 202, 203, 206, 244, 248, 250, 251, 275, 278, 282, 283, 284, 288, 289, 310, 311, 312, 313, 314, 315, 316, 317, 320, 321

Cultured Meat 11, 17, 22, 24, 26, 27, 29, 30, 34, 35, 36, 37, 39, 40, 43, 44, 46, 47, 48, 52, 53, 54, 58, 59, 60, 62, 63, 64, 65, 85, 91, 94, 98, 100, 101, 103, 104, 105, 107, 108, 110, 111, 112, 114, 115, 117, 118, 121, 122, 123, 124, 125, 126, 148, 149, 150, 151, 152, 153, 154, 156, 157, 158, 160, 161, 162, 166, 167, 168, 169, 170, 173, 174, 175, 176, 177, 178, 179, 180, 181, 182, 183, 184, 185, 186, 187, 188, 189, 190, 191, 192, 193, 194, 196, 197, 198, 199, 200, 201, 202, 203, 204, 205, 206, 210, 211, 214, 216, 220, 221, 226, 227, 228, 229, 235, 236, 237, 240, 243, 244, 245, 246, 247, 249, 250, 251, 252,

254, 264, 269, 270, 275, 278, 279, 280, 281, 282, 283, 285, 287, 311, 317, 318, 319, 320

E

Environmental Impact 34, 50, 51, 52, 58, 60, 73, 78, 79, 81, 101, 110, 128, 129, 132, 145, 148, 150, 204, 208, 209, 211, 212, 218, 221, 263, 264, 265
Ethical Considerations 26, 33, 34, 35, 70, 72, 73, 74, 79, 86, 119, 120, 150, 208, 211, 218, 239, 268
ethics 41, 47, 67, 69, 70, 71, 73, 74, 77, 78, 79, 80, 81, 82, 84, 86, 87, 88, 90, 91, 105, 109, 121, 147, 194, 213, 214, 217, 225, 226, 228, 229, 233, 242, 318

F

Fermentation 3, 26, 51, 52, 53, 59, 112, 130, 133, 134, 135, 136, 137, 138, 139, 140, 141, 210, 218, 219, 241, 256, 257, 261, 280, 293, 313, 321
food production 2, 3, 20, 24, 27, 33, 35, 36, 43, 51, 54, 58, 60, 61, 64, 67, 73, 74, 77, 78, 79, 81, 84, 85, 90, 97, 106, 128, 129, 130, 131, 134, 143, 148, 151, 180, 200, 208, 209, 211, 214, 218, 220, 221, 227, 231, 232, 234, 235, 236, 241, 242, 247, 250, 255, 259, 263, 264, 278, 289, 305, 318
Food Security 25, 56, 62, 74, 97, 98, 104, 121, 122, 123, 129, 130, 131, 141, 145, 149, 185, 186, 220, 225, 226, 231, 232, 233, 234, 235, 237, 238, 239, 240, 241, 242, 243, 245, 246, 248, 249, 250, 252, 253, 265, 266, 280, 282, 284, 288, 289, 292, 303, 316, 317
Food System 2, 61, 109, 129, 130, 132, 140, 141, 144, 146, 148, 154, 194, 208, 223, 228, 231, 232, 233, 240, 247, 248, 250, 252, 253, 255, 256, 272, 273, 278, 280, 284, 288, 289, 315

G

GHG Emission 25, 52, 53
Global Population 1, 35, 95, 185, 209, 231, 232, 234, 235, 250, 260, 262, 305

I

In Vitro Harvest 109, 111, 112

L

Labelling 31, 99, 118, 120, 121, 176, 185, 288, 291, 292, 294, 295, 296, 297, 299, 301, 305, 306, 307, 309, 320
Lab Grown 58, 62, 92, 93, 97, 204, 245, 252, 277
Lab-Grown Plate 92

M

Market Trends 208, 211

N

Naturalness 37, 120, 121, 179, 202, 217, 304, 308, 309, 313, 318, 320
Novel Foods 123, 171, 176, 177, 269, 284, 291, 292, 295, 296, 299, 307, 314, 318, 321

P

Plant Cells 1, 4, 6, 9, 20, 32, 33, 44, 210
Plant Tissue Culture 5, 25, 32, 33, 38, 41, 43, 45

R

Regulatory Coordination 121, 291, 299, 300, 307, 309, 310, 321
Regulatory Frameworks 74, 118, 119, 121, 144, 185, 208, 211, 221, 222, 240, 269, 270, 288, 290, 291, 298, 299, 301, 308, 321
research 3, 4, 10, 13, 19, 20, 22, 29, 30, 31, 34, 36, 37, 38, 40, 44, 45, 47, 52,

61, 67, 68, 69, 70, 71, 72, 73, 74, 75, 76, 78, 81, 82, 84, 85, 86, 87, 88, 89, 90, 91, 92, 98, 99, 100, 110, 114, 116, 118, 121, 124, 131, 132, 134, 136, 138, 139, 140, 141, 144, 145, 146, 147, 148, 149, 150, 151, 159, 161, 162, 164, 179, 181, 182, 185, 186, 190, 197, 201, 202, 204, 206, 211, 220, 221, 222, 226, 227, 228, 233, 235, 236, 237, 241, 243, 244, 245, 248, 250, 252, 253, 255, 256, 257, 258, 261, 264, 268, 269, 270, 272, 273, 276, 278, 280, 281, 282, 284, 286, 287, 289, 296, 303, 306, 309, 315

S

Scaffolding Biomaterial 162
SDGs 2, 25
Sustainability 2, 43, 48, 56, 63, 65, 85, 101, 104, 107, 109, 115, 120, 121, 126, 129, 130, 131, 134, 138, 139, 140, 141, 145, 146, 149, 150, 152, 162, 178, 193, 194, 195, 200, 208, 209, 210, 212, 216, 220, 221, 225, 228, 229, 242, 243, 244, 246, 247, 248, 249, 250, 251, 252, 254, 261, 262, 274, 278, 279, 280, 282, 283, 284, 285, 286, 305, 308, 309, 311, 320
Sustainable Food Production 24, 51, 60, 78, 81, 250

T

Technological Advancements 80, 110, 111, 208, 211, 255
Tissue Cultures 1, 2, 3, 4, 9, 101

W

Waste Management 54, 55, 56, 57, 64, 65, 277

Publishing Tomorrow's Research Today

Uncover Current Insights and Future Trends in
Scientific, Technical, & Medical (STM)
with IGI Global's Cutting-Edge Recommended Books

Print Only, E-Book Only, or Print + E-Book.
Order direct through IGI Global's Online Bookstore at www.igi-global.com or through your preferred provider.

ISBN: 9798369303689
© 2024; 299 pp.
List Price: US$ 300

ISBN: 9798369314791
© 2024; 287 pp.
List Price: US$ 330

ISBN: 9798369300442
© 2023; 542 pp.
List Price: US$ 270

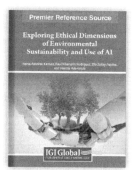

ISBN: 9798369308929
© 2024; 426 pp.
List Price: US$ 265

ISBN: 9781668489383
© 2023; 299 pp.
List Price: US$ 325

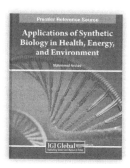

ISBN: 9781668465776
© 2023; 454 pp.
List Price: US$ 325

Do you want to stay current on the latest research trends, product announcements, news, and special offers? Join IGI Global's mailing list to receive customized recommendations, exclusive discounts, and more.
Sign up at: www.igi-global.com/newsletters.

Scan the QR Code here to view more related titles in STM.

www.igi-global.com   Sign up at www.igi-global.com/newsletters   facebook.com/igiglobal   twitter.com/igiglobal   linkedin.com/igiglobal

Ensure Quality Research is Introduced to the Academic Community

Become a Reviewer for IGI Global Authored Book Projects

  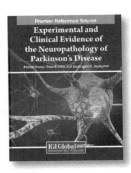 

The overall success of an authored book project is dependent on quality and timely manuscript evaluations.

Applications and Inquiries may be sent to:
development@igi-global.com

Applicants must have a doctorate (or equivalent degree) as well as publishing, research, and reviewing experience. Authored Book Evaluators are appointed for one-year terms and are expected to complete at least three evaluations per term. Upon successful completion of this term, evaluators can be considered for an additional term.

If you have a colleague that may be interested in this opportunity, we encourage you to share this information with them.

www.igi-global.com

Publishing Tomorrow's Research Today
IGI Global's Open Access Journal Program
Including Nearly 200 Peer-Reviewed, Gold (Full) Open Access Journals across IGI Global's Three Academic Subject Areas: Business & Management; Scientific, Technical, and Medical (STM); and Education

Consider Submitting Your Manuscript to One of These Nearly 200 Open Access Journals for to Increase Their Discoverability & Citation Impact

Web of Science Impact Factor **6.5**	Web of Science Impact Factor **4.7**	Web of Science Impact Factor **3.2**	Web of Science Impact Factor **2.6**
JOURNAL OF **Organizational and End User Computing**	JOURNAL OF **Global Information Management**	INTERNATIONAL JOURNAL ON **Semantic Web and Information Systems**	JOURNAL OF **Database Management**

Choosing IGI Global's Open Access Journal Program Can Greatly Increase the Reach of Your Research

Higher Usage
Open access papers are 2-3 times more likely to be read than non-open access papers.

Higher Download Rates
Open access papers benefit from 89% higher download rates than non-open access papers.

Higher Citation Rates
Open access papers are 47% more likely to be cited than non-open access papers.

Submitting an article to a journal offers an invaluable opportunity for you to share your work with the broader academic community, fostering knowledge dissemination and constructive feedback.

Submit an Article and Browse the IGI Global Call for Papers Pages

We can work with you to find the journal most well-suited for your next research manuscript.
For open access publishing support, contact: journaleditor@igi-global.com

Publishing Tomorrow's Research Today
IGI Global e-Book Collection

Including Essential Reference Books Within Three Fundamental Academic Areas

Business & Management
Scientific, Technical, & Medical (STM)
Education

- Acquisition options include Perpetual, Subscription, and Read & Publish
- No Additional Charge for Multi-User Licensing
- No Maintenance, Hosting, or Archiving Fees
- Continually Enhanced Accessibility Compliance Features (WCAG)

| Over **150,000+** Chapters | Contributions From **200,000+** Scholars Worldwide | More Than **1,000,000+** Citations | **Majority of e-Books Indexed in Web of Science & Scopus** | **Consists of Tomorrow's Research Available Today!** |

Recommended Titles from our e-Book Collection

Innovation Capabilities and Entrepreneurial Opportunities of Smart Working
ISBN: 9781799887973

Advanced Applications of Generative AI and Natural Language Processing Models
ISBN: 9798369305027

Using Influencer Marketing as a Digital Business Strategy
ISBN: 9798369305515

Human-Centered Approaches in Industry 5.0
ISBN: 9798369326473

Modeling and Monitoring Extreme Hydrometeorological Events
ISBN: 9781668487716

Data-Driven Intelligent Business Sustainability
ISBN: 9798369300497

Information Logistics for Organizational Empowerment and Effective Supply Chain Management
ISBN: 9798369301593

Data Envelopment Analysis (DEA) Methods for Maximizing Efficiency
ISBN: 9798369302552

Request More Information, or Recommend the IGI Global e-Book Collection to Your Institution's Librarian

For More Information or to Request a Free Trial, Contact IGI Global's e-Collections Team: eresources@igi-global.com | 1-866-342-6657 ext. 100 | 717-533-8845 ext. 100

Are You Ready to Publish Your Research?

IGI Global offers book authorship and editorship opportunities across three major subject areas, including Business, STM, and Education.

Benefits of Publishing with IGI Global:

- Free one-on-one editorial and promotional support.
- Expedited publishing timelines that can take your book from start to finish in less than one (1) year.
- Choose from a variety of formats, including Edited and Authored References, Handbooks of Research, Encyclopedias, and Research Insights.
- Utilize IGI Global's eEditorial Discovery® submission system in support of conducting the submission and double-blind peer review process.
- IGI Global maintains a strict adherence to ethical practices due in part to our full membership with the Committee on Publication Ethics (COPE).
- Indexing potential in prestigious indices such as Scopus®, Web of Science™, PsycINFO®, and ERIC – Education Resources Information Center.
- Ability to connect your ORCID iD to your IGI Global publications.
- Earn honorariums and royalties on your full book publications as well as complimentary content and exclusive discounts.

Join Your Colleagues from Prestigious Institutions, Including:

Learn More at: www.igi-global.com/publish
or by Contacting the Acquisitions Department at: acquisition@igi-global.com